Astrophysics
in the XXI Century
with Compact Stars

Astrophysics in the XXI Century with Compact Stars

Editors

César Augusto Zen Vasconcellos
Universidade Federal do Rio Grande do Sul (UFRGS), Porto Alegre, Brazil
International Center for Relativistic Astrophysics Network (ICRANet), Pescara, Italy

Fridolin Weber
San Diego State University (SDSU), San Diego, USA
University of California at San Diego (UCSD), San Diego, USA

World Scientific

NEW JERSEY • LONDON • SINGAPORE • BEIJING • SHANGHAI • HONG KONG • TAIPEI • CHENNAI • TOKYO

Published by

World Scientific Publishing Co. Pte. Ltd.
5 Toh Tuck Link, Singapore 596224
USA office: 27 Warren Street, Suite 401-402, Hackensack, NJ 07601
UK office: 57 Shelton Street, Covent Garden, London WC2H 9HE

Library of Congress Control Number: 2022945024

British Library Cataloguing-in-Publication Data
A catalogue record for this book is available from the British Library.

ASTROPHYSICS IN THE XXI CENTURY WITH COMPACT STARS

ISBN 978-981-122-093-7 (hardcover)
ISBN 978-981-122-094-4 (ebook for institutions)
ISBN 978-981-122-095-1 (ebook for individuals)

For any available supplementary material, please visit
https://www.worldscientific.com/worldscibooks/10.1142/11848#t=suppl

Desk Editor: Nur Syarfeena Binte Mohd Fauzi

Typeset by Stallion Press
Email: enquiries@stallionpress.com

We dedicate this book to our families

Fridolin Weber
Department of Physics, San Diego State University (SDSU), San Diego, USA and Center for Astrophysics and Space Sciences, University of California at San Diego, La Jolla, USA
César Augusto Zen Vasconcellos
Instituto de Física, Universidade Federal do Rio Grande do Sul (UFRGS), Porto Alegre, Brazil and International Center for Relativistic Astrophysics Network (ICRANet), Rome, Pescara, Italy
The Editors

https://doi.org/10.1142/9789811220944_fmatter

Foreword

There are many reasons to believe that the 21st Century will be the best and most exciting century yet for astrophysics and related fields of physics. The James Webb Space Telescope, for instance, has opened up a new window on infrared astronomy, allowing unprecedented insights into the early universe as well as observations of planets outside of our solar system; neutrino massiveness provides a new window into the exploration of dark energy and dark matter physics and is expected to shed light on the fate and evolution of the Universe; the Higgs boson could provide an understanding of the strength of gravity; and last but not least, gravitational waves generated in the very early Universe and by cataclysmic astrophysical physical events involving compact stellar objects (black holes, neutron stars, white dwarfs) have lead to a new branch of observational astronomy called gravitational wave astronomy. Given this complex and rich variety of topics, compact stars are proving to be unique astrophysical laboratories that can be used to study the structure of spacetime, the building blocks of matter and their interactions in physical regimes not accessible in terrestrial laboratories. The contributions compiled in this book provide an up-to-date overview of such research.

https://doi.org/10.1142/9789811220944_fmatter

Preface

As already mentioned in the preface, there are many reasons to believe that the 21st century will become an extraordinarily exciting century for astrophysics. This development is driven by unprecedented advances in observational astronomy together with exciting advances in nuclear and particle physics and a much better understanding of the structure and composition of ultra-dense and hot matter. All of this is occurring at a time when extremely powerful computers have become available to students and researchers to perform detailed numerical simulations of extremely complex physical systems.

Against this background, compact stars represent unique astrophysical laboratories where the structure of spacetime, the building blocks of matter and their interactions can be studied in physical domains not accessible in terrestrial laboratories. This is the main theme of the present book, which provides an up-to-date overview of compact star research and related physical phenomena.

In the following we briefly describe the contents of the book.

About the content of the book

Chapter 1. Authors: Jorge E. Horvath, Livia S. Rocha, Antonio L.C. Bernard, Marcio G.B. de Avellar, and Rodolfo Valentim.
Abstract: Birth events of neutron stars are tightly related with their observed masses, which in turn are also linked to the state of matter that constitutes them. Using a sample of 76 measured neutron star masses, we show that its distribution requires more than one mass scale, a result that is probably related with the existence of different formation and/or evolutionary channels. The highest value allowed for the mass of a neutron

star, inferred from the sample, is compatible with 2.5 $M_\odot$, suggesting that very heavy neutron stars may be present in Nature. The case of some "spider" systems and the enigmatic high mass object in the GW190408 event, as well as the highest measured values from other binaries may support a high M_{max}, that gets closer to the Rhoades–Ruffini limit. Now we have entered the multi-messenger astronomy era, gravitational waves observations can substantially contribute to all these open questions. In the following chapter, we review all these issues and suggest future directions.

Chapter 2. Author: Rachid Ouyed.
Abstract: The Bodmer–Terazawa–Witten hypothesis of absolutely stable quark matter made of up, down, and strange quarks ((u,d,s) matter) is of much astrophysical and fundamental interest. One consequence is a large binding energy release associated with the conversion of a neutron star to a quark star. A quantitative understanding of the dynamic aspect of the conversion is necessary in order to find out whether this energy is released quietly or in an explosive manner. We address numerically (i.e., solving the reaction–diffusion–advection equations for (u,d) to (u,d,s) combustion) the dynamic processes through which the conversion appears. We find fundamentally very different results from semi-analytic calculations, with front speeds that are several orders of magnitude higher for the former. Resolving the hadronic-quark-matter interface is necessary, since approximations like Coll's condition may quench the burning, while properly resolving the flame can make combustion always thermodynamically favourable if the hypothesis of absolutely stable strange quark matter is true. We find that lepton physics, including weak decays, electron equation-of-state (EoS), neutrino EoS, and neutrino transport are at the very least as important to the physics of the burning front as the EoS of highly dense matter. In concert, these effects induce novel wrinkling instabilities (such as the deleptonization instability) possibly leading to a deflagration-detonation and/or a quark-core collapse Quark-Nova.

Chapter 3. Authors: Chengjun Xia, Xiaoyu Lai, and Renxin Xu.
Abstract: What if normal baryonic matter is squeezed so that atomic nuclei come in close contact? This is a problem implying a broad interest in astrophysics, but is not solved because of non-perturbative issues of the strong interaction. The weak interaction also plays an important role there: although it is in the mainstream that neutronization occurs to make a state with asymmetric isospin, we are addressing that quark flavour-symmetry

from 2 (u and d quarks) to 3 (u, d and s quarks) could be restored during squeezing. The latter is called strangeonization. Strangeon is conjectured to be the constituent of bulk strong matter, as an analogy of nucleon for an atomic nucleus. We will explain strangeon and strangeon matter in this chapter, as well as discuss their observational manifestations in astrophysics.

Chapter 4. Authors: Edson Otoniel, Jaziel G. Coelho, Manuel Malheiro, and Fridolin Weber.
Abstract: As is well known, the existence of fast and ultra-massive white dwarfs has been discovered in recent years. We estimate the mass density thresholds for the onset of electron capture reactions and pycnonuclear reactions in the cores of fast, massive and highly magnetized white dwarfs (HMWDs) and discuss the impact of microscopic stability and rapid rotation on the structure and stability of different WD constitutions. We confirm that fast rotation increases the mass of a WD by up to 10%, while the central density may drop by one to two orders of magnitude, depending on stellar mass and rate of rotation. We also note that the central densities of the rotating WDs are smaller than those of the non-rotating stars, since less pressure is to be provided by the equation-of-state (EoS) in the rotating case, and that the maximum-mass limit slightly decreases when lattice contributions are taken into account, which soften the EoS mildly. The softening leads to WDs with somewhat smaller radii and therefore smaller Kepler periods. Overall, one sees that very massive and magnetic $^{12}C+^{16}O$ WDs have rotational Kepler periods on the order of 0.5 s. Pycnonuclear reactions are triggered in these WDs at masses that are markedly smaller than the maximum WD masses. The corresponding rotational periods turn out to be in the 5 s (~ 0.2 Hz) range.

Chapter 5. Authors: Armen Sedrakian, Jia-Jie Li, and Fridolin Weber.
Abstract: We review the covariant density functional approach to the equation-of-state (EOS) of the dense nuclear matter in compact stars. The main emphasis is on the hyperonization of the dense matter, and the role played by the Δ-resonances. The implications of hyperonization for the astrophysics of compact stars, including the EOS, composition, and stellar parametres are examined. The mass-radius relation and tidal deformabilities of static and rapidly rotating (Keplerian) configurations are discussed in some detail. We briefly touch upon some other recent developments involving hyperonization in hot hypernuclear matter at high- and low-densities.

Chapter 6. Author: David Ian Jones.
Abstract: Spinning neutron stars can emit long-lived gravitational waves. There are several mechanisms that can produce such *continuous wave* emission. These mechanisms relate to the strains in the elastic crust, the star's magnetic field, superfluidity of the neutron fluid, and bulk oscillations of the entire star. In this chapter, we describe how the frequency content of the gravitational wave signal, and its relation to any electromagnetically observed spin frequency, can be used to constrain the mechanism producing the gravitational waves. These ideas will be of use in the event of the first detections of such signals, and help convert a detection into useful physical insight.

Chapter 7. Authors: Marco Antonelli, Alessandro Montoli, and Pierre M. Pizzochero.
Abstract: The presence of superfluid phases in the interior of a neutron star affects its dynamics, as now neutrons can flow relative to the non-superfluid (normal) components of the star with little or no viscosity. A probe of superfluidity comes from pulsar glitches, sudden jumps in the observed rotational period of radio pulsars. Most models of glitches build on the idea that a superfluid component of the star is decoupled from the spin-down of the normal component, and its sudden re-coupling leads to a glitch. This transition in the strength of the hydrodynamic coupling is explained in terms of quantum vortices (long-lived vortices that are naturally present in the neutron superfluid at the microscopic scale). After introducing some basic ideas, we derive (as a pedagogical exercise) the formal scheme shared by many glitch studies. Then, we apply these notions to present some progress in understanding and modelling pulsar glitches, notably the possibility to constrain the nuclear physics input and the extraction of the coupling strength from observations of the elusive spin-up phase of a glitch.

Chapter 8. Authors: Andreas Bauswein, David Blaschke, and Tobias Fischer
Abstract: We outline a theoretical approach supporting strong phase transitions from normal nuclear matter to the deconfined quark-gluon plasma, in the equation of state (EOS) for compact star matter. Implications of this hypothesis are discussed for astrophysical applications. Special emphasis is devoted to potentially detectable signatures, which can be directly related

with the occurrence of a sufficiently strong phase transition. Therefore, simulations of core-collapse supernovae and binary compact star mergers are considered, including the subsequent emission of gravitational waves and, in the case of supernova, in addition the neutrinos play the role of messengers.

Fridolin Weber and César Zen Vasconcellos
San Diego, USA and Porto Alegre, Brazil, May 2021

https://doi.org/10.1142/9789811220944_fmatter

Contents

https://doi.org/10.1142/9789811220944_fmatter

List of Figures

Chapter 4

Chapter 5

Chapter 6

Chapter 7

Chapter 8

https://doi.org/10.1142/9789811220944_fmatter

List of Tables

https://doi.org/10.1142/9789811220944_0001

Chapter 1

Birth Events, Masses and the Maximum Mass of Compact Stars

Jorge E. Horvath*,‡, Livia S. Rocha*,§, Antonio L.C. Bernardo*,¶, Marcio G.B. de Avellar†,‖ and Rodolfo Valentim†,**

**Universidade de São Paulo (USP) Instituto de Astronomia, Geofísica e Ciências Atmosféricas (IAG) Rua do Matão 1226, 05508-090, São Paulo — SP, Brazil*

†Instituto de Ciências Ambientais, Químicas e Farmacêuticas (ICAQF) Departamento de Física, Universidade Federal de São Paulo (UNIFESP) R. São Nicolau no.210, Centro, 09913-030, Diadema — SP, Brazil

‡foton@iag.usp.br

§livia.silva.rocha@usp.br

¶alc.bernardo@gmail.com

‖marcio.de.avellar@gmail.com

***valentim.rodolfo@unifesp.br*

Abstract: Birth events of neutron stars are tightly related with their observed masses, which in turn are also linked to the state of matter that constitutes them. Using a sample of 76 measured neutron star masses, we show that its distribution requires more than one mass scale, a result that is probably related with the existence of different formation and/or evolutionary channels. The highest value allowed for the mass of a neutron star, inferred from the sample, is compatible with $2.5\,M_\odot$, suggesting that very heavy neutron stars may be present in Nature. The case of some "spider" systems and the enigmatic high mass object in the GW190408 event, as well as the highest measured values from other binaries may support a high $M_{\max}$, that gets closer to the Rhoades–Ruffini limit. Now we have entered the multi-messenger astronomy era, gravitational waves observations can substantially contribute to all these open questions. In the following chapter, we review all these issues and suggest future directions.

Keywords: compact stars; neutron stars; birth events.

1. Introduction

At the beginning of the current century a scientific magazine published an article mentioning 11 unanswered questions of Physics whose resolutions could provide a new era in science (Haseltine, 2002). Among them is the behaviour of matter at ultrahigh densities and temperatures, a challenging topic that is directly related with neutron stars (NSs) Physics.

Such extreme conditions make it impossible to deal with this type of matter in a laboratory, therefore the equation-of-state (EoS) governing these compact objects cannot be attained by experimental methods only. This is the reason why is so important to direct efforts to the observational, theoretical and computational fields of NSs, in order to extract their bulk properties, reveal how matter behaves in their interior and how related phenomenons are generated, such as supernovae explosions. For example, in the observational realm, macroscopic parameters like mass and radius are tightly related with microphysics inside the star and can help to set reliable constraints on it.

NSs are the densest and smallest stars observed in the Universe, with a mean density of $\langle \rho \rangle$ exceeding the ordinary nuclear density $\rho_{sat} = 2.8 \times 10^{14}\,\mathrm{g/cm^3}$. Such an extreme density corresponds to a baryon number density of $\mathrm{n}_{sat} \sim 0.16\,\mathrm{fm}^{-3}$. There is evidence that these stars are the final stage of the life of ordinary stars whose masses are in the range $\sim 9.0\,\mathrm{M}_\odot < M < 25.0\,\mathrm{M}_\odot$ (Heger *et al.*, 2003) (although this is far from established, see Section 5) or even an outcome of an Accretion Induced Collapse of a white dwarf (or a pair of them, see below) (Ruiter *et al.*, 2019). NSs are considered "dead" because they have stopped producing nuclear energy as remnants of the stellar evolution of massive stars. They can be observed in various astrophysical sources, from radio to X-ray pulsars, X-ray bursters, compact thermal X-ray sources in supernovae remnants and in binary systems (some of them relativistic), among many possibilities.

As a consequence of their compactness, NSs must be described in terms of General Relativity (GR), thus the introduction of an appropriate EoS in the Einstein Field Equations and its hydrostatic equilibrium consequence, the Tolman-Oppenheimer-Volkoff (TOV) equation, leads to the existence of a maximum value for the mass of the star (see Section 3). The observation of very massive NSs in the last decade had profound implications for the open question mentioned above, because it establishes a lower limit on the maximum mass, and thus helps to set realistic constraints for matter beyond nuclear saturation density ρ_{sat} ruling out models that do not support masses in the observed range, and also helps to classify an object as a black hole (heareafter BH) if its mass is above the maximum allowed value for NSs. All these issues will be briefly reviewed here to gain insight on the state-of-the-art of this problem and suggest future directions.

2. The Sample of Neutron Stars and its Mass Distribution

From the first observation of a pulsar in 1967 by Jocelyn Bell (Hewish *et al.*, 1968) until now, computational and instrumental developments, especially in the current century, helped to increase the available sample of objects and improved precision on measurements with new generations of high-quality telescopes. In particular, the masses of the NSs has been targeted to connect them with the birth events and the properties of superdense matter. We present now an overview of these issues.

2.1. *Types of mass measurements*

The vast majority of NSs are observed as pulsars, rapidly-rotating and high-magnetized NSs emitting beams of radiation along its magnetic axis that are seen on Earth as a pulse due to the lighthouse effect (Rezzolla *et al.*, 2018). Nowadays, more than 2,800 pulsars were observed (see ATNF Pulsar Catalogue (2020) for a catalogue of radio pulsars), but only a few aspects of them can be inferred from observations and only a tiny fraction of the total sample allows mass measurements because observed pulsars are mainly isolated stars and the calculation methods are generally based on orbital motions, although efforts are beginning to be made to measure features of isolated NSs (Horvath, 1996).

The extreme regularity of pulses, responsible for recognizing pulsars as the most stable clocks in the observable universe, makes *pulsar timing* the most accurate method to determine masses of NSs, as well as test fundamental physics. The procedure consists in monitoring the times-of-arrival (TOAs) of pulses over several years to determine the pulsar's rotation period. Thanks to the regularity, small deviations of TOAs are detectable with precision. Additional parameters of pulsar and its orbit (in binary systems) are also obtained from pulsar timing.

Orbital motion of binary systems is described by five Keplerian parameters: binary period P_b, projection of pulsar's semimajor axis (a_p) on the line of sight $x_p = a_p \sin i$, eccentricity e, time T_0 and longitude ω of periastron. The masses of components can be determined from a mass function in light of Kepler's Third Law:

$$f_1(m_1, m_2, i) = \frac{(m_2 \sin i)^3}{(m_1 + m_2)^2} = \frac{4\pi^2}{T_\odot} \frac{x_p^3}{P_b^2}, \tag{1}$$

where $T_\odot \equiv G M_\odot / c^3 = 4.925490947\ \mu\text{s}$ (G is the gravitational constant, c is the speed of light and $M_\odot$ is the solar mass) (see Chapter 9 of Shapiro and Teukolsky (2008) for a pedagogical discussion of the mass function).

Whenever mass functions are measurable, it is possible to obtain individual masses provided the inclination angle i is known, but this quantity is difficult to be inferred with accuracy and the mass function of the companion is only obtained in some cases where it is a detectable pulsar or a star with an observable spectrum. However, binary systems containing a pulsar are compact and thus some relativistic effects might be observed. The relativistic corrections for orbital motion for GR are described in terms of post-Keplerian (PK) parameters, which are functions of Keplerian parameters (Stairs, 2003):

(i) Orbital period decay, $\dot{P}_b$:

$$\dot{P}_b = -\frac{192\pi}{5} \left(\frac{P_b}{2\pi T_\odot}\right)^{-\frac{5}{3}} \left(1 + \frac{73}{24}e^2 + \frac{37}{96}e^4\right) \left(1 - e^2\right)^{-\frac{7}{2}} \frac{m_p m_c}{m^{1/3}}; \tag{2}$$

(ii) Range of Shapiro delay, r:

$$r = T_\odot m_c; \tag{3}$$

(iii) Shape of Shapiro delay, s:

$$s = \sin i = x_p \left(\frac{P_b}{2\pi}\right)^{-2/3} \frac{m^{2/3}}{T_\odot^{1/3} m_c}; \tag{4}$$

(iv) "Einstein delay", γ:

$$\gamma = e \left(\frac{P_b}{2\pi}\right)^{1/3} T_\odot^{2/3} \frac{m_c (m_p + 2m_c)}{m^{4/3}}; \tag{5}$$

(v) Advance of periastron, $\dot{\omega}$:

$$\dot{\omega} = 3 \left(\frac{P_b}{2\pi}\right)^{-5/3} \left(1 - e^2\right)^{-1} \left(m T_\odot\right)^{2/3}. \tag{6}$$

In Eqs. (2)–(6) the subindexes of Eq. (1) were changed from 1 to p standing for the pulsar component and from 2 to c for the companion, and $m = m_p + m_c$. If only one PK parameter is observed, constrains can be imposed on individual masses. The observation of any two PK parameters allows unique individual mass determinations, and if extra parameters are specified it is possible to test GR (Lyne *et al.*, 2004).

Millisecond pulsars (MSPs) are pulsars with very short spin periods, $1 < P < 30\,\text{ms}$ and $\dot{P} < 10^{-19}$, which were spun-up by its companion during a transfer of angular momentum to the NS, being *recycled*. These short periods make MSPs extremely precise for pulsar timing and they are now recognized as the most useful objects to test fundamental physics (Özel and Freire, 2016).

For a large number of binaries only one PK parameter is measured, but as mentioned before in cases where the companion is optically observed (main sequence, post-main sequence and white dwarfs stars) it is possible to provide additional information about the system. Phase-resolved spectroscopy provides orbital radial velocity amplitude (K_c), which together with Keplerian parameters yields the mass ratio ($q = m_p/m_c$). For white dwarfs, particularly, their radii can be estimated if the distance to Earth is known and the optical flux and effective temperature are measured. Thus, from a theoretical relation based on their spectrum its mass can be obtained, leading to the pulsars mass estimation through the relation of mass ratio. For NSs with high-mass companions, X-rays emitted by the pulsar are blocked by the companion during part of its orbit, allowing mass measurements also.

The Shapiro delay, in its turn, consists in an retardation of pulses TOA due to space-time curvature near a massive companion. This effect can remain unobserved, even after years of timing, if it is too small, but for almost edge-on MSP systems containing a high-mass WD, the effect is strong enough and allows extremely precise measurements of both masses. PSR J1614-2230 with $1.97 \pm 0.04\,M_\odot$ (Demorest *et al.*, 2010) (latter corrected for $1.928 \pm 0.017\,M_\odot$ by Fonseca *et al.* (2016)) and PSR J0740+6620 with $2.14^{+0.10}_{-0.09}\,M_\odot$ (Cromartie *et al.*, 2020) are two MSPs that became important references when talking about imposing a lower limit for the maximum mass of NSs where measured through Shapiro delay. However, there is more overall information in the whole distribution that can be exploited, even if the individual measurements are not that precise. We turn to this question in the rest of this section.

2.2. *Exploring the mass distribution of neutron stars*

Although pioneers works pointed for the possibility of a wide range of masses for NSs, the old idea of a unique-mass scale with a small dispersion around a central value was strengthened by the theoretical argument on the pre-supernova explosion, in which an iron core supported by electron degeneracy (thus with an almost invariant mass) gives place to a NS with a slightly lower mass. Observational works seemed to converge with the theoretical statement for decades. Finn (1994) employed for the first time a statistical analysis of a sample of four double NS systems (DNS) with the constrained masses of eight NS and found that masses should fall predominantly in the range $1.3 < m/M_{\odot} < 1.6$. Thorsett and Chakrabarty (1999) obtained a mean value of $1.35 \pm 0.04\,M_{\odot}$ for a sample of 19 NS masses, and no evidence for a significant dispersion around the single scale.

The rapid development of instrumental and computational fields, specially at the turn of the century, prompted a paradigm questioning and break. New observations put on stage old arguments about a different supernova mechanism leading to the formation of lighter pulsars. On the other hand, X-ray and optical observations provided evidence for the existence of masses above $2\,M_{\odot}$, despite large uncertainties, bringing up accretion histories and/or NSs that are born massive. As seen in (Van den Heuvel, 2004), a speculation of the possible existence of three NS classes already existed. The current sample of NS mass measurements and its uncertainties is shown in Fig. 1, with different types of binary systems distinguished by colors.

Technological advances has provided us with faster and higher-capacity computers, permitting the Bayesian approach to spread among scientific community through *Markov Chain Monte Carlo* (MCMC) simulations. Bayesian analyzes became an allied for Astronomy, embracing also the study of NS physics, including the formation channels and evolution histories, especially because it is quite suitable for small samples. Several works employed these techniques to extract information from the mass distribution (Antoniadis *et al.*, 2016; Farrow *et al.*, 2019; Kiziltan *et al.*, 2013; Linares, 2019; Özel *et al.*, 2012; Valentim *et al.*, 2011; Zhang *et al.*, 2011), each of them choosing their own *prior* models particularity. For example, Schwab *et al.* (2010) discussed the evidence of a bimodal distribution based on a sample of 14 masses, where a peak in $\sim 1.25\,M_{\odot}$ might be related with electron-capture supernovae (Nomoto, 1984; Podsiadlowski *et al.*, 2004) and a peak in $\sim 1.35\,M_{\odot}$ related with the Fe core-collapse supernovae.

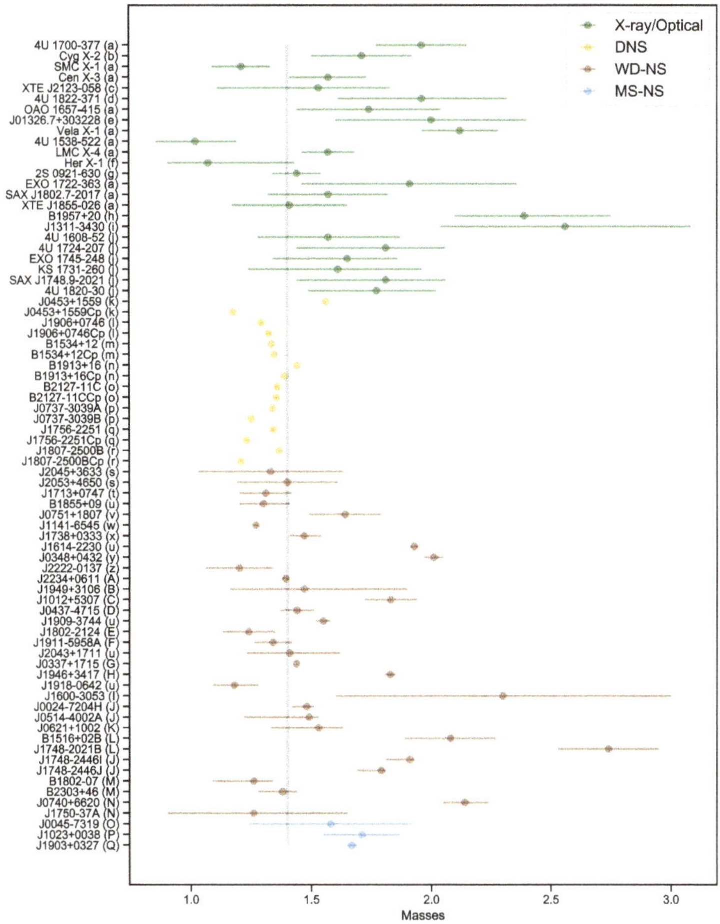

Fig. 1. Masses compiled by L.S. Rocha (October/2020). (a) (Falanga *et al.*, 2015), (b) (Casares *et al.*, 2010), (c) (Tomsick *et al.*, 2004), (d) (Munoz-Darias *et al.*, 2005), (e) (Bhalerao, 2012), (f) (Rawls *et al.*, 2011), (g) (Steeghs and Jonker, 2007), (h) (Van Kerkwijk *et al.*, 2011), (i) (Romani *et al.*, 2012), (j) (Özel *et al.*, 2016), (k) (Martinez *et al.*, 2015), (l) (Van Leeuwen, 2015), (m) (Fonseca *et al.*, 2014), (n) (Weisberg *et al.*, 2010), (o) (Jacoby *et al.*, 2006), (p) (Kramer *et al.*, 2006), (q) (Ferdman *et al.*, 2014), (r) (Lynch *et al.*, 2012), (s) (Berezina *et al.*, 2017), (t) (Zhu *et al.*, 2015), (u) (Fonseca *et al.*, 2016), (v) (Desvignes *et al.*, 2016), (w) (Bhat *et al.*, 2008), (x) (Antoniadis *et al.*, 2012), (y) (Antoniadis *et al.*, 2013), (z) (Kaplan *et al.*, 2014), (A) (Stovall *et al.*, 2019), (B) (Deneva *et al.*, 2012), (C) (Antoniadis *et al.*, 2016), (D) (Reardon *et al.*, 2016), (E) (Ferdman *et al.*, 2010), (F) (Bassa *et al.*, 2006), (G) (Ransom *et al.*, 2014), (H) (Barr *et al.*, 2017), (I) (Arzoumanian *et al.*, 2018), (J) (Kiziltan *et al.*, 2013), (K) (Kasian, 2012), (L) (Freire *et al.*, 2008), (M) (Thorsett and Chakrabarty, 1999), (N) (Cromartie *et al.*, 2020), (O) (Nice, 2003), (P) (Deller *et al.*, 2012), (Q) (Freire *et al.*, 2011).

This e-capture SN is supposed to occur in a very degenerate O–Ne–Mg core in progenitors, quenching the development of an iron core, being triggered by a sudden capture of electrons onto Ne nuclei when a critical mass is reached (see Section 5).

Generally speaking, the mass distribution of observed NS masses (the likelihood in the Bayesian language) can be modelled as a Gaussian mixture with n components:

$$\mathcal{L}(m_p|\theta) = \sum_{i}^{n} r_i \, \mathcal{N}(m_p|\mu_i, \sigma_i), \tag{7}$$

where μ_i and σ_i are the mean and standard deviation of *i-th* normal component $\mathcal{N}$ and r_i is it relative weight, satisfying $\sum_i^n r_i = 1$ to ensure normalization.

Frequentist inferences for the mass distribution Although Bayesian statistics have gained strength over the years, frequentist non-parametric hypothesis tests are also commonly employed by astronomers (about 500 refereed papers each year in astronomical literature), to help exploring tentative conclusions about different phenomena and to better understand the data. Kolmogorov–Smirnov (K–S) and Anderson–Darling (A–D) are two well-known suitable examples of hypothesis tests.

Both tests are based on the maximum distance between *empirical distribution function* (EDF) of the sample and the *cumulative distribution function* (CDF) of a specified distribution. The K–S statistic is most sensitive to large-scale differences in location and shape, whilst A–D statistic is sensitive for both large and small-scale differences and also weights the tails of distributions, presenting a more robust result (Babu and Feigelson, 2006). A p-value is extracted from tests and the model is rejected or not related to a significance level α. By default $\alpha = 0.05$, so the hypothesis is rejected if $p < 0.05$. We applied both K–S and A–D tests for the sample of NSs shown in Fig. 1 to compare unimodal, bimodal and trimodal distribution models. For the model with one Gaussian we set $\mu = 1.48$ and $\sigma = 0.35$, while for the 2-component model $\mu_i = \{1.38, 1.84\}$, $\sigma_i = \{0.15, 0.35\}$ and $r_i = \{0.6, 0.4\}$, with $i = \{1, 2\}$ respectively. Lastly, for the 3-component model we assume $\mu_i = \{1.25, 1.40, 1.89\}$, $\sigma_i = \{0.09, 0.14, 0.30\}$ and $r_i = \{0.10, 0.5, 0.4\}$ for $i = 1, 2, 3$. Resultant p-values are shown in Table 1. The simplest (unimodal) model is ruled out in both K–S and A–D tests. Both 2 and 3-component models are not rejected, and presents a similar p-value. Figure 2(a) shows the EDF of sample in black, and the CDF of

Table 1. P-value of two hypothesis test for three different models.

Model	K-S	A-D	Conclusion
Unimodal	0.025	0.032	Reject
Bimodal	0.971	0.990	Do not reject
Trimodal	0.974	0.953	Do not reject

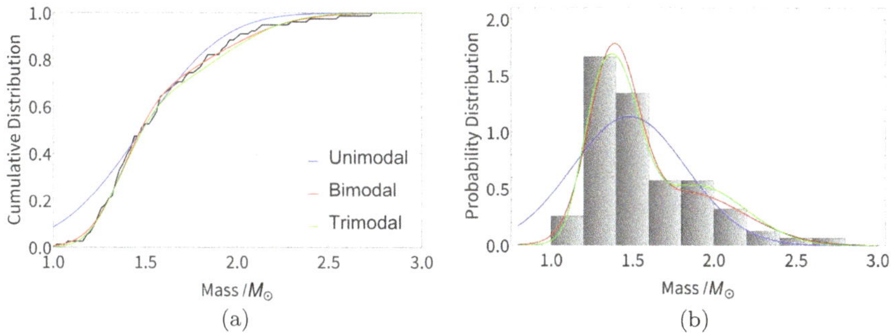

Fig. 2. (a) The black curve represents the EDF of sample, while the blue curve is the CDF of an unimodal distribution, the red curve is the CDF of a bimodal distribution and in green is the CDF of a trimodal distribution. As can be noted the blue curve provides a good fit to the mean position, but is quite bad in adjusting the tails. (b) The histogram of observed masses is shown, together with PDF of the 3 mixture models. Although the p-value of 2 and 3-component models are similar, the three peaks of the last one are not clearly identified, even if the binning is reduced.

three mentioned models used for tests calculations, showing that unimodal distribution (blue) fits well the position of mean but is bad in adjusting tails while the bimodal (red) and trimodal (green) curve provide better fits along all the sample distribution. Figure 2(b) shows the PDF of all models together with the sample histogram. As stated above, hypothesis tests help to gather a intuitive comprehension of the problem, but are sensitive to the parameter values and are not conclusive.

Bayesian Approach Bayesian statistics is based on *Bayes theorem* derived from axioms on conditional probabilities. In the case one is interested in the probability distribution of a parameter θ (or a set of parameters) that describes a model, based on observed data (D), the posterior distribution of it is set as:

$$P(\theta|D) = \frac{\mathcal{L}(D|\theta)P(\theta)}{P(D)}, \tag{8}$$

where $P(\theta)$ is called *prior distribution* and represents previous information about the parameter and $\mathcal{L}(D|\theta)$ is the *likelihood function* which describes the distribution of observed data, conditioned by the parameters. Depending on the adopted family distribution and on the dimension of parameter space, the main difficulty of calculating posterior distributions is in the normalization term $P(D) = \int P(\theta)P(D|\theta)d\theta$. MCMC methods provide an efficient way for sampling a given distribution, bypassing the problem of accounting for $P(D)$. We recommend (Sharma, 2017) and references therein for those who want to familiarize with MCMC algorithms. An extensive list of algorithms to perform MCMC sampling is available today, so the better choice for a given problem will depend on the specific purposes to be attained.

The features and shape of the distribution obtained from sampling methods helps to better comprehend the formation and evolution of NSs in different systems. The old model of a single-mass scale around $1.35-1.4\,M_{\odot}$ is expressed as unimodal, so $n = 1$ in order to represent a unique mood for NS formation. However, today is known that $m_p \geq 2.0\,M_{\odot}$ exist, therefore the mass distribution needs to be at least bimodal ($n = 2$) to accommodate these heavy objects, with the lower peak accounting for stars formed from the "normal" Fe core-collapse supernova and the higher peak ones may containg "directly formed" NSs, AIC remnants and NSs with a substantial accretion history. In order to analyze the existence of a different supernova mechanism forming (electron-capture) lighter NSs, as mentioned before, an extra component must be introduced in the model ($n \geq 3$).

Model comparison to determine which one is favourable respect to data can be done by many different techniques, depending on aspects of the model and chosen algorithm. Two of the most famous techniques are the Bayesian Information Criterion (BIC) and the Akaike Information Criterion, although some statisticians do not recommend them. Both methods are based on the calculation of posterior likelihood, penalizing the model due to the number of parameters and sample size (for BIC).

We have performed a Bayesian analysis to the sample shown in Fig. 1, assuming that the maximum mass allowed for NSs $m_{\max}$ to be located at 3σ to the right of the highest-value peak (referring to heavier objects). As we are dealing with Gaussian distributions, values within three standard deviations (to the left and right) account for $99,73\%$ of the distribution, so there is a very small probability of finding a NS with a mass higher than that.

The unimodal model, were all objects are assumed to belong to the same distribution peak with a likelihood given by:

$$\mathcal{L}(m_p|\theta) = \mathcal{N}(m_p|\mu, \sigma), \tag{9}$$

results unlikely, in agreement with previous works mentioned here and the frequentist result shown before.

Bimodal model results are the favoured one in our analysis. The likelihood of this model is:

$$\mathcal{L}(m_p|\mu_i, \sigma_i, r_i) = r_1\mathcal{N}(m_p|\mu_1, \sigma_1) + r_2\mathcal{N}(m_p|\mu_2, \sigma_2). \tag{10}$$

The prior distributions were assumed as a $Beta(2, 2)$ for the relative weights, $r = \{r_1, r_2\}$, a Gaussian $\mathcal{N}(0, 2)$ for $\sigma = \{\sigma_1, \sigma_2\}$, and a Gaussian $\mathcal{N}(1.45, 0.30)$ to $\mu = \{\mu_1, \mu_2\}$.

When trying to account for the peak of NSs formed by the electron-capture supernovae, in addition to the "commom" ones and the accretion history leading to heavy objects, with a likelihood:

$$\begin{aligned}\mathcal{L}(m_p|\mu_i, \sigma_i, r_i) = r_1\mathcal{N}(m_p|\mu_1, \sigma_1) + r_2\mathcal{N}(m_p|\mu_2, \sigma_2)\\ + r_3\mathcal{N}(m_p|\mu_3, \sigma_3),\end{aligned} \tag{11}$$

we found that we could not distinguish the peak around $1.25\,M_\odot$. Despite the fact that the progenitors of these objects are expected to be abundant in the Universe (as will be seen in Section 5), this result can be interpreted as if only a tiny fraction of observed NSs arise from these events (although the reason would be unknown), but it is not unlikely that improved statistical schemes can separate this peak.

In Bayesian analysis parameters are random variables, and not fixed values as in frequentist analysis, so they are expressed in terms of marginalized posterior distributions, which we summarized in Table 2. The mean value for each parameter is in the second column, followed by the corresponding standard deviation in the third column. Fourth and fifth columns presents the Highest Posterior Density from which we conclude that, for example, $1.305 < \mu_1 < 1.398$ with a 94% credible interval.

Although in this case $m_{\max}$ is a generated quantity and not a free parameter of the model, it depends on the values of μ_2 and σ_2 so it is also a random variable given by a distribution shown in Fig. 3. As a summary we have $m_{\max} = 2.616 \pm 0.196$ for the mean and standard deviation, with a 94%

Table 2. Parameter estimation of bimodal model trough MCMC methods in the Bayesian framework. The second column shows the mean value of each parameter, followed by the corresponding standard deviation. Fourth and fifth columns present the highest posterior density.

Parameters	Mean	sd	HPD (3%)	HPD (97%)
r_1	0.588	0.105	0.382	0.775
r_2	0.412	0.105	0.225	0.618
μ_1	1.351	0.025	1.305	1.398
μ_2	1.756	0.098	1.581	1.944
σ_1	0.087	0.023	0.047	0.133
σ_2	0.286	0.069	0.154	0.414

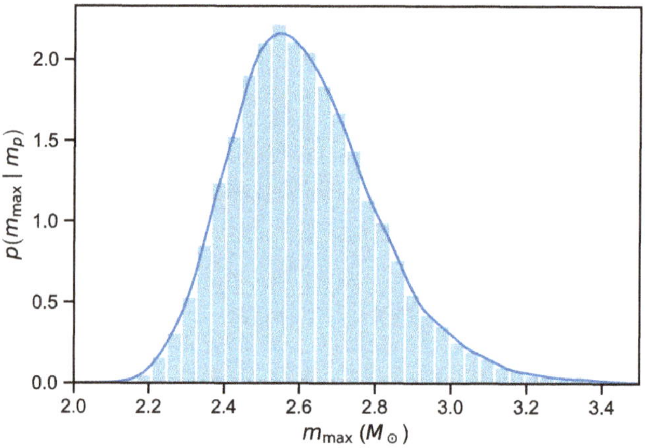

Fig. 3. Maximum mass distribution in the case where it is assumed to be located at $3\sigma_2$ to the right of μ_2.

credible interval in $\{2.273, 2.985\}$, a skewed distribution. This will be of relevance to understand the meaning of several high masses (and particularly the lighter one of the event GW190408) as will be seen below.

Recently, a Bayesian analysis (for a sample of 74 NSs) using Gaussian mixtures models with $n = \{1, \ldots, 4\}$ components and an additional investigation on $m_{\max}$ was presented (Alsing *et al.*, 2018). For each n two routines were implemented, one where the maximum mass is fixed at $m_{\max} = 2.9\,M_\odot$, and other where it is set as a free parameter bounded by $m_{\max} < 2.9\,M_\odot$. They found that $n = 2$ and $m_{\max} < 2.9\,M_\odot$ were favoured among all models. The posterior result also shows a sharp

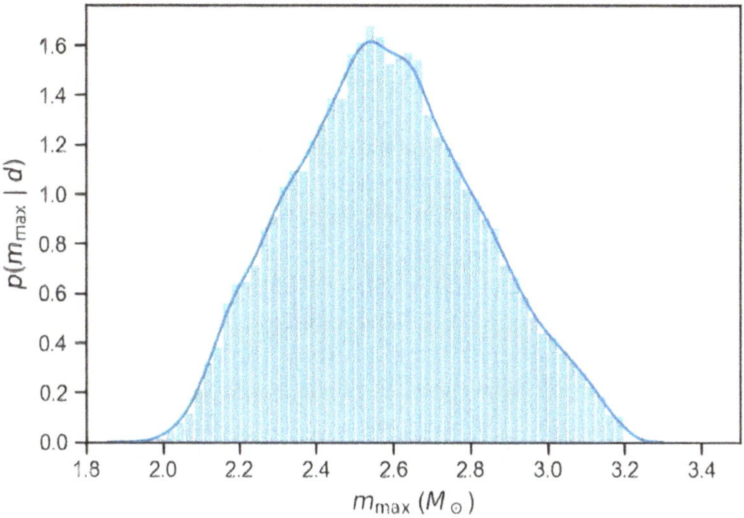

Fig. 4. Marginalized distribution of maximum mass as a free parameter of a truncated model.

cut-off on maximum mass distribution, with a peak in $2.12\,M_{\odot}$ as shown in Fig. 3 of their work that shows the marginalized posterior distribution of the maximum mass, with a 90% credible interval in $2.0 < m_{\max} < 2.6\,M_{\odot}$. Although the distribution reaches values $\sim 2.6\,M_{\odot}$, they have a low probability, so this result is disfavoured when considering a scenario in which heavy NSs are expected to exist.

We have conducted our own calculations leaving $m_{\max}$ as a free parameter to be determined (and penalized) by the Bayesian algorithm with the result given in Fig. 4. Even though there is a wide distribution of possible $m_{\max}$ the peak is precisely around the same value $m \sim 2.5\,M_{\odot}$, giving support to the idea that the "light" component of GW190408 may be a NS near the TOV limit. Studies along these lines are guaranteed to establish this important issue.

3. Maximum Mass of (Non-rotating, Isotropic) Neutron Stars

As final states of massive star evolution (and possibly additional formation channels as well), NSs are relativistic objects and as such, the pressure term becomes important for their structure as a source of gravity. Theory predicts the existence of a maximum mass for NSs resulting from the TOV equation, the General Relativistic version of the hydrostatic equilibrium discussed in Section 3.1. The exact value of $m_{\rm TOV}$ depends on the EoS, in any case, but the EoS itself is difficult to pin down and remains uncertain. Many of the EoS reflect mostly different compositions, from pure neutrons to baryon rich matter to even exotic matter composed by a soup of u, d and s asymptotically free quarks (Heiselberg and Pandharipande, 2000; Lattimer, 2012; Özel and Freire, 2016). The EoS of dense matter, i.e., the relation between pressure and (energy) density, is informative of the composition and the inner structure of a NS (Glendenning, 2012; Özel and Freire, 2016; Weber, 2017). Observational parameters like the mass or the radius have an intrinsic correlation with the EoS, albeit a complex one.

Irrespective of any experimental data, Rhoades and Ruffini (1974) derived a theoretical limit of 3.2 $M_\odot$ for the maximum mass of any compact object independent of the exact nature of the EoS. This is a result based on general relativistic arguments, together with Le Chatelier's principle. We discuss this limit in Section 3.2. Today, we know that the Rhoades and Ruffini limit is *not* an absolute mass limit in the general situation (for example, anisotropic models can violate the Rhoades–Ruffini limit), but it is still considered a milestone of the knowledge of the dividing line between NSs and black holes. As mentioned before, the observation of a very massive NS poses stringent constraints on the EoS of these objects, and also sets the lower limit for black holes masses.

3.1. *Tolman–Oppenheimer–Volkoff equations for relativistic stellar structure*

Oppenheimer and Volkoff (1939), based on the previous work on spherically symmetric metrics in GR by Tolman (1934), published a seminal paper where they derived the relativistic equations for hydrostatic equilibrium. This equation taken together with the continuity of the mass, written below, completely determine the structure of spherically symmetric, static NSs:

$$\frac{dP}{dr} = -\frac{G\rho m}{r^2}\left(1+\frac{P}{\rho c^2}\right)\left(1+\frac{4\pi P r^3}{mc^2}\right)\left(1-\frac{2Gm}{rc^2}\right)^{-1}, \tag{12}$$

where $m(r)$ is defined by:

$$\frac{dm}{dr} = 4\pi r^2\rho. \tag{13}$$

Solving the TOV equations for a compact star implies integrating the equations, usually from the centre ($r = 0$) to the surface ($r = R$) where the pressure vanishes, with the aid of an EoS, i.e., a function relating pressure and density. Besides, for solving the TOVs is necessary two boundary conditions: $m(0) = 0$ and $\rho(0) = \rho_c$. However, this procedure implies some assumptions throughout the calculations; inside the star, we must have (Chamel *et al.*, 2013):

(1) $\rho \geq 0$: the mass density should be positive;
(2) $dP/d\rho \geq 0$: the NS matter must remain in an equilibrium state (Le Chatelier's principle);
(3) $P \geq 0$: the pressure should be positive;
(4) $dP/dr \leq c^2$: the sound speed should not exceed c.

Conditions 1–4 imply that the pressure decreases from the centre to the star's surface, guaranteeing the integration of the TOVs given a value for the central pressure through the EoS. A general characteristic of these equations is that they naturally produce a maximum mass for the compact star. This feature is a purely general relativistic effect due to the denominator factor $\left(1 - 2Gm/rc^2\right)$ of the pressure gradient, and the presence of the pressure added to the energy density in the numerator of Eq. (12).

The nature of the resulting objects depends upon the exact properties of the EoS. Oppenheimer and Volkoff, for example, used an EoS for degenerate Fermi gas of neutrons, obtaining a maximum mass of only $0.7\,M_{\odot}$.

At the time of these writings the catalog of NS measured masses fills almost one hundred entries, and their distribution points to masses ranging from 1.1 to $2.7\,M_{\odot}$ as shown in Fig. 1. The higher mass confidently detected is $m = 2.14^{+0.10}_{-0.09}\,M_{\odot}$ for the millisecond pulsar MSP J0740+6620 (Cromartie *et al.*, 2020).

Theoretically, we learned a lot about the properties of matter at high and supranuclear densities, and now we know that EoS must allow masses above $2.14\,M_{\odot}$, and that other measurements and arguments may push this limit above $2.5\,M_{\odot}$, surprisingly closer to the original Rhoades–Ruffini value to be addressed below.

3.2. *The Rhoades–Ruffini limit of a neutron star mass*

Establishing the maximum mass of NSs is vital not only to understand the true nature of matter inside it but also to set the limit from where the formation of a black hole is unavoidable. As stated above, the exact nature of matter at supra-nuclear densities is a long-posed problem for which a solution is still nowadays elusive, this is why the availability of an "absolute" value for the limit mass is so important.

In fact, as Rhoades and Ruffini (1974) argued in 1974, it is not really necessary to know the exact EoS to obtain a valid theoretical limit. This theoretical limit on the mass comes from well-established physical principles that all the matter should obey, which are:

(1) The compact star must obey the General Relativistic equation for hydrostatic equilibrium;
(2) Matter must obey the Le Chatelier Principle, stating that a disturbance in a system in equilibrium will be opposed to restoring the equilibrium;
(3) Matter must obey the Principle of Causality, which implies that the sound speed must remain lower than the speed of light in the medium.

To comply with these three requirements, Rhoades and Ruffini (1974) assumed that the EoS is uncertain above a fiducial density of $\rho_\star = 4.6 \times 10^{14}\,\mathrm{g/cm^3}$, the region in which they assumed that the EoS is the stiffest possible, producing a sound velocity equal to the velocity of light. Thus, they found a maximum possible mass for a NS of about $3.2\,M_\odot$.

The Rhoades–Ruffini mass limit can be expressed as (Chamel *et al.*, 2013):

$$m_{RR} \simeq 3.0\Big(\frac{5 \times 10^{14}}{\rho_\star}\Big)^{1/2}[M_\odot]. \tag{14}$$

As a conclusion, all compact objects whose mass is measured to be higher than $3.2\,M_\odot$ must be a black holes provided it is non-rotating, isotropic and GR holds.

As a further development, allowing the violation of the causality, using an incompressible fluid, the mass limit can be expressed as (Chamel *et al.*, 2013):

$$m_{RR,non-causal} \simeq 5.09\Big(\frac{5 \times 10^{14}}{\rho_\star}\Big)^{1/2}[M_\odot]. \tag{15}$$

In Eqs. (14) and (15), $\rho_\star$ is the value assumed to be the fiducial one.

It is worth stressing again that the Rhoades–Ruffini limit does not take into account other effects that, as we will see below, allows matter inside the NS to support masses higher than $3.2\,M_\odot$. Examples of these effects are rotation and anisotropies in the fluid's pressure, discussed in Section 3.2.1.

3.2.1. *Types of neutron star models*

The several decades of work showed that, although the Rhoades–Ruffini limit meaning remains strictly unchallenged, realistic EoS hardly achieve the limit of causality. However, quite recently actual data is "pushing" the actual maximum mass towards the Rhoades–Ruffini extreme.

On the one hand, the maximum mass exact value is determined by the EoS. The better knowledge we now have about the behaviour of matter above supranuclear densities allows us to infer different compositions for the inner core of neutrons stars, and each one produces a unique value for the maximum mass.

A NS is usually depicted as an onion-like object whose structure is something like: a fragile gaseous plasma atmosphere $\sim$ centimeters thick; a $\sim$ few-metre deep liquid plasma "ocean"; a few hundred meters thick atomic *outer crust* followed by a 1–2 kilometers crystal lattice of neutrons and protons immersed in an electron gas and neutron-rich liquid *inner crust*. All these layers altogether, however, add up from 0.01 to 0.1 of the star's total mass, and their composition is relatively well-known from nuclear physics (Chamel *et al.*, 2013; Haensel *et al.*, 2007).

The EoS of the crust up to densities of $\sim 2\rho_{\rm sat}$ (where $\rho_{\rm sat} = 2.8 \times 10^{14}\,{\rm g/cm^3}$ is the nuclear saturation density) is relatively well-known, and common EoS used in calculations include the Baym–Pethick–Sutherland (Baym *et al.*, 1971) EoS and the HP (Haensel and Pichon, 1994).

Things begin to get complicate at the core of the star, divided into two regions: the outer core, with densities ranging from $\rho_{\rm sat}/2$ to $2\rho_{\rm sat}$, and the inner core with densities ranging from $2\rho_{\rm sat}$ to around $10\rho_{\rm sat}$ (Chamel *et al.*, 2013; Haensel *et al.*, 2007). Neutrons (mostly), protons, electrons, and muons compose the NS's outer core. The composition of the inner core is, actually, the long-posed mystery about NSs. The inner core is also the dominant structure for the NS's mass budget. Depending on the approach to the behaviour of matter at densities higher than $\sim 5 \times 10^{14}{\rm g/cm^3}$, in the inner core region, one builds a sequence of stars with a unique value of the maximum mass.

Figure 5 shows examples of several sequences of stars, each built by a specific EoS. Each EoS produces a unique value of maximum mass. In this figure we also show, represented by the magenta horizontal stripes, some observed values for the mass of the most massive NSs ever detected, and a candidate to the uttermost massive NS: PSR J1614-2230 ($m = 1.928 \pm 0.016\, M_\odot$ (Fonseca *et al.*, 2016)), PSR J0740+6620

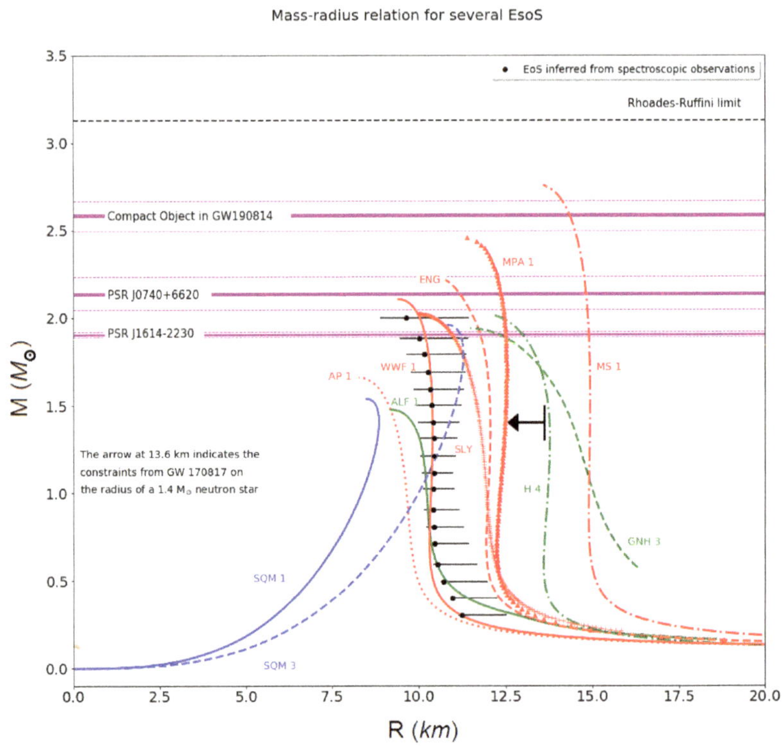

Fig. 5. Mass–radius relation for several equations of state (EoS), named in the figure, illustrating the "families" of EoSs. The hadronic EoSs are the AP 1, ENG, MS 1, MPA 1, SLY and WWF1; the hybrid ones are the ALF 1, GNH 3 and H4; the ones with strange quark matter are the SQM 1 and SQM 3. The black dots represent an EoS inferred from spectroscopic observations. The data for this $M - R$ diagram and the observational dots were taken from The University of Arizona (2020). (Earlier compilations and naming conventions are from Lattimer and Prakash (2001) and Read *et al.* (2009). The full list included above is from Özel and Freire (2016)). In this figure, we also plot: PSR J1614-2230 ($m = 1.928 \pm 0.016\,M_{\odot}$), PSR J0740+6620 ($m = 2.14^{+0.10}_{-0.09}\,M_{\odot}$), and the compact object in GW190814 ($m = 2.59^{+0.08}_{-0.09}\,M_{\odot}$). The sequences were calculated from the TOV, i.e., they do not include rotation.

($m = 2.14^{+0.10}_{-0.09}\,M_{\odot}$ (Cromartie *et al.*, 2020)), and the compact object in GW190814 ($m = 2.59^{+0.08}_{-0.09}\,M_{\odot}$ (Abbott *et al.*, 2020b)).

That said, there are "types" of EoS to describe the inner core (see Chamel *et al.* (2013)). The first of these types is the hadronic one, where matter is assumed as a liquid of nucleons only, i.e., protons and neutrons, the whole structure being neutralized by a lepton gas such electrons and, at very high densities, muons. This nuclear matter has been studied since the '50s, as the many approaches in many-body calculations were becoming available. Also, heavy-ion experiments are used to gauge the models, making the EoSs more realistic. Some representatives of this family are the following equations: AP 1 (Akmal and Pandharipande, 1997), ENG (Engvik *et al.*, 1994), MPA 1 (Müther *et al.*, 1987), MS 1 (Müller and Serot, 1996), SLY (Douchin and Haensel, 2001), and WFF 1 (Douchin and Haensel, 2001).

Although such calculations produce similar results up to $5 \times 10^{14}\,\mathrm{g/cm^3}$, as the density becomes high the predictions open a fan of different results, producing maximum masses from 1.8 to 2.5 $M_\odot$, the first case revealing that the EoSs are too soft for supporting recent determinations of NS masses.

In a sense, the recent detection of the very massive MPSs, or the successive LIGO-VIRGO reports of the GW170817 coalescence of two compact objects into an object with $\geq 2.4\,M_\odot$ rule out soft hadronic EoSs. On the theoretical side, it has been known that the possible appearance of kaon and pion condensates at high densities implies that the EoS should be even softer, and some exotic mechanisms have been suggested to circumvent this softening to match the observations. Some kind of repulsion must be introduced to match high masses, otherwise the sequences fall short to produce room for $\geq 2\,M_\odot$ stars.

The second type of model comes from the formation of *hyperons* or another species that will populate the core of the star at the high-density inner core. Brueckner–Hartree-Fock (BHF) calculations using realistic two- and three-body forces suggest to be unavoidable the appearance of hyperons, and this makes the EoS containing them extremely soft, producing a maximum mass of at most $1.4\,M_\odot$. This situation has been called the *Hyperon Puzzle* (Horvath and De Souza, 2017). However, it is still possible to circumvent the hyperon puzzle at densities above $\sim 5 \times 10^{14}\,\mathrm{g/cm^3}$ with the relativistic mean-field calculation approach by "tuning" hyperonic coupling interactions. Some representatives of this family are the following EoS: ALF 1 (Alford *et al.*, 2005), GNH 3 (Glendenning, 1985), and H4 (Lackey *et al.*, 2006).

A third type comes with the deconfinement of nucleons into quarks. By this, we mean a soup of u and d quarks in asymptotic freedom that can produce NSs with masses in agreement with observations. However, this agreement is only achievable if the EoS for the quark matter has a stiffness so that the sound velocity is close to the velocity of light (however, never achieving the Rhoades–Ruffini mass limit discussed before). If quarks appear at high density, at some critical density, we may have quark cores for the so-called "hybrid models".

However, there is an extreme version of this deconfinement scenario, the *strange star* picture. This hypothesis of the *absolute stability* of a quark matter with strangeness originated in 1971 with Bodmer (1971) and his speculation that the true ground state of matter was not the one we know from ordinary physics. Then, Witten (1984) calculated that this ground state would be a mixture of u, d and s quarks attained at supranuclear densities $\rho \sim 10^{15}\,\mathrm{g/cm^3}$.

If deconfinement is achieved in compact objects, then NSs would be almost certainly "strange stars" (so named because of the almost equal proportion between the upper, down, and strange quarks). Strange stars show significant differences with normal neutrons stars: to begin with, the absence of an onion-like structure. Another critical difference is that the surface density is as huge as $\sim 10^{15}\,\mathrm{g/cm^3}$, which is not that different from the centre's density. It is not that difficult to produce maximum masses larger than $2\,M_\odot$ (Horvath, 2007), around the values recently observed, provided the interactions represented in the vacuum energy and/or related quantities have suitable values, contrary to the earliest views about the softness of a relativistic quasi-massless gas of quarks (see, for instance, Franzon *et al.* (2012)).

Some representatives of the third type, strange stars, are the following EoSs: SQM 1 and SQM 3 (Prakash *et al.*, 1995). See also (Bodmer, 1971), (Witten, 1984) and (Alcock *et al.*, 1968). A complete survey of SQM EoS previous to pairing and other issues is given in (Benvenuto and Horvath, 1989).

In the end, the quest for the exact EoS of NSs is still far from being closed, but recent observations and measurements of NSs masses seem to push the theory to stiff EoS. To view the heat of the debate, Shibata *et al.* (2019) constrained the maximum mass of a NS to less than 2.3 $M_\odot$ using the detection of the event GW 170817. Soon after the paper of Wu *et al.* (2020) showed that NSs could have masses greater than $2.3\,M_\odot$ if the EoS could become stiffer in some regime. To study the stiffness of the EoS, the authors employed polytropic models dependent upon two free parameters in two scenarios, with a similar spirit than the Rhoades–Ruffini approach. For hadronic NSs, there would be a transition density above which the stiffening occurs. In this case, the transition would occur at $\rho_t/\rho_{\mathrm{sat}} = 0.5$ if the polytropic index is $\gamma = 2.65$. In the scenario of strange stars, the compact object would support $m > 2.3\,M_\odot$ if the two free parameters are in the range $(\rho_s/\rho_{\mathrm{sat}}; \gamma) = (1.0 - 1.58; 1.40 - 2.0)$, where ρ_s is the surface density of these self-bound systems.

Still, using polytropic EoS, O'Boyle *et al.* (2020) presented a generalized piecewise polytropic parametrization for NS. The innovation of their method was to impose, despite the continuity in pressure and energy density, the continuity in the sound speed in the fluid. This generalized parametrization has advantages over the old one in hydrodynamic simulations of NS mergers since the tidal deformability is sensitive to the sound speed c_s, which is discontinuous in old approaches. The tidal deformability, in turn, strongly influences the gravitational waves (GW) produced at merger events.

Application of the generalized piecewise polytropic parametrization goes in the sense of recover the correct EoS parameters from observables in a secure way, providing strong constraints on the maximum mass of NSs, shedding light on the fundamental physics of black holes formation and the nature of dense matter.

Anisotropies in the stellar fluid are also an exciting approach for calculating the stellar structure and have been widely used to include a new degree of freedom, especially when dealing with exact solutions of the Einstein Field Equations (de Avellar and Horvath, 2010; Rocha *et al.*, 2020; Sharma and Maharaj, 2007). The extra degree of freedom allows one to impose an analytical EoS, as the MIT Bag Model, and an *ansatz* for a functional form of the metric elements.

Anisotropies inside compact stars can be significant for high densities, e.g., $\geq 10^{15}$ (Ruderman, 1972). One consequence is that they allow a higher central density, which augments the pressure support against gravitational collapse, leading the star to higher masses. Using the simple MIT Bag Model (Alcock *et al.*, 1968; Witten, 1984) and Sharma *ansatz* (Sharma and Maharaj, 2007), (de Avellar and Horvath, 2010) reached a maximum mass compatible with the observations of PSR J0740+6460. Rocha *et al.* (2020), on the other hand, reached maximum masses ranging from 3 to 5 $M_{\odot}$ using the Thirukkanesh-Ragel-Malaver *ansatz* (Malaver, 2014; Thirukkanesh and Ragel, 2012) and an exact form for a CFL-type EoS (Alford *et al.*, 2005; Lugones and Horvath, 2002, 2003; Rajagopal and Wilczek, 2001).

On the other hand, studies on (quasi-)universal relations (involving, for example, the moment of inertia or the quadrupole moment, among others, known as "I-Love-Q"), and the role of rotation allowed for making new and improved predictions, i.e., place new constraints almost independently of the EoS.

Recently, Breu and Rezzolla (2016) derived, from universal relations, that uniform rotation would increase the maximum mass up to 20% for all the EoS they considered: they found $m_{\max} = (1.203 \pm 0.022)m_{TOV}$, where m_{TOV} is the maximum mass from solving the non-rotating TOV equations. Applying this result on the maximum mass resulting from some EoSs, one can get a value compatible with recent observations such as GW190814.

Finally, we point out that compact objects are also studied within so-called *braneworld* models with extra dimensions (Lugones and Arbañil, 2017). By solving the stellar structure equation in the braneworld and using two types of EoSs (a hadronic one and a strange quark matter one), the authors found that a new branch of stellar configurations arises for masses above two in which the causal limit can be violated. Not only this implies that there is no maximum mass in a braneworld NS but also that these very massive stars are stable against (small) radial perturbations, being supported by non-local effects of the bulk on the brane.

Thus, as we see from Fig. 5, although some EoS can support masses compatible with recent observations, the detection of the compact object in GW190814, whose mass is $\sim 2.5\,M_\odot$, poses a new challenge to the theory of dense matter provided it is indeed a NS. Not even rotation would allow a straightforward match. Is the compact object in GW190814 the lower limit for a black hole?

The general conclusion of this discussion is that there is no "absoluteness" in the Rhoades–Ruffini maximum mass limit (see, for example, Fig. 6). It can be violated in some circumstances, and the maximum mass could be higher than $3.2\,M_\odot$ within these models.

3.3. *Redback/black widow binary systems and the maximum mass/lightest black hole problem*

As explained in the Section 2, from the first determinations to the beginning of the 21st century, the idea that NS masses cluster around $1.4\,M_\odot$, called the "canonical mass", was settled in favour of a wider distribution. As newer accurate data from double NS systems and other binaries became available, it was revealed that NSs with masses as high as $2.14\,M_\odot$ exist, suggesting that, at the very least, another higher mass scale should be considered.

"Spider" systems — black widow and redbacks — are a class of relativistic binary systems in which the NS strongly interacts with its ordinary companion stars (Fruchter *et al.*, 1988; Hui and Li, 2019; Roberts, 2012). This interaction is possibly different from what we see in other binary systems, such as low-mass X-ray binaries, since two new ingredients come into

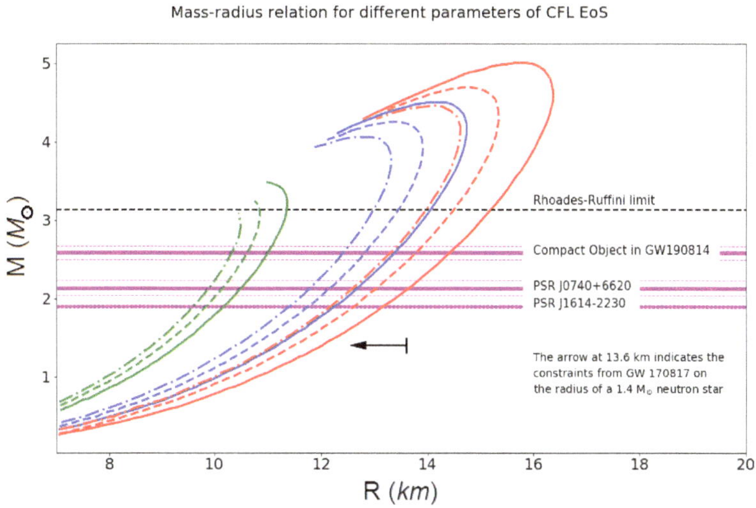

Fig. 6. Mass–radius relation for different parameters of the CFL equation-of-state, using Thirukkanesh-Ragel-Malaver *ansatz* for the metric element $e^{-2\lambda(r)}$. For solid lines, we assume $\Delta = 100\,MeV$ and $m_s = 150\,MeV$, while for the dashed lines, we assume $\Delta = m_s = 0\,MeV$ (resembling the MIT bag model), and for the dash-dotted lines $\Delta = 150\,MeV$ and $m_s = 150\,MeV$. Also, for the red lines we set $B = 57.5\,MeV/\mathrm{fm}^3$, for the blue lines $B = 70\,MeV/\mathrm{fm}^3$, and for the green lines $B = 115\,MeV/\mathrm{fm}^3$. Figure adapted from (Rocha *et al.*, 2020).

play: the back illumination onto the donor and the ablation of the donor by the pulsar wind in the later stages.

The evolutionary history of these systems suggests a connection between redbacks and black widow through accretion, going back until the formation of the NS itself (Benvenuto *et al.*, 2012, 2014). The systems at formation time should have initial orbital periods of about 1 day. On a nuclear timescale of the secondary, the accretion sets in, but at a highly variable rate, due to a variety of effects. The authors in (Benvenuto *et al.*, 2012) present a complete model calculation of this scenario, showing the oscillations in the rate of accretion due to the so-called *quasi-Roche Lobe overflow* state.

A critical observation for the scenario mentioned above, of relevance to our discussion here, is that it takes a long time until the donor star enters the degenerate regime, allowing the NS to accrete a substantial fraction of a solar mass. Therefore, if the efficiency of the accretion is not too low, the scenario may produce very massive NSs. In Fig. 7, we show the measured masses of 17 redbacks/black widow systems.

Because of this feature, we may expect that the spider systems can shed light on the value of the maximum mass of NSs, as soon as new observations become available. If, contrary to widespread beliefs, nature produces very massive objects they will be a challenge to the physics of dense matter and to the actual mass difference between the $m_{\max}$ and the Rhoades–Ruffini limit itself. Several NSs in spiders shown in Fig. 7 have been reported to be above the $2\,M_{\odot}$ mark.

Within this evolutionary scenario, Redback/black widow NSs may harbor the highest masses in nature, and can be an alternative channel to form low-mass BHs by pushing some NS masses over the TOV limit. This new

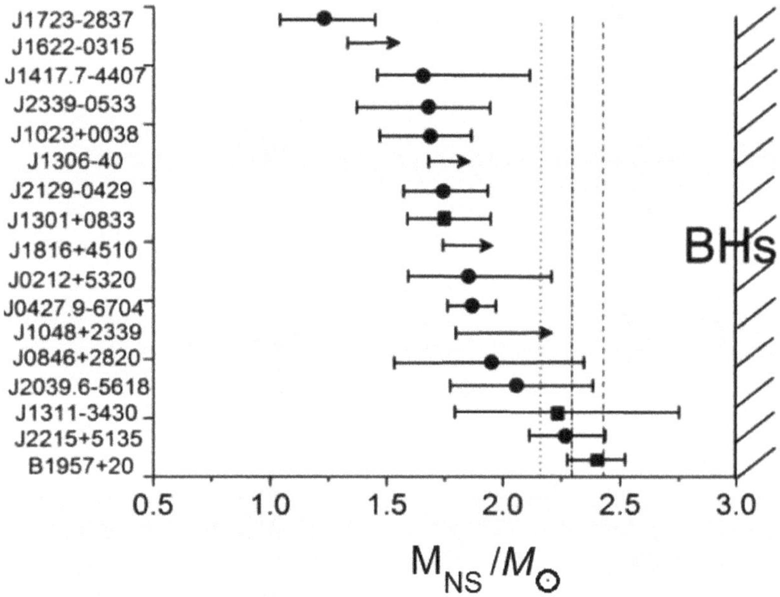

Fig. 7. The masses of Redback/black widow NSs. The circles denote a Redback system and the squares a Black Widow one. Lower limits on the masses of J1622-0315, J1306-40, J1816+4510 and J1048+2339 are indicated with the arrows. The dotted vertical line is the maximum mass derived by Margalit and Metzger (2017), the dashed-dotted line the upper limit of the range quoted by Ruiz *et al.* (2018) and the dashed line the value of Ai *et al.* (2020) for the SMNS formation case in the GW170817 merger. It is apparent that the lower numbers begin to collide with direct observations unless significant systematic errors are present. Other inferences are available and do better in this sense (see Table 3), allowing higher maximum masses (full data of the direct observations and references can be found in Özel and Freire (2016) and Linares (2019)). The hatched region on the right marks the region of BHs, although strictly speaking all object pushed above will be BHs before the vertical line. Figure from Horvath *et al.* (2020).

channel is not subject to the constraints of gravitational collapse involved in the existence of the mass gap between high-mass NSs and low-mass BHs. However, the number of low-mass BHs formed in spider systems must be small due to the restrictive conditions, as suggested by Benvenuto *et al.* (2012, 2014). The detection strategy must rely on the existence of accretion in binary systems where a compact object is present and the orbital periods are of the order of 1 day or less. Of course, typical requisites of redback and black widow donors must also be fulfilled. An interesting candidate has been studied by Tetarenko *et al.* (2016). The binary system VLA J2130+12 is at a distance of $2.2^{+0.5}_{-0.3}$ kpc. Its compact object is a stellar black hole whose mass is *assumed* to be $10\,M_{\odot}$ in the analysis. Other features of this system are a 0.1–$0.2\,M_{\odot}$ companion star and an orbital period of about 1–2 hours. We suggest that it may be still possible to identify the compact object with a black hole with a smaller mass descendant of NS pushed over the TOV limit if one does not assume anything about the mass *a priori*.

There are other methods to detect the existence of low-mass BHs other than by the spider channel. For instance, the LIGO event detection S200316bj was reported to have a probability of 0.9957 in favour of a black hole with 3–$5\,M_{\odot}$ (see LIGO/VIRGO (2020)). Other channels cannot be discarded yet, and new methods such the eLISA, when it becomes operational, will improve our knowledge of the many channels, finally exploring the actual structure of the mass gap between NSs and black holes.

3.4. *GW detection events and the possibility of NS masses above* $\sim$**2.5** $M_{\odot}$

The gravitational wave detection events have opened a window of opportunity for studying the physics of dense matter. Initially, the LIGO/VIRGO experiments detected GW from a BH merger, one with $\sim 36\,\mathrm{M}_{\odot}$ and other with $\sim 29\,\mathrm{M}_{\odot}$ resulting in a BH with $\sim 62\,\mathrm{M}_{\odot}$.

The first detected NS–NS merger event was GW170817, with individual masses between 1.17–1.60 $M_{\odot}$ and with a total mass of $2.74^{+0.04}_{-0.01}\,\mathrm{M}_{\odot}$ (see some references in Table 3). GW170817 was the first GW event confirmed by non-gravitational counterparts, proving that it was the result of the collision of two NSs.

Later works on GW170817 inferred values for the maximum mass of the resulting NS, using different assumptions. As we see in Table 3, the mass inferences are dangerously close to the Rhoades–Ruffini mass limit,

Table 3. Inferred value of M_{TOV} from the behaviour of GW170817 according to (a) (Shao *et al.*, 2020); (b) (Ai *et al.*, 2020); (c) (Shibata *et al.*, 2019); (d) (Ruiz *et al.*, 2018); (e) (Rezzolla *et al.*, 2018); (f) (Margalit and Metzger, 2017); (g) (Alsing *et al.*, 2018).

Value [$M_{\odot}$]	Type of the resulting object/ assumption
$2.13\pm^{0.08}_{0.07}$ (a)	SMNS
$2.09\pm^{0.11}_{0.09}$ (b)	SMNS
$2.43\pm^{0.10}_{0.08}$ (b)	SMNS
$2.43\pm^{0.10}_{0.08}$ (b)	SNS
2.3 (c)	Conserv. + simulation
2.16 – 2.28 (d)	Minimal assumption
$2.16\pm^{0.17}_{0.15}$ (e)	—
2.17 (f)	—
2.6 (g)	(3σ)

revealing that not only the EoS of NSs must be stiff, but also that other effects can be at play, like rotation and anisotropies.

One final remark is motivated by the recent release of a complete analysis of the GW population (Abbott *et al.*, 2020c), which found a "gap" between the lightest GW190408 component at $\sim 2.6\,M_{\odot}$ and the second lightest object with $\sim 6\,M_{\odot}$, surely a black hole. If true this analysis reinforces the idea of a "mass gap", but also suggests that the controversial $\sim 2.6\,M_{\odot}$ should be a NS. When the number of GW mergers increases in the next years this "gap" will be explored thoroughly. This is a clear example of the interplay between GW measurements and local NS Astrophysics previously stated.

In the end, it seems likely that the solution for the long-standing quest for the correct EoS of NSs and, consequently, for the maximum mass of these unusual compact objects, will be settled by Gravitational Wave Astronomy, explored in Section 4.

4. Gravitational Waves from Merging NS

The discovery of the system PSR B1913+16 by Hulse and Taylor in 1974 exposed the first detected pulsar in a binary system around another NS. The system has its orbit shrinking in precise agreement with Einstein General Theory of Relativity. The shrinkage of the orbit is due to the emission of GW. This was the first indirect measurement of GW, granting Hulse and Taylor the Nobel Prize of Physics in 1993 for that contribution.

Gravitational Waves are perturbations on spacetime curvature, resulting from accelerated masses that propagate at the speed of light outwards from the source, decaying with the inverse of the distance. After decades of technological developments and tests, the first *direct* detection of GW was made in 2015 by the LIGO/VIRGO collaboration (Abbott *et al.*, 2016a, 2016b), although since the 1960s attempts were made using resonant mass antennas, large masses of metal that would oscillate with the passage of GWs. Detections using this method were claimed, but never validated or consensual.

The new generation of GW detectors are very large (4 km) interferometers of high precision that are able to observe relative changes in length as small as 10^{-21} cm. The active ones are the two interferometers from the LIGO and VIRGO collaboration. These observatories are placed in a way that allow the triangulation of the event with a fairly reasonable precision (usually around 20 deg^2) and in conjunction with the wave analysis is possible to estimate the distance and the so-called chirp mass of the system $\mathcal{M}$ for the case of merging objects.

The chirp mass is an observable quantity that can be expressed as

$$\mathcal{M} = M \left[q/(1+q)^2 \right]^{3/5}, \tag{16}$$

where q is the quotient of masses and M the total mass of the merging system. Note that it is not possible to determine both masses independently without further assumptions as, e.g., the distribution of masses or spin of the objects.

4.1. *Inferring the masses of NS from GW events*

The determination of the chirp mass $\mathcal{M}$ is straightforward and very precise, being obtained from the GW frequency evolution as the objects coalesce. The main source of uncertainty is the redshift from the event to the geocentric frame, but still less than 1%. In contrast, the mass quotient q is degenerate with the alignment of the spins with the orbital angular momentum, imposing a dependency between the two quantities.

Another way to approach the problem of individual masses determination is to use the distribution of *observed* NS masses. It is generally assumed that merging NSs have a higher probability to have their masses within the most populated places of the diagram of masses (Valentim *et al.*, 2011). For example, for $\mathcal{M} = 1.18\,M_\odot$ it is more plausible to have two NSs around $1.36\,M_\odot$, since it is in the most populated region of the diagram, than one with (say) $2\,M_\odot$ and the other with $1\,M_\odot$ where no NSs in DNSs are found.

We cannot, however, assume that NSs masses in binaries are strictly independent from each other. There are several mass transfer events on the stellar evolution that leads to such systems that will create a *correlation* between the two masses and other parameters, such as orbit periods and spins. Figure 8 suggest that in a general way, the greater m_1, the lower m_2. This shows that it would not be reasonable to use the same distribution as prior for both masses. Moreover, NSs in NS-WD systems for example, tend to be more massive than NSs in DNSs. We should not therefore treat them equally when estimating masses on DNS merging events, but focus the DNSs themselves. However, recent evidence of asymmetric DNS has raised a flag about the old belief in fully symmetric systems. And it is also important to remark that simulations obtained masses at birth in binary systems that are even higher than the ones determined for the two asymmetric DNS systems. The formation channel of the latter is still unknown and the small sample may contain observational biases.

With all these caveats, the time of a system to merge due to GW, τ, can be written as:

$$\tau = \frac{5}{264} \frac{a^4}{m^3} \frac{(1+q)^2}{q} \frac{1}{f(e)}, \tag{17}$$

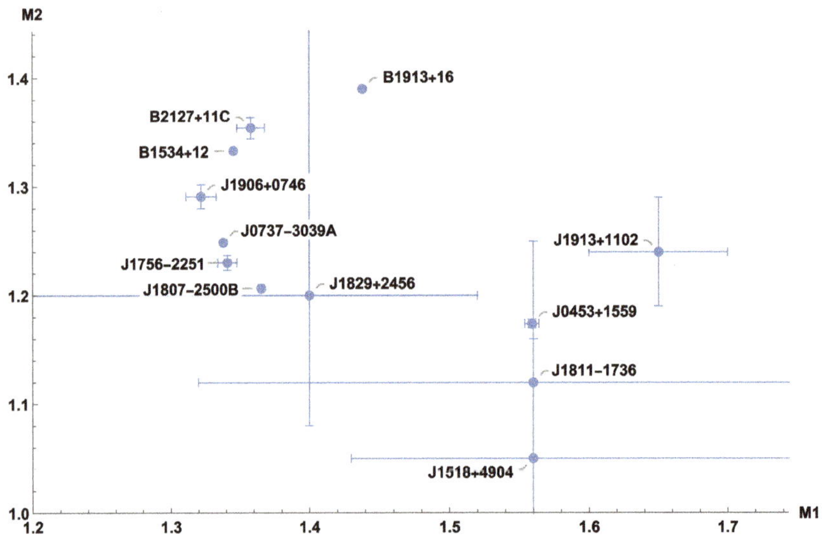

Fig. 8. Individual masses of Double NS systems.

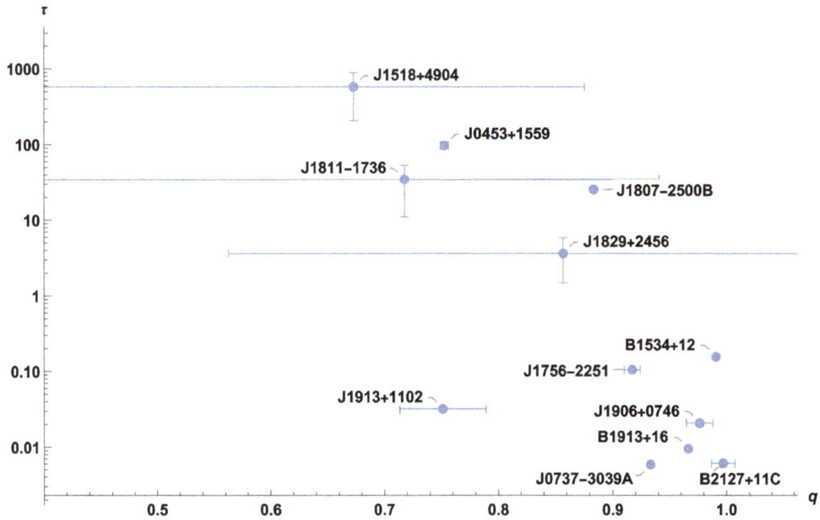

Fig. 9. Time to merge versus mass-ratio. The graphic shows the correlation between the time to merger due to gravitation waves τ and the mass quotient q. Systems with low q tend to have a higher τ.

where:

$$f(e) = \left[1 + \frac{73}{24}e^2 + \frac{37}{96}e^4\right]\left[1 - e^2\right]^{-\frac{7}{2}}. \tag{18}$$

The parameter a is the semi major axis, m the total mass, and e the eccentricity of the orbit. Looking to Eq. (17) we can expect that systems with higher q will have a smaller τ, around 5%, which falls very shy from the observed on Fig. 9, where systems with higher q systematically have a smaller τ by several orders of magnitude.

It is still unclear why the observed DNSs display these correlations, and whether they can be generalized to all systems. As we discover new DNSs the questions about the veracity of these correlations will be answered and further studies on binary evolution should explain the mechanisms behind them.

4.2. *Asymmetry in the systems GW170817 and GW190425*

Since the operation of the new interferometers, two events compatible with NS mergers had been observed. The first event, GW170817 had a luminous counterpart detected in various wavelengths from radio to gamma-rays.

The observations of the electromagnetic spectrum was important for the determinations such as ejected mass, details of the nucleosynthesis and others. The gamma rays detection connected to the merger helped consolidate NS mergers as a major candidate to the source of gamma-ray bursts — although the detected GRB was longer and weaker than most sGRBs-. The chirp mass of GW170817 was found to be $\mathcal{M} = 1.186\,M\odot$, which is consistent with parameters close to most of the known binary systems.

The second event of this class was GW190425, albeit without a luminous counterpart and with a chirp mass of $M = 1.44\,M\odot$. This chirp mass indicates a higher total mass than any known galactic DNS. The analysis of the spin suggested that the most probable total mass lies around $3.35\,M\odot$, $0.45\,M\odot$ above the most massive observed system (PSR J1913+1102). This suggests that the source had an ultra-tight orbit origin, invisible to radio pulsar surveys due to high Doppler smearing effects. Alternatively, it is possible that the event was *not* the merger of a DNS, but rather of a NS-BH system. This alternative would require a BH in the apparent mass gap between NSs and black holes, with 2–3 $M\odot$. These objects however were never observed and lack a solid Stellar Evolution understanding.

Abbott *et al.* (2019, 2020a) presented a Bayesian analysis for the individual masses of both events given two different priors for the spins, one of them considering the maximal spin allowed by any known EoS as the upper limit, and the other considering the maximal spin observed in galactic NSs.

The interval of masses obtained by these two different priors differ a lot from each other in both events. For the GW170817 the high-spin prior yields $m_1 \in (1.36, 1.89)\,M_{odot}$ & $m_2 \in (1.00, 1.36)\,M_{odot}$ while the low-spin prior yields $m_1 \in (1.36, 1.60)\,M_{odot}$ & $m_2 \in (1.16, 1.36)\,M_{odot}$ in the 90% confidence interval as shown in Fig. 10.

Figure 11(a) shows the individual masses determination by the spin chirp mass analysis for the event GW190425 and Fig. 11(b) shows the comparison of the total mass estimation of the event with the known galactic DNSs. For the low spin prior we have $m_1 \in (1.60, 1.87)\,M\odot$ and $m_2 \in (1.46, 1.69)\,M\odot$, while for the high spin: $m_1 \in (1.61, 2.52)\,M\odot$ and $m_2 \in (1.12, 1.68)\,M\odot$.

From the above description, it is quite clear that GWs have opened a new era for NS physics, especially when combined with multi-wavelength data as in the case of the merger GW170817. This "unseen" population is in many senses complementary to the information gathered from the known and yet-to-be-discovered nearby systems, and the joint study will produce a positive feedback already apparent in the works that explore this issues.

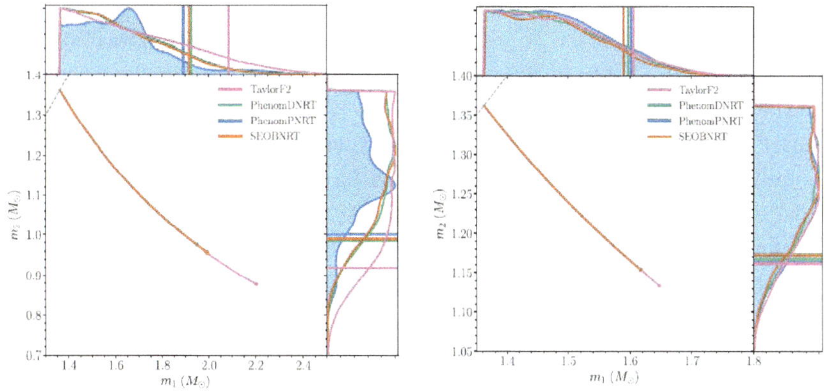

(a) High spin prior.

(b) Low spin prior.

Fig. 10. Allowed masses interval from event GW170817 given different spin priors. Figure extracted from Abbott *et al.* (2019).

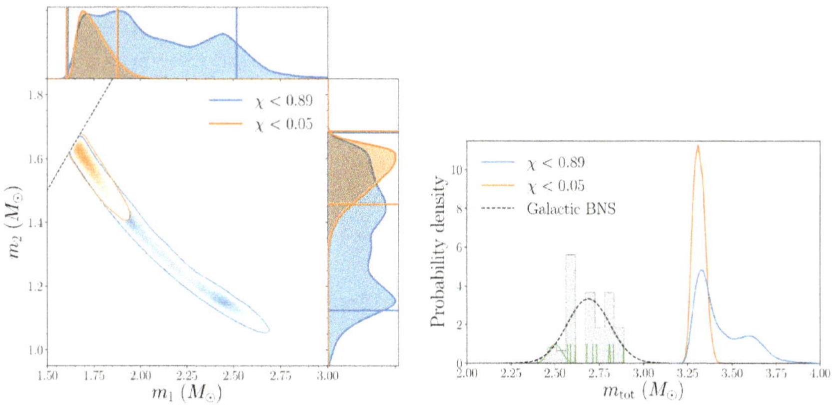

(a) Individual masses determination.

(b) Total mass comparison.

Fig. 11. (a) Shows the joint probability of individual masses of GW190425 for to different prior assumptions on the spin. (b) Shows the comparison of the total mass of the event with the observed Galactic DNSs. Figure extracted from Abbott *et al.* (2019).

5. Neutron Star Birth Events

The birth of NSs has been generally associated to explosions of massive stars. This conviction stems from the pioneer work of (Baade and Zwicky, 1934), the link of the Crab pulsar with the 1054 A.D explosion (the latter itself not yet fully understood) and later association of several pulsars with supernova remnants (Tian and Leahy, 2004). While there are several reasons supporting this picture, it is not obvious that the formation of NSs cannot have contributions from other channels. The evidence regarding the mass distribution presented in Section 2 can help, when complemented with theoretical studies of massive stars and their very explosions (or failures) and alternative events which have been around for years, and that have been characterized much better in recent times (binary evolution and Accretion-Induced Collapse (AIC)). We shall address these scenarios below and relate them to the empirical evidence as much as possible, in an attempt to complete with this topic an appraisal of the NS population origin as we understand today.

5.1. *Neutron stars formed in single explosions*

The evolution of single stars has produced a body of knowledge about their final fate which is relatively robust. In spite that several ingredients can modify somewhat the final numbers, it is generally agreed that the initial (ZAMS) mass and metallicity suffice to evolve the models and predict what the final outcome of the structure is, an important step towards the issue of NS formation itself. Figure 12 depicts the broad-brush picture of single star evolution. Atop the axis, the numbers indicate the masses of the stars at the ZAMS which produce the features indicated below the axis. For solar metallicity, assumed to be $Z = 0.01$, stars above $\sim 7.5\, M_{\odot}$ are the ones believed to be heavy enough to ignite carbon at their centres. This threshold is referred as M_{up} (or alternatively M_{CO}) in the literature. The minimum mass for the production of a NS after an explosive event is

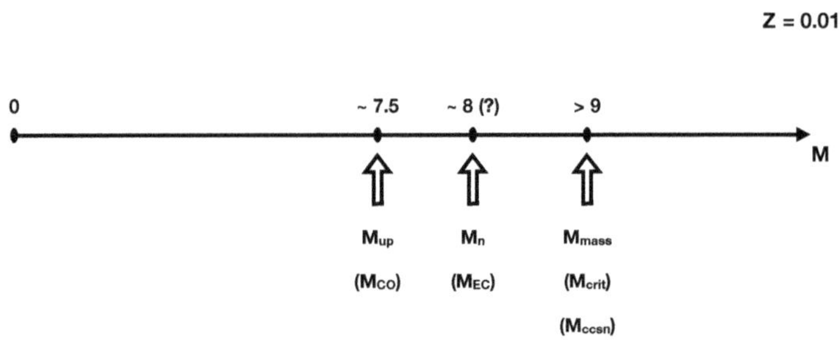

Fig. 12. The boundaries of mass (ZAMS) separating the different regimes for solar metallicity. The numbers above the axis denote the approximate locations of the separation between regimes, with their names and alternatives indicated below. See text for details.

though to be slightly higher, at $M_n \sim 8\,M_\odot$ but with a higher uncertainty denoted with a question mark (M_{EC} is an alternative name for this quantity, stemming from "electron capture"). Finally, above $M_{\rm mass} \geq 9\,M_\odot$ or so, depending on metallicity and input physics, all possible nuclear reactions are ignited and we enter the regime of "true" massive stars, developing an iron core and exploding as a core collapse supernova (note the alternative names $M_{\rm crit}$ and $M_{\rm ccsn}$ also found in the literature for the same mass (Doherty *et al.*, 2017)). If the limits can not be made more precise is due to two main factors (although other ones also exist): the uncertainties in the mass loss rate in the advanced stages and the true nature of convection inside the progenitors. As mentioned, the metallicity is important for the location of the boundaries, and as a general trend we know that the values are lower for lower metallicities, and even so a substantial spread in the calculations still exist, as shown in Figs. 13 and 14 (Doherty *et al.*, 2017).

While the actual problems above $M_{\rm mass}$ are mainly related to the collapse-implosion and launch of the supernova explosion, the fate of massive stars below $M_{\rm mass}$ presents a series of difficulties even before their final fate. The latest stage of these objects, characterized by the off-centre ignition of degenerate carbon before the thermal pulses (TPs), is known as the "super-AGB" phase. Stars just below the M_n limit are thought to leave *O-Ne-Mg* white dwarfs, and those slightly less massive a *C-O-Ne* white dwarf, but without explosions involved. A true unsolved problem is that the super-AGB stars have never been identified observationally with confidence, and indeed they do not stand out clearly from neighbours at nearby positions in the HR diagram. Consequently, we must rely on the accurateness of theoretical calculations for the determination of $M_{\rm up}$ and related quantities.

A second important mass scale starts at a value M_n, above which the cores may undergo electron captures and explode as a class of supernovae, forming NSs. The electron-capture supernovae start with the reduction of the electron fraction per baryon Y_e because of capture reactions onto Mg and Ne, followed by pressure support loss and collapse of the core having a quite *fixed value* of mass, $1.375 M_\odot$. The explosion receives a massive contribution from the energy released by oxygen burning in nuclear statistical equilibrium, and therefore the event is actually close to a Type Ia thermonuclear event (a recent claim of the identification of an electron-capture supernova showing all the expected features has been presented by Hiramatsu *et al.* (2020)). Because of this features, it is believed that an almost-fixed mass NS is formed, with $m \sim 1.25 M_\odot$ which results from the

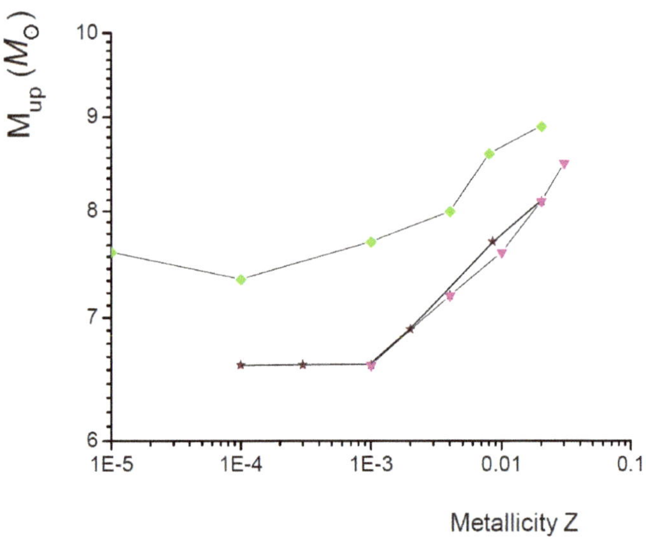

Fig. 13. Sample calculations of the quantities M_{up} as a function of the metallicity Z. The three curves correspond to the calculations of Siess (2007) (diamonds, upper), Doherty *et al.* (2015) (stars, middle) and Suwa *et al.* (2018) (triangles, lower).

emission of the binding energy of the fixed core. Since this mass range is the most abundant among the exploding progenitors, it is expected that they make a large fraction of the full supernova rate, as it is easily seen from the form of the IMF function. Because of these features, it is expected that a group of "light" NSs stands out in the mass distribution (Schwab *et al.*, 2010), although confirming its presence may still be statistically tricky and would require a larger sample (Rocha *et al.*, 2019), as stated in Section 1.

The discussion above implicitly requires that if even *lighter* ($< 1.25\,M_\odot$) NSs exist, they must be formed in the explosion of small *iron* cores, not the ones undergoing the electron-capture supernovae. In other words, it is important to determine both the smallest NS mass and the lightest iron core resulting from the evolution of a $M \sim M_{\rm mass}$ star. From the observation's diagram shown in Section 2, we identify the masses of PSR J1453+1559 companion, with $1.174 \pm 0.004\,M_\odot$ as the lightest exactly measured NS. Other sources such as 4U1538-522 and Her X-1 may also be considered, but their error bars are larger and must be refined. With this figure for the gravitational mass M_G, the iron core progenitor of this low-mass NS should have been no heavier than $\sim 1.28 M_\odot$. Generally speaking, it is easier to find the baryonic masses of the remnants M_B than the gravitational mass M_G in the literature. The difference of both quantities, related to the

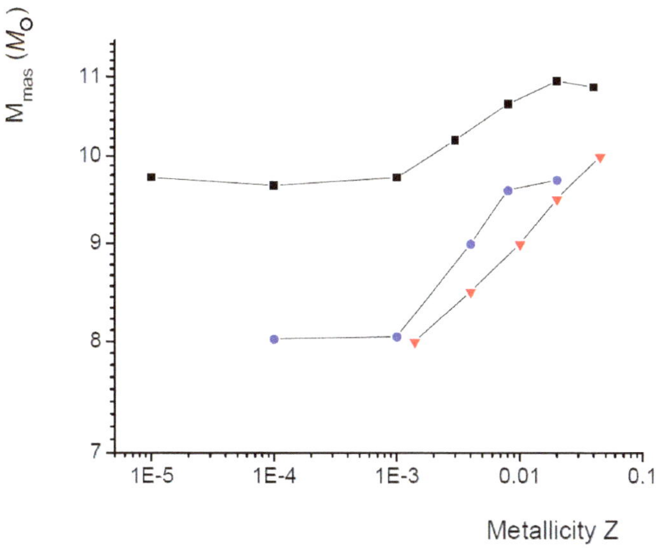

Fig. 14. Sample calculations of the quantities $M_{\rm mass}$ as a function of the metallicity Z. The curves are due to Siess (2007) (squares, upper), Doherty *et al.* (2015) (dots, middle) and Eldridge and Tout (2004) (triangles, lower).

binding energy, has to be calculated for each underlying theory of gravitation. However, a simple approximate expression for the latter quantity has been found by Lattimer and Prakash (2001) in terms of the quantity $\beta = GM_G/c^2 R_0$, where G is Newton's constant, c the speed of light and R_0 a fiducial radius (safely set to $12km$) to relate both quantities quite accurately, namely:

$$\frac{M_B - M_G}{M_G} = 0.6\frac{\beta}{1 - 0.5\beta}, \qquad (19)$$

and can be used quite safely if an extreme accuracy is not required.

The simulation of explosions of single stars have found confronting results for the minimum iron core forming the lightest NSs. For instance, Timmes *et al.* (1996) found a small number of progenitors that can produce a $M_G \leq 1.2\,M_\odot$. A more recent work by Sukhbold *et al.* (2016), calibrated for two progenitors, does not produce any NS below $M_G = 1.2\,M_\odot$. A dedicated study of the iron core at the onset of collapse (Suwa *et al.*, 2018), formed by low-mass CO cores has obtained "light" NSs in the mass range of the observed sources, and even close to $\sim 1 M_\odot$. This is consistent with the results of Ugliano *et al.* (2012), in which a minimum baryonic mass of $\sim 1.2\,M_\odot$ would render suitable gravitational mass after applying

Eq. (19) (although for a quite narrow mass range of the progenitors). All these works and a few others are quite difficult to compare because of different prescriptions for the stellar physics, different pre-supernova models and different numerical codes. In any case, it is entirely possible that a single star explosion does not constitute a valid evaluation for this lower limit, since the systems are found in binaries and binary stellar evolution is likely important. We shall return to this point in the next section.

With all the above caveats, it is fair to state that theoretical model explosions are not at odds with the empirical mass distribution found in Section 2. However, all the simulations produce stellar remnants which could be associated with the "second peak" around $1.8\,M_{\odot}$, but fall short of explaining the higher masses detected in actual systems. An example of this statement can be seen in Fig. 15, and similar features are present in other works. Naively one could have expected that the jump in the size of iron core above $\sim$19 $M_{\odot}$ could be the reason for the production of heavy NSs, and even if this is true, the final values of the NSs M_G does not go above $1.9\,M_{\odot}$. The reason for these large iron cores, often overlooked in general works, is that there is a finite entropy inside them, making the "effective" Chandrasekhar mass $M_{\mathrm{Ch,eff}}$ to grow from its cold value $M_{\mathrm{Ch},0}$

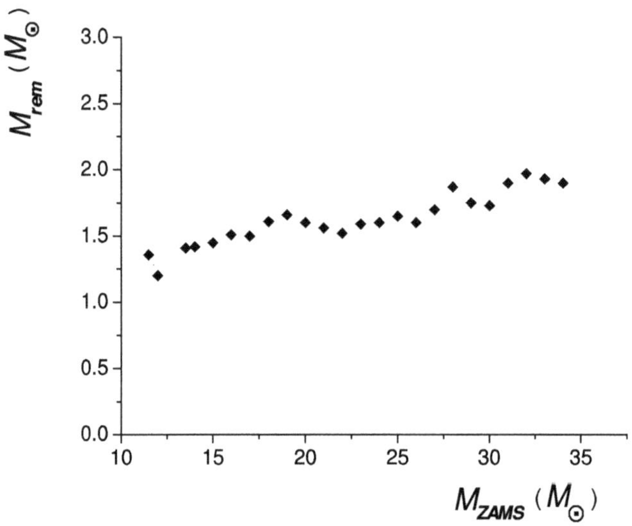

Fig. 15. The compact remnants of single star explosions (baryonic mass) as a function of the ZAMS progenitor mass (Ugliano *et al.*, 2012).

according to Timmes *et al.* (1996):

$$M_{\rm Ch,eff} \simeq M_{Ch,0}\left(1+\left(\frac{s_e}{\pi Y_e}\right)^2\right), \tag{20}$$

where, inserting a rough average for the electronic entropy per baryon $\langle s_e \rangle = 1$ and given that $Y_e \geq 0.4$ in general, produces collapsing cores of $\sim 1.8\,M_\odot$. However, in spite of this growth, the iron cores never reach values well above $2\,M_\odot$, that would be necessary to reproduce the highest measured NS masses. A contrasting view has been recently presented by Burrows and Vartanyan (2021), which in a summary of their present and former simulations were able to obtain not only explosions, but also gravitational masses up to $\sim 2\,M_\odot$ for remnant NSs. There is some important factor(s) not yet understood in the systematics of collapse simulations obtained by different groups to firmly assert whether high masses above $\sim 2\,M_\odot$ can be produced by this formation channel at birth, or and/or subsequent accretion is needed to reach the highest observed values. This statement is also important for massive pulsars with $\geq 2\,M_\odot$, which have been suggested to be born "as is" (Deng *et al.*, 2020), without suffering substantial accretion: if this happens to be true, current models of the explosions should obtain them.

As a general trend we see that the simulations do well to reproduce the "first" peak around $1.35 - 1.4\,M_\odot$ and can populate the "second peak" as well. It is the high-mass tail definitely present in the distribution that is not obtained. In addition, the presence of a "low-mass" peak is quite interesting and worth mentioning. We have stated that since light-mass ZAMS progenitors are very abundant, it is expected that presence of $\sim 1.25\,M_\odot$ can be easily seen. There are definitely a group of NSs in the histogram that were identified as such (Schwab *et al.*, 2010), but they are often "merged" with the main "first peak" using statistical discriminators (Section 2). In other words, even though their presence is possible, it has a low probability from this point of view (Horvath and Valentim, 2017). Moreover, it is intriguing that, even *ignoring* the electron-capture events, a peak at that position has been obtained by Sukhbold *et al.* (2016). Therefore, we find somewhat counter-intuitive that the statistical significance of this group is not higher than the presently obtained one.

Finally, we would like to point out that a substantial advance in the knowledge of the connection between events and NS masses has a long road ahead. It is clear that the attack to this problem produced a lot of advances, and revealed an extreme complexity that is still being deciphered. One of the major issues, in our view, will be to establish whether the formation of NS and BH is monotonic (Burrows and Vartanyan, 2021) or "intermittent" (Sukhbold *et al.*, 2016), and which are the conditions for that behaviour.

5.2. *Neutron stars formed in binaries*

NSs are known to be present in a variety of systems. Some of them required quite sophisticated models for the evolution of a binary system featuring several phases along its life. Therefore the question of a *binary* NS origin, in contrast with the isolated single-star evolution discussed above arises.

The full evolution of stars in binary systems is not yet calculated until the *CO* core forms. This means that the pre-supernova structure is not really known, but rather assumed using reasonable prescriptions. It is generally assumed that the whole consequence of binary interaction is to promptly remove the entire hydrogen envelope of the star at helium ignition (Ertl *et al.*, 2020). The exact amount of accurateness of this simplification is not known, although it is a very reasonable first attempt and does not seem to imply any obvious misbehaviour. An attempt to improve the situation can be seen in the work by Patton and Sukhbold (2020).

Works using some version of this simple stripping hypothesis have been presented, with exploding "helium star" masses ranging from ≥ 1 to $40\,M_{\odot}$ since the removal of the envelope is a common feature of almost all close binary system evolution, the relation of the final iron core mass with the initial $M_{\rm ZAMS}$ is uncertain, because it depends on the mass loss rate, and even the specific code results need to be validated. Nevertheless, the NS distribution presents some features which are not present in the single-star explosions. For example, Woosley *et al.* (2020) obtain some objects that, because of larger fallback, lie above the $2\,M_{\odot}$ mark for all the mass-loss prescriptions. Ertl *et al.* (2020) also obtain a small fraction of heavy objects. In both cases the synthesis of a bimodal (or even trimodal) population presented in Section 2 remains to be demonstrated in detail. The production of stars with $M_G \leq 1.2\,M_{\odot}$ does not appear to be favoured in these simulations. However, the work of Fortin *et al.* (2016) claims to produce NSs as light as $\sim 1\,M_{\odot}$, although their framework is difficult to compare with the former.

Given that there are many systems in which the interaction of both components inevitably leads to consider the issue of the explosion(s) themselves, it is likely that substantial advances can be made in the near future, once the pre-collapse structure can be determined (Patton and Sukhbold, 2020). One prime example is the well-known path towards binary NSs (see, for example, Tauris *et al.* (2017); Van den Heuvel (2018)) in which a detailed evolution implying "ultra-stripped" supernovae is needed. To connect with the issue of the NS distribution, we should state that even if the binary NSs that originated the "$1.4\,M_{\odot}$ paradigm" have not accreted substantially after formation, no real reason to expect a "fixed mass" exists. In fact, recent observations of *asymmetric* binary NS systems (Ferdman *et al.*, 2020) that can merge on less than a Hubble time may call for a reanalysis of the masses at birth, in line with the simulation work just described. On the other hand, BH production with a supernova explosion has been also found possible, and in fact a recent report Maxted *et al.* (2020) claims one of these identifications.

5.3. *The Accretion-Induced Collapse channel for NS formation*

The process by which a white dwarf can collapse and form a NS, known as AIC, has been present for more than 40 yrs and invoked to solve a large variety of problems in stellar astrophysics. The basic original idea is that for white dwarfs near their own Chandrasekhar limit, a high accretion rate $\dot{M}$ (but below the Eddington limit) from a MS or RG companion can make the electron captures more efficient than the thermonuclear ignition of carbon or even oxygen, triggering the collapse of the star by removing the degenerate electron pressure (see, for example, Nomoto and Kondo (1991)). Theoretically, AIC events could be quite frequent and originate NSs in many binary systems. In fact, the detection of a large number of low-mass binary pulsars in globular clusters was one of the features in which AIC was suggested to operate (Grindlay and Bailyn, 1988). AIC has been also invoked in the magnetar formation problem (Margalit *et al.*, 2019), expecting a flux-conservation amplification of an initial magnetic field of the white dwarf, and many other systems like intermediate-mass and MSPs. A "double degenerate" AIC resulting from the merging of two white dwarfs with a short orbital period was later discussed as a complementary possibility to the "single degenerate" channel, analogously to the problem of Type Ia supernova progenitors to which they have a kinship.

Since we are concerned with the NS formation from these events only, we just point out what are the results and expectations relevant for our subject. The first important observation is that the very events have never been positively identified, although it is expected that their luminosity remains low, and therefore this is not a big surprise. The first detailed calculation of the AIC (Fryer *et al.*, 1999) obtained an important output of exotic isotopes in the $\sim 0.1\,M_{\odot}$ ejecta, and proceed to deduce an upper limit to the occurrence in the galaxy based on the measured abundance of them. This exotic isotope production has been challenged by Qian and Wasserburg (2007), and makes the issue of the upper limit uncertain. On the other hand, population synthesis have yielded the expectation of $\sim 10^7$ pulsars formed by AIC in the single-degenerate channel and a few times this figure coming from the double-degenerate channel (Wang and Liu, 2020). This are high numbers and may be in tension with the estimated rate of $\leq 0.1\%$ of the total NS population estimated by Fryer *et al.* (1999). If the overproduction of exotic isotopes can be avoided, high rates could be eventually accepted. The alternative way out of this quandary is that the number of suitable progenitor systems is overestimated, and that AIC NSs will not be produced in many of the current merging candidates, but which of these solutions is viable remains unsolved for the present.

The production of NSs in the single-degenerate channel would be in a narrow range around $1.25\,M_{\odot}$ naively, but models envisage the accretion to resume after their formation. Therefore, the actual range of the NS masses may extend all the way up to the highest measured masses if the accretion conditions allows it. For the double-degenerate channel, the expected range is different, and is believed to span the range $1.4 - 2.8\,M_{\odot}$, with a slight variation according to the chemical composition of the white dwarfs (Wang and Liu, 2020). We note that this may be an efficient way to form extremely heavy NSs "at birth", a feature that may be required if more heavy objects are detected. The contribution of GW observations to this task is very important, as suggested by the detection of a $2.6\,M_{\odot}$ object in the merge GW190814 (Abbott *et al.*, 2020b), although the true nature of this component remains to be confirmed, as stated in Section 4.

6. Conclusions

The study of NS origins and masses has entered a mature age, with a growing body of reliable data and perspectives to clarify the basic facts of these amazing objects. The main points discussed above can be summarized as follows:

- Contrary to earlier beliefs, there is no "canonical" NS mass. The most abundant NSs in the sample do have masses around this value, but a substantial fraction of the sample is either smaller or much larger than that, possibly up to $\sim 1\,M_\odot$ heavier.
- The situation with the so-called "mass gap", closely related to the M_{TOV} issue and the mechanism(s) of formation, is not clear yet. A "gap" has been reported in the merging GW events. The presence of "light" (i.e., $3\,M_\odot$) black holes is not certain, although this may still be due to detection biases and could be associated to the light components in a few GW merging processes. It is important to keep looking for them in local systems.
- Theoretical models of NS must produce high masses, maybe closer to the Rhoades–Ruffini limit than we use to think a few years ago.
- The study of merging NS–NS and possibly NS–BH mergers could prove crucial for the determination of M_{TOV} and the mass ratio of the binaries, while the studies of local systems are complementary and amenable to more direct analysis.
- Supernova simulations are now on a full different level of sophistication, and had achieved some important goals. However, there are still differences between the results of different groups that are important for our main focus. It is still not clear whether heavy NS with $M \geq 2\,M_\odot$ can be formed directly, and the issue of monotonicity of the NS-BH remnants with the remnant mass and other relevant factors (binarity, treatment of pre-SN physics, etc.) does not seem fully established.
- The existence of NSs with mass significantly below $1.25\,M_\odot$ may be interpreted as empirical evidence for small iron cores. It is important for Stellar Evolution to determine this low-end of the distribution of masses.

Acknowledgements

JEH wishes to acknowledge the financial support of the FAPESP Agency (São Paulo) through the grant 13/26258-4 and the CNPq (Federal Government) for the award of a Research Fellowship. The CAPES Agency (Federal Government) is acknowledged for financial support in the form of Scholarships.

References

Abbott, B. P., Abbott, R., Abbott, T. D., *et al.* (2016). Observation of gravitational waves from a binary black hole merger, *Physical Review Letters*, **116**, p. 061102.

Abbott, B. P., Abbott, R., Abbott, T. D., *et al.* (2016). Properties of the binary black hole merger GW150914, *Physical Review Letters*, **116**, p. 241102.

Abbott, B. P., Abbott, R., Abbott, T. D., *et al.* (2019). Properties of the binary neutron star merger GW170817, *Physical Review X*, **9**, p. 011001.

Abbott, B. P., Abbott, R., Abbott, T. D., *et al.* (2020). GW190425: Observation of a compact binary coalescence with total mass $3.4\,M_\odot$, *The Astrophysical Journal Letters*, **892**, p. L3.

Abbott, R., Abbott, T. D., Abraham, S., *et al.* (2020). GW190814: Gravitational waves from the coalescence of a 23 $M_\odot$ black hole with a 2.6 $M_\odot$ compact object, *The Astrophysical Journal Letters*, **896**, p. L44.

Abbott, R., Abbott, T. D., Abraham, S., *et al.* (2020). Population Properties of Compact Objects from the Second LIGO-Virgo Gravitational-wave Transient Catalog. arXiv preprint arXiv:2010.14533v1.

Ai, S., Gao, H., and Zhang, B. (2020). What constraints on the neutron star maximum mass can one pose from GW170817 observations?, *The Astrophysical Journal*, **893**, p. 146.

Akmal, A. and Pandharipande, V. R. (1997). Spin-isospin structure and pion condensation in nucleon matter, *Physical Review C*, **56**, p. 2261.

Alcock, C., Farhi, E., and Olinto, A. (1968). Strange stars, *The Astrophysical Journal*, **310**, p. 261.

Alford, M., Braby, M., Paris, M., and Reddy, S. (2005). Hybrid stars that masquerade as neutron stars, *The Astrophysical Journal*, **629**, p. 969.

Alsing, J., Silva, H. O., and Berti, E. (2018). Evidence for a maximum mass cut-off in the neutron star mass distribution and constraints on the equation of state, *Monthly Notices of the Royal Astronomical Society*, **478**, p. 1377.

Antoniadis, J., Freire, P. C. C., Wex, N., *et al.* (2013). A massive pulsar in a compact relativistic binary, *Science*, **340**, p. 6131.

Antoniadis, J. (2015). Multi-wavelength Studies of Pulsars and their Companions, *Springer International Publishing, Switzerland AG.*

Antoniadis, J., Tauris, T. M., Ozel, F., *et al.* (2016). The millisecond pulsar mass distribution: Evidence for bimodality and constraints on the maximum neutron star mass, arXiv preprint arXiv:1605.01665.

Antoniadis, J., Van Kerkwijk, M. H., Koester, D., Freire, P. C. C., *et al.* (2012). The relativistic pulsar — white dwarf binary PSR J1738+ 0333–I. Mass determination and evolutionary history, *Monthly Notices of the Royal Astronomical Society*, **423**, p. 3316.

Arzoumanian, Z., Brazier, A., Burke-Spolaor, S., *et al.* (2018). The NANOGrav 11-year data set: High-precision timing of 45 millisecond pulsars, *The Astrophysical Journal Supplement Series*, **235**, p. 37.

ATNF pulsar catalogue (2020). Neutron Star Data, (https://www.atnf.csiro.au/research/pulsar/psrcat/).

Baade, W. and Zwicky, F. (1934). On Supernovae, *Proceedings of the National Academy of Sciences*, **20**, p. 254.

Babu, G. J. and Feigelson, E. D. (2006). Astro-statistics: Goodness-of-fit and all that!, *Astronomical Data Analysis Software and Systems XV*, **351**, p. 127.

Barr, E. D., Freire, P. C. C., Kramer, M., *et al.* (2017). A massive millisecond pulsar in an eccentric binary, *Monthly Notices of the Royal Astronomical Society*, **465**, p. 1711.

Bassa, C. G., Van Kerkwijk, M. H., Koester, D., and Verbunt, F. (2006). The masses of PSR J1911–5958A and its white dwarf companion, *Astronomy & Astrophysics*, **456**, p. 295.

Baym, G., Pethick, C., and Sutherland, P. (1971). The ground state of matter at high densities: Equation of state and stellar models, *The Astrophysical Journal*, **170**, p. 299.

Benvenuto, O. G. and Horvath, J. E. (1989). On the structure of strange stars and bagged QCD parameters, *Monthly Notices of the Royal Astronomical Society*, **241**, p. 43.

Benvenuto, O. G., De Vito, M. A., and Horvath, J. E. (2012). Evolutionary trajectories of ultracompact black widow pulsars with very low mass companions, *The Astrophysical Journal Letters*, **753**, p. L33.

Benvenuto, O. G., De Vito, M. A., and Horvath, J. E. (2014). Understanding the evolution of close binary systems with radio pulsars, *The Astrophysical Journal Letters*, **786**, p. L7.

Berezina, M., Champion, D. J., and Freire, P. C. C. (2017). The discovery of two mildly recycled binary pulsars in the northern high time resolution universe pulsar survey, *Monthly Notices of the Royal Astronomical Society*, **470**, p. 4421.

Bhalerao, V. B. (2012). Neutron stars and NuSTAR: A systematic survey of neutron star masses in high mass X-ray binaries and characterization of CdZnTe detectors for NuSTAR, Universal-Publishers, Irvine, USA.

Bhat, N. D. R., Bailes, M., and Verbiest, J. P. W. (2008). Gravitational-radiation Losses from the Pulsar–white-dwarf Binary PSR J1141–6545, *Physical Review D*, **77**, p. 124017.

Bodmer, A. R. (1971). Collapsed nuclei, *Physical Review D*, **4**, p. 1601.

Breu, C. and Rezzolla, L. (2016). Maximum mass, moment of inertia and compactness of relativistic stars, *Monthly Notices of the Royal Astronomical Society*, **459**, p. 646.

Burrows, A. and Vartanyan, D. (2021). Core-collapse supernova explosion theory, *Nature*, **589**, p. 29.

Casares, J., Hernández, J. I. G., Israelian, G., and Rebolo, R. (2010). On the mass of the neutron star in Cyg X-2, *Monthly Notices of the Royal Astronomical Society*, **401**, p. 2517.

Chamel, N., Haensel, P., Zdunik, J. L., and Fantina, A. F. (2013). On the maximum mass of neutron stars, *International Journal of Modern Physics E*, **22**, p. 1330018.

Cromartie, H. T., Fonseca, E., Ransom, S. M., *et al.* (2020). Relativistic shapiro delay measurements of an extremely massive millisecond pulsar, *Nature Astronomy*, **4**, p. 72.

de Avellar, M. G. B. and Horvath, J. E. (2010). Exact and quasi-exact models of strange stars, *International Journal of Modern Physics D*, **19**, p. 1937.

Deller, A. T., Archibald, A. M., and Brisken, W. F. (2012). A parallax distance and mass estimate for the transitional millisecond pulsar system J1023+0038, *The Astrophysical Journal Letters*, **756**, p. L25.

Demorest, P. B., Pennucci, T., Ransom, S. M., Roberts, M. S. E., and Hessels, J. W. T. (2010). A two-solar-mass neutron star measured using shapiro delay, *Nature*, **467**, p. 1081.

Deneva, J. S., Freire, P. C. C., Cordes, J. M., *et al.* (2012). Two millisecond pulsars discovered by the PALFA survey and a shapiro delay measurement, *The Astrophysical Journal*, **757**, p. 89.

Deng, Z. L., Gao, Z. F., Li, X. D., and Shao, Y. (2020). On the formation of PSR J1640+2224: A neutron star born massive?, *The Astrophysical Journal*, **892**, p. 4.

Desvignes, G., Caballero, R. N., Lentati, L., *et al.* (2016). High-precision timing of 42 millisecond pulsars with the European pulsar timing array, *Monthly Notices of the Royal Astronomical Society*, **458**, p. 3341.

Doherty, C. L., Gil-Pons, P., Siess, L., and Lattanzio, J. C. (2017). Super-AGB stars and their role as electron capture supernova progenitors, *Publications of the Astronomical Society of Australia*, **34**, p. e056.

Doherty, C. L., Gil-Pons, P., Siess, L., Lattanzio, J. C., and Lau, H. H. B. (2015). Super-and massive AGB Stars-IV. Final fates-initial-to-final mass relation, *Monthly Notices of the Royal Astronomical Society*, **446**, p. 2599.

Douchin, F. and Haensel, P. (2001). A unified equation of state of dense matter and neutron star structure, *Astronomy & Astrophysics*, **380**, p. 151.

Eldridge, J. J. and Tout, C. A. (2004). The progenitors of core-collapse supernovae, *Monthly Notices of the Royal Astronomical Society*, **353**, p. 87.

Engvik, L., Hjorth-Jensen, M., Osnes, E., Bao, G., and Østgaard, E. (1996). Asymmetric nuclear matter and neutron star properties, *Physical Review Letters*, **73**, p. 2650.

Ertl, T., Woosley, S. E., Sukhbold, T., and Janka, H.-T. (2020). The explosion of helium stars evolved with mass loss, *The Astrophysical Journal*, **890**, p. 51.

Falanga, M., Bozzo, E., Lutovinov, A., *et al.* (2015). Ephemeris, orbital decay, and masses of ten eclipsing high-mass X-ray binaries, *Astronomy & Astrophysics*, **577**, p. A130.

Farrow, N., Zhu, X.-J., and Thrane, E. (2019). The mass distribution of galactic double neutron stars, *The Astrophysical Journal*, **876**, p. 18.

Ferdman, R. D., Stairs, I. H., Kramer, M., *et al.* (2010). A precise mass measurement of the intermediate-mass binary pulsar PSR J1802–2124, *The Astrophysical Journal*, **711**, p. 764.

Ferdman, R. D., Stairs, I. H., Kramer, M., *et al.* (2014). PSR J1756- 2251: A pulsar with a low-mass neutron star companion, *Monthly Notices of the Royal Astronomical Society*, **443**, p. 2183.

Ferdman, R. D., Freire, P. C. C., Perera, B. B. P., *et al.* (2020). Asymmetric mass ratios for bright double neutron-star mergers, *Nature*, **583**, p. 211.

Finn, L. S. (1994). Observational constraints on the neutron star mass distribution, *Physical Review Letters*, **73**, p. 1878.

Fonseca, E., Pennucci, T. T., Ellis, J. A., *et al.* (2016). The NANOGrav nine-year data set: Mass and geometric measurements of binary millisecond pulsars, *The Astrophysical Journal*, **832**, p. 167.

Fonseca, E., Stairs, I. H., and Thorsett, S. E. (2014). A comprehensive study of relativistic gravity using PSR B1534+12, *The Astrophysical Journal*, **787**, p. 1878.

Fortin, M., Bejger, M., Haensel, P., and Zdunik, J. L. (2016). Progenitor neutron stars of the lightest and heaviest millisecond pulsars, *Astronomy & Astrophysics*, **586**, p. A109.

Franzon, B., Fogaça, D. A., Navarra, F. S., and Horvath, J. E. (2012). Self-bound interacting QCD matter in compact stars, *Physical Review D*, **86**, p. 065031.

Freire, P. C. C., Wolszczan, A., van den Berg, M., and Hessels, J. W. T. (2008). A massive neutron star in the globular cluster M5, *The Astrophysical Journal*, **679**, p. 1433.

Freire, P. C. C., Bassa, C. G., Wex, N., *et al.* (2011). On the nature and evolution of the unique binary pulsar J1903+0327, *Monthly Notices of the Royal Astronomical Society*, **412**, p. 2763.

Fruchter, A. S., Stinebring, D. R., and Taylor, J. H. (1988). A millisecond pulsar in an eclipsing binary, *Nature*, **333**, p. 237.

Fryer, C., Benz, W., Herant, M., and Colgate, S. A. (1999). What can the accretion-induced collapse of white dwarfs really explain?, *The Astrophysical Journal*, **516**, p. 892.

Glendenning, N. K. (1985). Neutron stars are giant hypernuclei?, *The Astrophysical Journal*, **293**, p. 470.

Glendenning, N. K. (2012). *Compact Stars: Nuclear Physics, Particle Physics and General Relativity*, Springer Science & Business Media, Berlin, Germany.

Grindlay, J. E. And Bailyn, C. D. (1988). Birth of millisecond pulsars in globular clusters, *Nature*, **336**, p. 48.

Haensel, P., and Pichon, B. (1994). Experimental nuclear masses and the ground state of cold dense matter, *Astronomy & Astrophysics*, **283**, p. 313.

Haensel, P., Potekhin, A. Y., and Yakovlev, D. G. (2007). Neutron Stars 1: Equation of state and structure, Springer Science & Business Media, Berlin, Germany.

Haseltine, E. (2006). The 11 greatest unanswered questions of physics, *Discover-New York*, **23**, p. 36.

Heger, A., Fryer, C. L., Woosley, S. E., Langer, N., and Hartmann, D. H. (2003). How massive single stars end their life, *The Astrophysical Journal*, **591**, p. 288.

Heiselberg, H. and Pandharipande, V. (2000). Recent progress in neutron star theory, *Annual Review of Nuclear and Particle Science*, **50**, p. 481.

Hewish, A., Bell, S. J., Pilkington, J. D. H., Scott, P. F., and Collins, R. A. (1968). Observation of a rapidly pulsating radio source, *Nature*, **217**, p. 709.

Hiramatsu, D., Howell, D. A., Van Dyk, S. D., *et al.* (2020). The Electron-capture Origin of Supernova 2018zd, arXiv preprint arXiv:2011.02176v1.

Horvath, J. E. (1996). Possible determination of isolated pulsar masses with gravitational microlensing, *Monthly Notices of the Royal Astronomical Society*, **278**, p. L46.

Horvath, J. E. (2007). *What Do Exotic Equations of State Have to Offer?*, *Isolated Neutron Stars: From the Surface to the Interior*, Springer, Berlin, Germany, p. 431.

Horvath, J. E. and De Souza, R. A. (2017). The stifness of the supranuclear equation of state (once again), *Journal of Physics: Conference Series*, **861**, p. 012010.

Horvath, J. E. and Valentim, R. (2017). The Masses of Neutron Stars, *Handbook of Supernovae*, Springer International Publishing AG, Switzerland AG, p. 1317.

Horvath, J. E. (2019). The binaries of the NS-NS merging events, *AIP Conference Proceedings*, **2127**, p. 020015.

Horvath, J. E., Bernardo, A., Rocha, L. S., Valentim, R., Moraes, P. H. R. S., and de Avellar, M. G. B. (2020). Redback/black widow systems as progenitors of the highest neutron star masses and low-mass black holes, *SCIENCE CHINA Physics, Mechanics & Astronomy*, **63**, p. id.129531.

Hui, C. Y. and Li, K. L. (2019). High energy radiation from spider pulsars, *Galaxies*, **7**, p. 93.

Hulse, R. A. and Taylor, J. H. (1975). Discovery of a pulsar in a binary system, *The Astrophysical Journal*, **195**, p. L51.

Jacoby, B. A., Cameron, P. B., Jenet, F. A., *et al.* (2006). Measurement of orbital decay in the double neutron star binary PSR B2127+11C, *The Astrophysical Journal Letters*, **644**, p. L113.

Kaplan, D. L., Boyles, J., Dunlap, B. H., *et al.* (2014). A 1.05 $M_{\odot}$ companion to PSR J2222–0137: The coolest known white dwarf?, *The Astrophysical Journal*, **789**, p. 119.

Kasian, L. E. (2012). *Radio Observations of Two Binary Pulsars*, University of British Columbia, Vancouver, Canada.

Kiziltan, B., Kottas, A., De Yoreo, M., and Thorsett, S. E. (2013). The neutron star mass distribution, *The Astrophysical Journal*, **778**, p. 66.

Kramer, M., Stairs, I. H., and Manchester, R. N. (2006). Tests of general relativity from timing the double pulsar, *Science*, **314**, p. 97.

Lackey, B. D., Nayyar, M., and Owen, B. J. (2006). Observational constraints on hyperons in neutron stars, *Physical Review D*, **73**, p. 024021.

Lattimer, J. M. (2012). The nuclear equation of state and neutron star masses, *Annual Review of Nuclear and Particle Science*, **62**, p. 485.

Lattimer, J. M. and Prakash, M. (2001). Neutron star structure and the equation of state, *The Astrophysical Journal*, **550**, p. 426.

LIGO/VIRGO (2020). Neutron Star Data (https://gracedb.ligo.org/superevents/S200316bj/view/).

Linares, M. (2019). Super-Massive Neutron Stars and Compact Binary Millisecond Pulsars, arXiv preprint arXiv:1910.09572.

Lugones, G. and Arbañil, J. D. V. (2017). Compact stars in the braneworld: A new branch of stellar configurations with arbitrarily large mass, *Physical Review D*, **95**, p. 064022.

Lugones, G. and Horvath, J. E. (2002). Color-flavor locked strange matter, *Physical Review D*, **66**, p. 074017.

Lugones, G. and Horvath, J. E. (2003). High-density QCD pairing in compact star structure, *Astronomy & Astrophysics*, **403**, p. 173.

Lynch, R. S., Freire, P. C. C., Ransom, S. M., and Jacoby, B. A. (2012). The timing of nine globular cluster pulsars, *The Astrophysical Journal*, **745**, p. 109.

Lyne, A. G., Burgay, M., Kramer, M., *et al.* (2004). A double-pulsar system: A rare laboratory for relativistic gravity and plasma physics, *Science*, **303**, p. 1153.

Malaver, M. (2014). Strange Quark Star Model with Quadratic Equation of State, arXiv preprint arXiv:1407.0760.

Margalit, B., Berger, E., and Metzger, B. D. (2019). Fast radio bursts from magnetars born in binary neutron star mergers and accretion induced collapse, *The Astrophysical Journal*, **886**, p. 110.

Margalit, B. and Metzger, B. D. (2017). Constraining the maximum mass of neutron stars from multi-messenger observations of GW170817, *The Astrophysical Journal Letters*, **850**, p. L19.

Martinez, J. G., Stovall, K., Freire, P. C. C., *et al.* (2015). Pulsar J0453+1559: A double neutron star system with a large mass asymmetry, *The Astrophysical Journal*, **812**, p. 143.

Maxted, N. I., Ruiter, A. J., Belczynski, K., Seitenzahl, I. R., and Crocker, R. M. (2020). A Supernova Remnant Associated with a Nascent Black Hole Low-mass X-ray Binary, arXiv preprint arXiv:2010.15341.

Müller, H. and Serot, B. D. (1996). Relativistic mean-field theory and the high-density nuclear equation of state, *Nuclear Physics A*, **606**, p. 508.

Muñoz-Darias, T., Casares, J., and Martínez-Pais, I. G. (2005). The K-correction for irradiated emission lines in LMXBs: Evidence for a massive neutron star in X1822–371 (V691 CrA), *The Astrophysical Journal*, **635**, p. 502.

Müther, H., Prakash, M., and Ainsworth, T. L. (1987). The nuclear symmetry energy in relativistic Brueckner-Hartree-Fock calculations, *Physics Letters B*, **199**, p. 469.

Nice, D. J. (2003). *The International Astronomical Union (IAU) Proceedings*, Sydney, Australia.

Nomoto, K. (1984). Evolution of 8–10 solar mass stars toward electron capture supernovae. I-formation of Electron-degenerate O+ NE+ MG cores, *The Astrophysical Journal*, **277**, p. 791.

Nomoto, K. and Kondo, Y. (1991). Conditions for accretion-induced collapse of white dwarfs, *The Astrophysical Journal*, **367**, p. L19.

O'Boyle, M. F., Markakis, C., Stergioulas, N., and Read, J. S. (2020). A parametrized equation of state for neutron star matter with continuous sound speed, *Physical Review D*, **102**, p. 083027.

Oppenheimer, J. R. and Volkoff, G. M. (1939). On massive neutron cores, *Physical Review*, **55**, p. 374.

Özel, F. and Freire, P. (2016). Masses, radii, and the equation of state of neutron stars, *Annual Review of Astronomy and Astrophysics*, **54**, p. 401.

Özel, F., Psaltis, D., Narayan, R., and Villarreal, A. S. (2012). On the mass distribution and birth masses of neutron stars, *The Astrophysical Journal*, **757**, p. 55.

Özel, F., Psaltis, D., Güver, T., *et al.* (2016). The dense matter equation of state from neutron star radius and mass measurements, *The Astrophysical Journal*, **820**, p. 28.

Patton, R. A. and Sukhbold, T. (2020). Towards a realistic explosion landscape for binary population synthesis, *Monthly Notices of the Royal Astronomical Society*, **499**, p. 2803.

Podsiadlowski, P., Langer, N., Poelarends, A. J. T., *et al.* (2004). The effects of binary evolution on the dynamics of core collapse and neutron star kicks, *The Astrophysical Journal*, **612**, p. 1044.

Prakash, M., Cooke, J. R., and Lattimer, J. M. (1995). Quark-hadron phase transition in protoneutron stars, *Physical Review D*, **52**, p. 661.

Qian, Y.-Z. and Wasserburg, G. J. (2007). Where, oh where has the R-process gone?, *Physics Reports*, **442**, p. 237.

Rajagopal, K. and Wilczek, F. (2001). The Condensed Matter Physics of QCD, *At The Frontier of Particle Physics: Handbook of QCD (In 3 Volumes)*, p. 2061.

Ransom, S. M., Stairs, I. H., Archibald, A. M., *et al.* (2014). A millisecond pulsar in a stellar triple system, *Nature*, **505**, p. 520–524.

Rawls, M. L., Orosz, J. A., McClintock, J. E., *et al.* (2011). Refined neutron star mass determinations for six eclipsing X-ray pulsar binaries, *The Astrophysical Journal*, **730**, p. 25.

Read, J. S., Lackey, B. D., Owen, B. J., and Friedman, J. L. (2009). Constrains on a phenomenologically parameterized neutron-star equation of state, *Physical Review D*, **79**, p. id. 124032.

Reardon, D. J., Hobbs, G., Coles, W., *et al.* (2016). Timing analysis for 20 millisecond pulsars in the parkes pulsar timing array, *Monthly Notices of the Royal Astronomical Society*, **455**, p. 1751.

Rezzolla, L., Most, E. R., and Weih, L. R. (2018). Using gravitational-wave observations and quasi-universal relations to constrain the maximum mass of neutron stars, *The Astrophysical Journal Letters*, **852**, p. L25.

Rezzolla, L., Pizzochero, P., Jones, D. I, Rea, N., and Vidaña, I. (2018). *The Physics and Astrophysics of Neutron Stars*, Springer, Berlin, Germany.

Rhoades, C. E. and Ruffini, R. (1974). Maximum mass of a neutron star, *Physical Review Letters*, **32**, p. 324.

Roberts, M. S. E. (2012). Surrounded by spiders! new black widow and redbacks in the galactic field, *Proceedings of the International Astronomical Union*, **8**, p. 127.

Rocha, L. S., Bernardo, A., Horvath, J. E., Valentim, R., and De Avellar, M. G. B. (2019). The distribution of neutron star masses, *Astronomische Nachrichten*, **340**, p. 957.

Rocha, L. S., Bernardo, A., De Avellar, M. G. B., and Horvath, J. E. (2020). Exact solutions for compact stars with CFL quark matter, *International Journal of Modern Physics D*, **29**, p. id. 2050044.

Romani, R. W, Filippenko, A. V., and Silverman, J. M. (2012). PSR J1311-3430: A heavyweight neutron star with a flyweight helium companion, *The Astrophysical Journal Letters*, **760**, p. L36.
Ruderman, M. (1972). Pulsars: Structure and dynamics, *Annual Review of Astronomy and Astrophysics*, **10**, p. 427.
Ruiter, A. J., Ferrario, L., Belczynski, K., *et al.* (2019). On the formation of neutron stars via accretion-induced collapse in binaries, *Monthly Notices of the Royal Astronomical Society*, **484**, pp. 698–711.
Ruiz, M., Shapiro, S. L., and Tsokaros, A. (2018). GW170817, general relativistic magnetohydrodynamic simulations, and the neutron star maximum mass, *Physical Review D*, **97**, p. 021501.
Schwab, J., Podsiadlowski, P., and Rappaport, S. (2010). Further evidence for the bimodal distribution of neutron-star masses, *The Astrophysical Journal*, **719**, p. 722.
Shao, D.-S., Tang, S.-P., Sheng, X., *et al.* (2020). Estimating the maximum gravitational mass of non-rotating neutron stars from the GW170817/GRB 170817A/AT2017gfo observation, *Physical Review D*, **101**, p. 063029.
Shapiro, S. L. and Teukolsky, S. A. (2008). *Black Holes, White Dwarfs, and Neutron Stars: The Physics of Compact Objects*, John Wiley & Sons, Hoboken, New Jersey, USA.
Sharma, R. and Maharaj, S. D. (2007). A class of relativistic stars with a linear equation of state, *Monthly Notices of the Royal Astronomical Society*, **375**, p. 1265.
Sharma, S. (2017). Markov chain monte carlo methods for bayesian data analysis in astronomy, *Annual Review of Astronomy and Astrophysics*, **55**, p. 213.
Shibata, M., Zhou, E., Kiuchi, K., and Fujibayashi, S. (2019). Constraint on the maximum mass of neutron stars using GW170817 event, *Physical Review D*, **100**, p. 023015.
Siess, L. (2007). Evolution of massive AGB stars-II model properties at non-solar metallicity and the fate of super-AGBStars, *Astronomy & Astrophysics*, **476**, p. 893.
Stairs, I. H. (2003). Testing general relativity with pulsar timing, *Living Reviews in Relativity*, **6**, p. 5.
Steeghs, D. and Jonker, P. G. (2007). On the mass of the neutron star in V395 Carinae/2S 0921–630, *The Astrophysical Journal Letters*, **669**, p. L85.
Stovall, K., Freire, P. C. C., Antoniadis, J., *et al.* (2019). PSR J2234+0611: A new laboratory for stellar evolution, *The Astrophysical Journal*, **870**, p. 74.
Sukhbold, T., Ertl, T., Woosley, S. E., Brown, J. M., and Janka, H.-T. (2016). Core-collapse Supernovae from 9 to 120 solar masses based on neutrino-powered explosions, *The Astrophysical Journal*, **821**, p. 38.
Suwa, Y., Yoshida, T., Shibata, M., Umeda, H., and Takahashi, K. (2018). On the minimum mass of neutron stars, *Monthly Notices of the Royal Astronomical Society*, **481**, p. 3305.
Tauris, T. M., Kramer, M., and Freire, P. C. C. (2017). Formation of double neutron star systems, *The Astrophysical Journal*, **846**, p. 170.

Tetarenko, B. E., Bahramian, A., and Arnason, R. M., *et al.* (2016). The first low-mass black hole X-ray binary identified in quiescence outside of a globular cluster, *The Astrophysical Journal*, **825**, p. 10.

The university of Arizona (2020). Neutron Star Data (xtreme.as.arizona.edu/).

Thirukkanesh, S. and Ragel, F. C. (2012). Exact anisotropic sphere with polytropic equation of state, *Pramana*, **78**, p. 687.

Thorsett, S. E. and Chakrabarty, D. (1999). Neutron star mass measurements. I. Radio pulsars, *The Astrophysical Journal*, **512**, p. 288.

Tian, W. and Leahy, D. (2004). 20 pairs of real pulsar/supernova remnant associations, *Symposium-International Astronomical Union*, **218**, p. 137.

Timmes, F. X., Woosley, S. E., and Weaver, T. A. (1996). The neutron star and black hole initial mass function, *The Astrophysical Journal*, **457**, p. 834.

Tolman, R. C. (1934). *Relativity, Thermodynamics, and Cosmology*, Clarendon, Oxford, UK.

Tomsick, J. A., Gelino, D. M., Halpern, J. P., and Kaaret, P. (2004). The low quiescent X-ray luminosity of the neutron star transient XTE J2123–058, *The Astrophysical Journal*, **610**, p. 933.

Ugliano, M., Janka, H.-T., Marek, A., and Arcones, A. (2012). Progenitor-explosion connection and remnant birth masses for neutrino-driven supernovae of iron-core progenitors, *The Astrophysical Journal*, **757**, p. 69.

Valentim, R., Rangel, E., and Horvath, J. E. (2011). On the mass distribution of neutron stars, *Monthly Notices of the Royal Astronomical Society*, **414**, p. 1427.

Van den Heuvel, E. P. J. (2004). X-ray Binaries and their Descendants: Binary radio pulsars; evidence for three classes of neutron stars?, *5th INTEGRAL Workshop on the INTEGRAL Universe*, **552**, p. 185.

Van den Heuvel, E. P. J. (2018). High-Mass X-ray Binaries: Progenitors of double compact objects, *Proceedings of the International Astronomical Union*, **14**, p. 1.

Van Kerkwijk, M. H., Breton, R. P., and Kulkarni, S. R. (2011). Evidence for a massive neutron star from a radial-velocity study of the companion to the black widow pulsar PSR B1957+20, *The Astrophysical Journal*, **728**, p. 95.

Van Leeuwen, J., Kasian, L., and Stairs, I. H. (2015). The binary companion of young, relativistic pulsar J1906+0746, *The Astrophysical Journal*, **798**, p. 118.

Wang, B. and Liu, D. (2020). The formation of neutron star systems through accretion-induced collapse in white-dwarf binaries, *Research in Astronomy and Astrophysics*, **20**, p. 135.

Weber, F. (2017). *Pulsars as Astrophysical Laboratories for Nuclear and Particle Physics*, Routledge, London, UK.

Weisberg, J. M., Nice, D. J., and Taylor, J. H. (2010). Timing measurements of the relativistic binary pulsar PSR B1913+16, *The Astrophysical Journal*, **722**, p. 1030.

Witten, E. (1984). Cosmic separation of phases, *Physical Review D*, **30**, p. 272.

Woosley, S. E., Sukhbold, T., and Janka, H.-T. (2020). The birth function for black holes and neutron stars in close binaries, *The Astrophysical Journal*, **896**, p. 56.
Wu, X. H., Du, S., and Xu, R. X. (2020). What if the neutron star maximum mass is beyond $\sim 2.3\,M_{\odot}$?, *Monthly Notices of the Royal Astronomical Society*, **499**, p. 4526.
Zhang, C. M., Wang, J., and Zhao, Y. H. (2011). Study of measured pulsar masses and their possible conclusions, *Astronomy & Astrophysics*, **527**, p. A83.
Zhu, W. W., Stairs, I. H., Demorest, P. B., *et al.* (2015). Testing theories of gravitation using 21-year timing of pulsar binary J1713+0747, *The Astrophysical Journal*, **809**, p. 41.

https://doi.org/10.1142/9789811220944_0002

Chapter 2

The Micro-physics of the Quark-nova: Recent Developments

Rachid Ouyed
Department of Physics and Astronomy, University of Calgary, Calgary, Alberta, Canada
rouyed@ucalgary.ca

Abstract: The Bodmer-Terazawa-Witten hypothesis of absolutely stable quark matter made of up, down, and strange quarks ((u,d,s) matter) is of much astrophysical and fundamental interest. One consequence is a large binding energy release associated with the conversion of a neutron star to a quark star. A quantitative understanding of the dynamic aspect of the conversion is necessary in order to find out whether this energy is released quietly or in an explosive manner. We address numerically (i.e. solving the reaction-diffusion-advection equations for (u,d) to (u,d,s) combustion) the dynamic processes through which the conversion appears. We find fundamentally very different results from semi-analytic calculations, with front speeds that are several orders of magnitude higher for the former. Resolving the hadronic-quark-matter interface is necessary, since approximations like Coll's condition may quench the burning, while properly resolving the flame can make combustion always thermodynamically favourable if the hypothesis of absolutely stable strange quark matter is true. We find that lepton physics, including weak decays, electron equation-of-state (EoS), neutrino EoS, and neutrino transport are at the very least as important to the physics of the burning front as the EoS of highly dense matter. In concert, these effects induce novel wrinkling instabilities (such as the deleptonization instability) possibly leading to a deflagration-detonation and/or a quark-core collapse Quark-Nova.

Keywords: neutron stars; nuclear matter; quark matter; quark deconfinement; phase-transition; combustion.

1. Introduction

Quark deconfinement appears at extremely high temperatures or densities. The reason is due to the property of asymptotic freedom where high momentum exchange between quarks weakens the attractive interaction between them. In Earth-based experiments, particle accelerator experiments tap into the high temperature regime by triggering very high energy collisions. In neutron stars, deconfinement would appear at low temperature but high ($\sim 10^{15}\,\mathrm{g/cm^3}$) densities. Given the high Fermi energies and therefore high momentum exchanges in the cores of compact stars, one can expect the nucleation of quark matter inside compact stars. A very consequential possibility to entertain is that there can exist stable deconfined quark matter at zero pressure. The reason is that although empirically we do not find two flavoured (up (u), down (d)) quark matter at zero pressure, the addition of an extra degree of freedom such as strange quarks (s) could decrease the Fermi energy for the same baryon number density, lowering the quark matter's free energy below the free energies of both hadronic and two flavoured quark matter. This hypothesis is referred as the Bodmer–Witten–Terasawa hypothesis (BWTH) (Bodmer, 1971; Terazawa, 1979; Witten, 1984).

Specifically, the BWTH requires that the state of matter with the lowest binding energy is (u,d,s) quark matter. The main reason is that the existence of a third degree of freedom in the form of s-quarks lowers the Fermi energy of the matter in general. Using a simple approximation of quark matter in the form of a Fermi gas with a constant B that acts as the confinement pressure, using the MIT Bag Model (Johnson, 1975), one can derive the EoS:

$$P_q = P_{fermi} - B; \tag{1}$$

$$e_q = 3P_{fermi} + 4B, \tag{2}$$

where P_q is the pressure of the quark gas, P_{fermi} is an ordinary Fermi pressure, and e_q is the energy density of quark matter. One can find a range of bag constants where (u,d,s) quark matter is lower than the hadronic binding energy of $\sim$930 MeV but at the same time, where (u,d) matter has a higher binding energy than hadronic matter. This hypothesis therefore argues that macroscopic objects made of (u,d,s) matter are thermodynamically plausible.

Since this hypothesis was posed, many interesting scenarios have been postulated in both astrophysics and particle physics. For example, the existence of pure strange quark stars and fragments of (u,d,s) matter called strangelets. The search for strangelets is currently an active research program, with experiments both trying to detect incoming strangelets in the form of cosmic rays, or attempting to produce strangelets in particle accelerators (Weber, 2005). An important aspect of strange quark stars is that they can be very phenomenologically distinct to more ordinary neutron stars, which is very interesting from an observational and theoretical perspective (Weber, 2005). An interesting argument put forward by Drago and Pagliara (2016) is that (u,d,s) stars hold the key to solve the so-called "hyperon puzzle". Burning of a hadronic star into a quark or a hybrid star, where current hadronic EoS are soft because of the appearance of hyperons at high nuclear densities, leads to maximum masses of hadronic stars of less than $2\,M_\odot$, which is incompatible with current observations. Drago and Pagliara (2016) argue that massive stars of $\sim 2\,M_\odot$ could be (u,d,s) stars, and therefore hadronic stars coexist with quark stars.

The conversion of hadronic to quark matter could appear in the following way: Once two flavoured quark matter is nucleated in the core of neutron stars, the weak interaction can turn some of the d quarks into s-quarks, lowering the Fermi energy of the quark matter. Because at this point, the free energy of (u,d,s) matter is lower than the free energy of hadrons, the hadrons accreted by the quark core would find it energetically favourable to deconfine into lower energy quark matter. Eventually, the quark core would grow, engulfing the whole compact star, turning it into a pure (u,d,s) star. There are other alternate scenarios for conversion of a whole compact star to a (u,d,s) star, for example, through "seeding" of cosmic strangelets (e.g., Olinto (1987)), or dark-matter annihilation in neutron stars heating up parcels of neutron star matter which would make it favourable for the creation of quark bubbles (Perez-Garcia *et al.*, 2010).

Beyond the existence of macroscopic objects like strange quark stars, another interesting consequence of the BTWH is the release of large amounts of energy when hadronic matter converts to (u,d,s) matter. Assuming a bag constant of $B = 145\,\text{MeV}$, using the above model the energy per baryon becomes $\sim 840\,\text{MeV}$ which is roughly 100 MeV less than for ordinary hadronic matter (930 MeV) (Weber, 2005). This implies that a conversion from hadronic to (u,d,s) matter should release about 100 MeV per converted baryon. Assuming a neutron star has about 10^{57} baryons, conversion of every baryon into (u,d,s) matter would generate $\sim 10^{53}$ erg in radiation and kinetic energy, which is in the same order of magnitude of typical explosive events in astrophysics such as core collapse supernovae.

Although, the 10^{53} erg of energy release predicted by energetics compares favourably to explosive events like supernovae, it does not pinpoint to whether this energy is released explosively or in a slow simmer. Since the 1980s, different groups have tried to figure out the phenomenology of this energy release. Olinto (1987) pioneered a hydrodynamic formalism for exploring the conversion of hadronic to (u,d,s) quark matter as hydrodynamic combustion. They modelled the conversion as a "combustion front" that "burns" hadronic fuel into (u,d,s) ash. However, the exact equations that govern this reaction zone, the reaction-diffusion-advection equations cannot be solved in analytic form since they are nonlinear. So Olinto (1987) had to linearize the equations and impose mechanical equilibrium and come up with semi-analytic, steady state solutions. This study yielded timescales of conversion for the whole compact star of minutes to days, which would imply a slow simmer, since the timescale of a supernova explosion is about 1 s.

Another pioneering paper was published by Benvenuto and Horvath (1989) in the late 1980s. In contrast to Olinto (1987), they assumed an initial shock and solved the relativistic jump conditions. This model leads to a steady state solution but without the mechanical equilibrium assumed by Olinto (1987). This approach yields a supersonic detonation, which Benvenuto and Horvath (1989) argue can impact enough kinetic energy to make a core collapse supernova explode. Although this explosive solution contrasts to Olinto (1987)'s much slower timescales, the reason is that a shock is assumed aprioristically, with arguments that the initial deconfined bubble of quark matter creates a sharp pressure discontinuity. Drago *et al.* (2007) also solved the jump conditions without assuming mechanical equilibrium, but in their case they found that the combustion takes the form of subsonic deflagration.

All of the literature on hadron-quark combustion before the 2010s can be roughly categorized as following either the mechanical equilibrium approach (Olinto, 1987) or the jump condition approach (Drago *et al.*, 2007), and always assuming a steady state. Because of the variety of the assumptions, such as whether pressure equilibrium is assumed or not, or whether a shock is hypothesized as an initial condition or not, the timescales of conversion predicted for the compact star varied by various orders of magnitude, from milliseconds to days.

However, Niebergal *et al.* (2010) published for the first time a study that solved the reaction-diffusion-advection equations for hadron-quark combustion numerically. This study coupled transport, chemical, and entropic processes into a numerical simulation. Not only was the burning velocity much faster than many of the previous estimates, but also Niebergal *et al.* (2010) hinted that leptons may trigger feedbacks that may accelerate to burning front into supersonic detonation or quench it. Their argument was based on solving the jump conditions and parameterizing the cooling behind the front.

Later on, Ouyed *et al.* (2018a) solved the reaction-diffusion-advection equations and coupled them to neutrino transport using a flux limited diffusion scheme, and added an electron EoS. Ouyed *et al.* (2018a) confirmed numerically that leptons can trigger extreme feedbacks, with the burning halting completely for certain choices of initial conditions. Ouyed *et al.* (2018a)'s study was important in so far that it showed that due to nonlinear couplings between lepton physics and hydrodynamics, the simulation is extremely sensitive to the details of neutrino transport. This hinted that the system is genuinely a nonlinear dynamical process, and that simplifying it by imposing mechanical equilibrium or steady state conditions was extremely inaccurate.

1.1. *Quark-nova*

The usual picture of stellar evolution for massive ($>8\,M_\odot$) stars is that after they have burnt their core into iron through fusion, nuclear fusion becomes energetically unfavourable and therefore there is not sufficient energy that is being released to counter gravitational contraction. At some point, the star suffers gravitational collapse, triggering a spectacular core collapse supernova, leaving behind a neutron star or a black hole.

The quark-nova hypothesis argues that this traditional picture of stellar evolution is not the whole story. There is the possibility that the neutron star would experience further collapse into a (u,d,s) star. Such a phenomenon would be of similar energetics as a core collapse supernova, with about $\sim 10^{53}$ erg released of both chemical and gravitational binding energy. Such an addition to the stellar evolution picture has immense phenomenological consequences. For one, this hypothesis essentially argues that the neutron star explodes. The neutron-rich ejecta released from the outer layers of the exploding neutron star is also a very suitable site for nuclEoSynthesis and r-process elements, since there is a very low proton fraction (Y_e).

This neutron star explosion adds both an extra source of energy and an extra source of nuclEoSynthesis elements (Jaikumar *et al.*, 2007; Keränen *et al.*, 2005; Ouyed *et al.*, 2002).

There's been various observational phenomena that point to the explosion of a neutron star. Beyond the raw energetic and nuclEoSynthesis budget issues, were traditional models have a hard time accounting for all the "sources" of energy and nuclEoSynthesis in the galaxy, we find that there are phenomenological consequences. For example, most models that try to explain the large luminosities and kinetic energy of super-luminous supernovae do so by using a "point source" that injects energy into an envelope of 1-20 $M_{\odot}$, whether this point source is a core collapse supernova, or a magnetar (Sukhbold and Woosley, 2016). However, in the case of transforming a core-collapse supernova's energy into luminosity by shocking it with a 1 $M_{\odot}$ envelope or "wall", one has to explain the source of the envelope itself, which is a non-trivial problem. For the case of the magnetar, one has to assume almost 100% efficiency of conversion between rotational energy to luminosity/kinetic energy of the envelope (Sukhbold and Woosley, 2016). Furthermore one has to explain the source of the large magnetic field.

The necessity of low electron fraction environments for r-process nuclEoSynthesis may also suggest the explosion of a neutron star. A common model that makes neutron stars "explode" is through the neutron star merger, where tidal forces rip apart some of the neutron rich material, making this an ideal site for nuclEoSynthesis. However, it's been pointed out that the coalescence rate of neutron stars is too low to explain the r-process abundances in old, metal-poor stars (Jaikumar *et al.*, 2007). Recently, a study "exploded" neutron stars by arguing that primordial, microscopic black holes may rip neutron stars apart, generating copious r-process materials (Sasaki *et al.*, 2018). This model was suggested partly to solve the timescale constraints of neutron star mergers.

Given the observational evidence, and the nuclEoSynthetic constraints that suggest that a neutron star should explode, we find that a very natural way of triggering the detonation of a neutron star is simply postulating a "quark core collapse" where the neutron star simply collapses into a more compact (u,d,s) configuration, releasing massive amounts of energy.

There are multiple ways that a quark-nova could be triggered. There are two "mechanisms" for initiating the combustion of a neutron star into a quark star. One mechanism relates to the core of a neutron star somehow reaching sufficiently high densities that favour the deconfinement of quark matter. These nucleated (u,d) quark bubbles in turn would beta equilibrate into (u,d,s) matter, and then in accordance to the BWTH, grow, engulfing the whole neutron star (Lugones, 2016). However, the density at which quark matter deconfines is very uncertain. It could be that most neutron stars achieve the deconfinement density and therefore turn into quark stars. However if the deconfinement density is higher than the one of the average core of a neutron star, then one could achieve the sufficiently high density through other processes such as accretion, fall-back from supernovae, or spin-down evolution, leading to a two family compact star scenario, where quark stars and neutron stars coexist (Drago and Pagliara, 2016).

The other mechanism for triggering a quark-nova could be through "seeding". According to the BWTH, a strangelet that interacts with hadronic matter can convert the latter provided there is not an electrostatic barrier preventing the interaction. Since neutrons do not have a charge, neutron stars are an ideal site for strangelet contamination. The source of strangelets can be arbitrary — for example, cosmic strangelets that are released by a (u,d,s) star merger, or strangelets formed through the annihilation of dark matter in the core of neutron stars.

Although the quark-nova hypothesis is very phenomenologically rich and it could have immense astrophysical consequences, its physical realization requires that the conversion of hadronic to (u,d,s) matter is violent enough, which ultimately depends on the micro physics and the hydrodynamics of the conversion front.

2. Microphysics

2.1. *Particle physics*

The theory that ultimately treats the strong force, which governs the behaviour of quarks, is quantum chromodynamics (QCD). The Lagrangian is

$$\mathcal{L} = \overline{\psi}^a_f (i\gamma_\mu D^\mu_{ab} - m_f)\psi^b_f - \frac{1}{4} F^i_{\mu\nu} F^{\mu\nu}_i, \tag{3}$$

where D^μ_{ab} is the color gauge-covariant derivative, ψ^a_f are the quark fields for each flavour f, and m_f the quark masses. $F^u_{\mu\nu}$ is the gluon field tensor. This Lagrangian is symmetric under SU(3).

The equations of motion are:

$$\left(i\gamma_\mu \partial^\mu - m_f\right)\psi^a_f = -g_s\gamma_\mu \left(\frac{\lambda}{2}\right)\psi^b_d G^\mu_i; \tag{4}$$

and

$$\partial^\mu F^i_{\mu\nu} + g_s f_{ijk} G^{i\mu} F^k_{\mu\nu} = -g_s \overline{\psi}^a_f \gamma_\mu \left(\frac{\lambda}{2}\right) \psi^b_f. \tag{5}$$

Solving these equations of motions yields the behaviour of the quark fields. These equations are not analytically solvable unless computed at very high momentum exchanges, where the coupling constant is low enough to allow for the use of perturbation theory. These high momentum exchanges correspond to really large densities or temperatures, and the technique used to solve the equations of motion for those thermodynamic environments is called perturbative QCD. It corresponds to a very useful tool for high energy physics experiments, as often the energies are so high that the strong coupling constant becomes small enough to allow for QCD perturbation theory.

However in the regime of compact stars, the densities are not high enough to allow for perturbation theory. Instead, one is forced to solve the equations of motion directly through numerical-computational techniques. A way to solve these equations of motion numerically is through a technique called Lattice QCD, where the space-time is discretised into a computational lattice. However Lattice QCD faces a very extreme computational challenge called the sign problem. The solutions to the equations oscillate very quickly between negative and positive values, so numerical techniques tend to erroneously average into zero, making the authentic solution intractable.

Thus, for the case of compact star, we are forced to use what are referred as phenomenological models. Those models are not derived directly from the QCD equations of motions but instead capture certain symmetries and behaviours in a posteriori way. For example, two processes that are very important in QCD is confinement and chiral symmetry breaking. Confinement is the process that holds quarks together in nuclei. Symmetry breaking is the process that makes quarks massive in nuclei. For example, a proton has a mass of about 938 MeV while up and down quarks have bare masses in the order of few MeV, which means about 99% of quark masses is gained through "chiral symmetry" breaking.

Most phenomenological models either include confinement or chiral symmetry breaking. A very popular phenomenological model that we should explore is the MIT Bag Model (Johnson, 1975), which models confinement but not chiral symmetry breaking. We superficially touched upon this model in the Introduction.

The MIT Bag Model (Johnson, 1975) assumes that nucleons can be modelled as a bag of free flowing quarks. The reason is that asymptotical freedom dictates that quarks are weakly interacting at short distances, so one can model quarks at high densities as a free quark gas. Confinement is modelled as a bag, where the bag "confines" the free flowing quarks into a sphere. In the case of a nucleon, one can model it as a bag of three quarks. For example, a bag representing a proton would have two u quarks, and one d quark.

Now one can take this model and construct a quark star. A quark star would be simply a large nucleon, a bag of about $\sim 10^{57}$ quarks.

One can model mathematically this bag of free quarks through the following way. Because we assume we are in the regime of asymptotic freedom, where the coupling constant between quarks is negligible, one can model the quarks as a gas of free fermions. This will give you a Fermi–Dirac distribution:

$$f(p) = \frac{1}{\exp(\frac{E-\mu}{T} + 1)}. \tag{6}$$

For a pure Fermi–Dirac gas, the EoS is.

$$n = \frac{g}{(2\pi)^3} \int f(p) d^3p; \tag{7}$$

$$e = \frac{g}{(2\pi)^3} E(p) \int f(p) d^3p; \tag{8}$$

$$p = \frac{g}{(2\pi)^3} \int \frac{p^2}{3E} f(p) d^3p, \tag{9}$$

where n is the baryon number density, e is the energy density, p is the pressure and g is a degeneracy constant. The bag constant simply acts as a negative pressure that confines the gas into a bag. If we make a Sommerfield approximation for degenerate matter, that is, with the Fermi momentum $p_f >> T$, where T is the temperature, we get:

$$n = \frac{gT^3}{6\pi} \left[\pi^2 \left(\frac{\mu}{T}\right) + \left(\frac{\mu}{T}\right)^3 \right]; \tag{10}$$

$$e = \frac{7g\pi^2 T^4}{120} \left[1 + \frac{30}{7\pi^2} \left(\frac{\mu}{T}\right)^2 + \frac{15}{7\pi^4} \left(\frac{\mu}{T}\right)^4 \right]. \tag{11}$$

Finally we can use these expressions for a fermi-gas and derive the MIT quark EoS by simply adding a correction to the bag constant B:

$$n_f = n; \tag{12}$$

$$e_f = e + B; \tag{13}$$

$$P_f = 3e - 4B. \tag{14}$$

where f stands for flavour.

What is the astrophysical connection to these equations? It depends on the value of the bag constant. Thermodynamics may allow for the existence of pure strange stars, which follows from the existence of stable strange quark matter at zero pressure. However, if $B^{1/4}$ is above about 145 MeV, then quark matter may not exist at zero pressure, and it is only thermodynamically favourable at the higher densities in the core of a compact star. This gives rise to a hybrid star with a core of quark matter and an outer layer of hadronic matter.

We can use the EoS to model strange quark stars. If we solve the Tolman–Oppemheimer–Volkoff (Oppenheimer and Volkoff, 1939; Tolman, 1934) (TOV) equations, which are the hydrostatic equilibrium equations for a general relativistic object, we can find the relationship between radius and mass of a star:

$$\frac{dr}{da} = \frac{1}{4\pi r^2 n_B e^{\lambda}}; \tag{15}$$

$$\frac{dm}{da} = \frac{\epsilon}{n_B e^{\lambda}}; \tag{16}$$

$$\frac{d\phi}{da} = \frac{e^{\lambda}}{4\pi r^4 n_B}(m + 4\pi r^2 P); \tag{17}$$

$$\frac{dP}{da} = -(\epsilon + P)\frac{e^{\lambda}}{4\pi r^4 n_B}(m + 4\pi r^3 P); \tag{18}$$

$$e^{-\lambda} = \sqrt{1 - \frac{2m}{r}}, \tag{19}$$

where m the gravitational mass enclosed at radius r, a is the enclosed baryon number at radius r, ϵ is the energy density, and n_B is the baryon number density.

However, it turns out that the simplest version of the MIT Bag Model equations cannot sustain compact stars above two solar masses, which contradicts current observations (Demorest *et al.*, 2010). On the other hand, it turns out that we can add a strong coupling constant a_s, which stiffens the EoS. A stiffer EoS implies there is more pressure for a given density. Doing so we can get 2 solar mass stars:

$$P_f(a_s) = P_f - \left[\frac{7}{60}T^4\pi^2\frac{50a_s}{21\pi} + \frac{2a_s}{\pi}\left(\frac{1}{2}T^4\mu^2 + \frac{\mu^4}{4\pi^2}\right)\right], \tag{20}$$

where P_f is the EoS of quark matter for flavour f without perturbative corrections.

What is the mechanism through which a neutron star converts to a quark star or a hybrid star? Most of the studies concerning the appearance of quark matter in compact stars depend on hydrostatic equilibrium approximations, which are computed using the TOV equations. The timescale of hydrostatic equilibrium should be larger than $\sim R/v_s$, where R is the radius of the star and v_s is the sound speed. For a degenerate matter sound speed of $v_s \sim 1\sqrt{3}c$ where c is the speed of light, and for $R \sim 10\,\text{km}$, the timescale should be larger than about $10^{-4}\,\text{s}$. For steady states, one can assume hydrostatic equilibrium equations with longer timescales and a phase transition modelled through the Gibbs criterion:

$$G_Q = G_H; \quad T_Q = T_H; \quad P_Q = P_H, \tag{21}$$

where the subscript Q stands for quark matter and H for hadronic matter. G is the Gibbs free energy, T is the temperature, and P is the pressure. In other words, a phase transition appears if the two phases are in chemical, thermal, and mechanical equilibrium.

However, the idea of mechanical equilibrium only makes sense if the hydrodynamic processes are too fast compared to the timescales of the whole system. When there is a pressure gradient, hydrodynamic processes will produce sound waves that may even destroy the pressure gradient, leading the system to mechanical equilibrium. On the other hand, for hydrodynamic processes to destroy the pressure gradient, the system under study must change at a much slower rate than the hydrodynamic timescales. In case of a steady state compact star in hydrodynamic equilibrium, it makes sense to assume both phases are in mechanical equilibrium by considering the timescale is large enough so that all the transient effects subsided. In order to track the dynamical physical processes that lead to the phase transition which converts hadronic to (u,d,s) quark matter, one cannot use the Gibbs condition, because the assumption of mechanical equilibrium is too stringent. We will explain this statement further in the following sub-section.

2.2. *Thermodynamics of phase transitions*

There is a long tradition of using Gibbs condition to model the phase transition of hadronic quark matter. As mentioned before, however, much of this work has to do with modelling steady states for hydrostatic compact stars, where the timescales are long enough to assume mechanical equilibrium. However, many studies that look at the phase transition as a dynamic process, rather than just a steady state system, still use the Gibbs condition (e.g., Furusawa *et al.* (2016)). Therefore, we will comment in this section about the different free energy functionals and how they are related to the physics of the problem and to the second law of thermodynamics.

The second law of thermodynamics states that the entropy, S, of the universe must increase:

$$dS > 0; \tag{22}$$

this law implies that a physical process will spontaneously happen if it raises the entropy of the Universe.

So, in order to predict if a process will spontaneously appear, we would have to calculate whether the process increases the entropy of the whole universe. However this is obviously unfeasible, so we instead approximate the part of the Universe we are studying as an isolated system. For example, a volume of gas in contact with a thermal bath. Since we cannot calculate the entropy in the whole universe, we must make approximations of the "second law" that are applicable to isolated systems. This is where the free energy functional comes about.

The two major free energy functions are the Gibbs Free Energy:

$$G = U + PV - TS, \tag{23}$$

and the Helmholtz Free Energy:

$$F = U - TS, \tag{24}$$

where U is the internal energy, T is the temperature, and P is the pressure.

The key aspect is that under specific conditions, the change of entropy is equal to the change of a free energy function, provided certain constraints are met. These constraints are related to fixing certain thermodynamic variables as constant. For example, let's find the conditions where the differential change of Gibbs Free Energy, dG, is equal to the differential increase of entropy, as predicted by the second law of thermodynamics. We can start with with G and taking its differential:

$$dG = dU + PdV + VdP - TdS - SdT, \tag{25}$$

from the first law of thermodynamics we know that:

$$dU = dQ - PdV. \tag{26}$$

Inputting the first law into dG we get:

$$dG = dQ + VdP - TdS - SdT. \tag{27}$$

We know that the second law implies $TdS > dQ$. In order to make the change dG equivalent to the second law, we must therefore fix T and P as constants, which would mean $dP = 0$ and $dT = 0$; then we find:

$$dG = dQ - TdS. \tag{28}$$

Because TdS must be larger than dQ, then according to the second law, $dG < 0$.

In other words if we fix T and P as constants, then a spontaneous process must decrease the Gibbs Free Energy. Therefore, only matching dT and dP to zero makes $dG < 0$ obey the second law of thermodynamics. We can make a similar derivation for the case of F. If V and T are constants, we would find that $dF < 0$ is equivalent to the second law.

Now, since these functionals are only equivalent to the second law if we fix the right variables, one must use the appropriate functional for the specific problem at hand. In other words, using dG is only appropriate if the change in pressure and temperature is very small that can be approximated away as zero. Similarly, we can only use dF if we can assume the volume and the temperature don't change much.

Let's get back to the problem of hadron-quark phase transition. We know from the previous section that dG is generally used in hadron-quark phase transitions, which implies that dP and dT are equal to zero. However, we believe this choice is inappropriate for modelling the phase transition as a dynamical process. In order to understand what free energy functional we must use, we should look closely at the timescales present in the hadron-quark process. Because the quark-nova makes use of the BWTH, our calculations will be based under that hypothesis.

In the case of the BTWH, conversion is spontaneously possible if there are enough s-quarks in the vicinity of the (u,d,s) matter's interface with hadronic matter. This is because s-quarks lower the Fermi surface of strange quark matter, making (u,d,s) matter more entropically favourable than hadronic matter. Since we are dealing with fluids, we are dealing with a hydrodynamic problem.

First let's analyze whether using dG is accurate for this model. The Gibbs condition assumes that pressure and temperature are changing slowly enough to be considered constants. What are the processes that govern the evolution of P and T? In the case of a hadron-quark phase transition, we are dealing with a transition governed by the strong interaction. The strong interaction is extremely fast, as once a phase transition from hadronic to quark matter is energetically favourable, for example, in the presence of sufficient s-quarks, or due to a sufficiently high density, the strong interaction can deconfine hadrons into quarks in a timescale of $\tau_s = 10^{-24}$ s. In other words, at that timescale, assuming high enough s-quark fraction or baryon number density, a fluid parcel will undergo an abrupt change in pressure since the strong interaction will change the EoS from hadronic to quark matter, which would lead to a large dP. Assuming the hydrodynamic approximation, during that fast timescale, the baryon number density will not change but pressure will. This is because in hydrodynamics, the mass density must be conserved, and therefore it will only change through transport processes. The counteracting process that would smoothen the pressure gradient is hydrodynamics, through sound waves. However, hydrodynamic processes are much slower that the strong interaction. Assuming the interface between hadrons and quarks is at most a centimeter long (e.g., Niebergal *et al.* (2010) and Ouyed *et al.* (2018a)), then a sound wave going at a speed $1/\sqrt{3}c$ would cross the interface in about $\tau_h = 10^{-10}$ s. Since $\tau_h \gg \tau_s$, dP cannot be modelled as zero in a hydrodynamic simulation of hadron-quark phase transition, therefore we must use Helmholtz free energy.

So for phase transitions, instead of the Gibbs conditions, we must use the Helmholtz conditions. In this case, we know $dV = 0$ and $dT = 0$, so the conditions for a phase transition using the Helmholtz conditions are:

$$F_Q = F_H; \quad T_Q = T_H; \quad n_{B,Q} = n_{B,Q}. \tag{29}$$

These conditions follow from the equivalent ones behind the Helmholtz free energy, where V and T are held fixed, since baryon number density is equivalent to the inverse of the volume. Helmholtz's free energy represents a more precise framework than Gibbs's condition not only to describe the hadron-quark phase transition but also for hydrodynamic simulations. The main reason is that pressure is not a "conserved" quantity in hydrodynamic simulations but density is. Because density is automatically outputted by most hydrodynamic numerical schemes, then fixing the density (volume) as required for F is fairly easy, as the free energy of the phases can be calculated explicitly from the density. In contrast, using G would be much more computationally intensive. In hydrodynamic simulations, modelling the phase transition through the Gibbs condition would require to invert the EoS using root-finding algorithms that check each computational zone for whether the other phase at the same pressure has a lower free energy in order to trigger a transition. This is because the only way to check whether a phase is energetically favourable is to "root-find" at what density does the other phase have the same pressure, since pressure is not a conserved quantity, and then check whether the free energy at this density is lower for the alternative phase.

What are the phenomenological differences between choosing a Helmholtz free energy functional over the corresponding Gibbs one? A common framework used to model the phase transition from hadronic to quark matter is using the Coll's condition (Coll, 1976). Coll's condition states that a phase transition from hadronic to quark matter appears if the energy density of quark matter is lower than the energy density of hadronic matter at a fixed pressure, in other words:

$$e(P)_H > e(P)_Q, \tag{30}$$

where e is the energy density, P is the pressure, H denotes the hadronic matter and Q the quark matter. This is almost equivalent to the Gibbs condition, as often a lower energy density would entail a lower free energy, as

$$G = e + P - Ts, \tag{31}$$

where s represents the entropy density. However it turns out that this condition is extremely stringent, since this condition only allows this phase transition at very high densities, of the order the saturation nuclear density, ρ_0. This in turn leads to a situation where the neutron star cannot convert completely into a quark star, since the Coll's condition is violated at lower densities. However this is analytically contradictory since by design, the BTWH should allow the conversion of the whole compact star.

Finally we should focus on the thermodynamics of the weak interaction. As mentioned above, hadronic matter will convert to quark matter if it is energetically favourable, or if in other words, the Helmholtz free energy is lower for quark matter than for hadronic matter. This transformation will be a phase transition triggered by the strong interaction. However, the most interesting dynamical part of the transition lies in the BTWH assumption that strange quark matter has the lowest binding energy. So once two flavour quark matter starts equilibrating through the weak interaction, the chemical process will be very exothermic. This equilibration is done through the following weak interaction reactions:

$$u + e^- \leftrightarrow s + \nu_e, \tag{32}$$

$$u + e^- \leftrightarrow d + \nu_e, \tag{33}$$

$$u + d \leftrightarrow u + s. \tag{34}$$

The weak interaction will equilibrate (u,d,s) matter in a timescale of 10^{-8} s. If we assume u,d and s quarks are massless, we can show analytically about how much energy is released. Following the methodology from Weber (2005) we can calculate the energy released per baryon for the equilibration of a two flavoured quark parcel into a three flavoured parcel. For a cold Fermi gas we can use the following relations:

$$n_f = \mu_f^3/\pi^2, \tag{35}$$

$$e_f = 3\mu_f^4/(4\pi^2) + B, \tag{36}$$

$$P_f = \mu_f^4/(4\pi^2) - B. \tag{37}$$

We will begin by calculating the binding energy of two flavoured quark matter. First, electric charge neutrality requires that $n_d = 2n_u$. Using the charge neutrality relations and equating pressure into zero we find that:

$$e = 4Bn_B = (n_u + n_d)/3, \tag{38}$$

where n_B is baryon number density. One can obtain the energy per baryon for two flavour quark matter:

$$4\frac{4B}{n_B} = 934\,\text{MeV} \times B_{145}^{1/4}, \tag{39}$$

where $B_{145}^{1/4} = B^{1/4}/145\,\text{MeV}$.

We can make a similar exercise for three flavour quark matter. In the three flavour quark matter case, $n_s = n_d = n_u$. Using a similar approach to two flavour quark matter, we will find that:

$$\frac{4B}{n_B} = 829 \times B_{145}^{1/4}\,\text{MeV}. \tag{40}$$

This difference arises from the fact that there is an extra degree of freedom in strange quark matter. What this calculation shows is that weak equilibration will release about 100 MeV in heat in a timescale of 10^{-8} s. Assuming cooling is slow compared to the timescale of weak equilibration (10^{-8} s), we can roughly calculate the temperature of quark matter by equating the average thermal energy of each baryon $\sim 3T$ where T is the temperature, to the energy released. For $3T = 100$ MeV, the temperature of the three flavoured quark matter ash will be about $T \sim 33$ MeV! This is an incredibly large temperature, only surpassed by the early quark gluon plasma universe, particle accelerator experiments, and comparable to the temperatures found in the heated cores of supernova explosions. If we take the 100 MeV of heat released per baryon, we can calculate the energy release of a whole neutron star converted into a quark star, we will get about 10^{53} erg which is comparable to the total energy released by a core collapse supernova explosion.

Although these thermodynamic calculations point to a powerful release of energy, the thermodynamic arguments are not enough. One has to study how the burning front will propagate. For example, if the propagation of the burning front is too slow, the energy released may not be able to be transformed into kinetic energy in a sufficiently efficient way, with the energy simply leaking out slowly as transparent neutrinos, which would not be very interesting from an observational standpoint. Therefore the next task is to include the thermodynamics outlined above into a hydrodynamic simulation, that consistently calculates how the weak interaction gives rise to particle and temperature spatial gradients that in turn trigger pressure gradients. A pressure gradient acts as a source of momentum density in the fluid, transforming some of the energy released into mechanical energy. Since the quark-nova hypothesis is based on the detonation of a neutron star to a quark star, finding the hydrodynamic processes that transform the heat released into mechanical energy is necessary. We will explore the hydrodynamics of this event in the following subsection.

2.3. *Hydrodynamics*

As mentioned in the subsection above, we know that assuming the BTWH, the conversion of two flavoured quark matter into three flavoured quark matter will release about 100 MeV in heat per baryon, where part of it can be transformed into mechanical energy. The appropriate framework to study this generation of mechanical energy is through hydrodynamics, since at the temperatures and densities that are characteristic of the core of a compact star, both quark and hadronic matter behave like fluids.

There are two possible avenues for the conversion of hadronic to quark matter under the BTWH. One of them is sudden deconfinement of an extended volume of hadronic matter. Under certain conditions, such as supersonic core-collapse of a supernova, the central density can rise very rapidly to sufficiently high densities so that quark deconfinement is energetically favourable. In that case, an extended hadronic volume of the order of kilometers can be rapidly converted into quark matter. The sudden nucleation of quark matter would lead to a discontinuous pressure gradient, which could generate a powerful shock induced conversion that would compress hadronic matter into quark matter supersonically.

The other possible avenue is a diffusion induced conversion. As mentioned before, under the BWTH, a large fraction of strangeness could make quark matter more stable than hadronic matter, given the former's reduction of the fermi energy. In this scenario, a small bubble of stable strange quark matter could be nucleated somehow, and then the bubble would expand, creating a well defined conversion front. The conversion would be ultimately decided by the transport of s-quarks into the interface, as the hadronic matter would convert to strange quark matter if a sufficiently high s-quark fraction is present.

The width of the reaction zone is imposed by the weak interaction reactions, which regulate the dynamics behind the interface. These weak reactions are shown in Eqs. (32)–(34). These reactions are coupled to the reaction-diffusion-advection equations that govern the burning in one dimension are given as (Niebergal *et al.*, 2010; Ouyed *et al.*, 2018a):

$$\frac{\partial n_i}{\partial t} = -\nabla \cdot (n_i v - D_i \nabla n_i) + R_i, \tag{41}$$

$$\frac{\partial (hv)}{\partial t} = -\nabla (hv^2) - \nabla P, \tag{42}$$

$$\frac{\partial s}{\partial t} = -\nabla \cdot (sv) - \frac{1}{T}\sum_i \mu_i \frac{dn_i}{dt} + \frac{1}{T}\frac{d\epsilon_{\nu_e}}{dt}. \tag{43}$$

Here, the variables represent the following parameters: i runs through all the particle species (u, d, s, e, ν), v is the velocity, h is the enthalpy density, n_i is the number density for species i, s is the entropy density, T is the temperature, P is the pressure, R_i is a reaction source term for species i, D is the diffusion coefficient.

Given the richness in particles and processes, the reaction zone is fairly complex. Figure 1, which is a snapshot of a simulation, is included in order

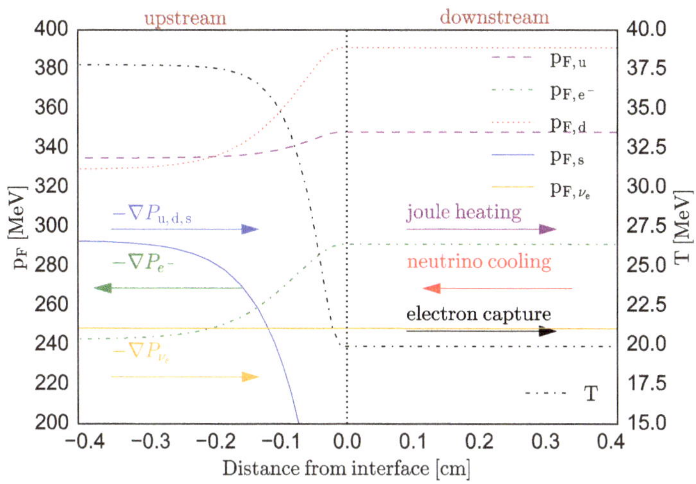

(a) Simulation snapshot of the burning interface.

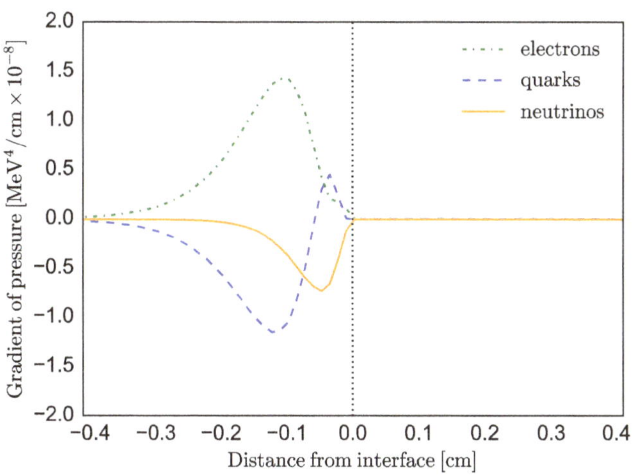

(b) Pressure gradients for the leptons and quarks shown in the upper panel.

Fig. 1. (a) $p_{F,i}$ are the Fermi momenta for particles i, and T is the temperature. In both panels, the interface lies at position zero depicted by the vertical line. The arrows represent the directions of the force vectors and their labels depict the processes that caused them. (b) Upstream is the side behind (left side of the vertical line) the interface, and downstream is the side in front of it (right side of the vertical line). Figures and caption are taken from Ouyed *et al.* (2018a).

to exemplify this complexity. Various processes, such as weak reactions that create and destroy particles, entropy generation through weak decay equilibration, hydrodynamic and diffusion processes that transport the s-quarks, and the evolution of pressure due to changes in particle numbers and temperature, are coupled nonlinearly in the above equations. For example, particle densities are coupled to the entropy terms, as their reaction rates affect the temperature gradients, and the entropy evolution affects the reaction rates as well as these rates are functions of temperature. Furthermore particle creation and destruction is in turn coupled to the very nonlinear momentum Eq. (42), as the ∇P term is affected by the entropy generation and particle change through modification of the EoS of quark matter, since pressure is both a function of temperature and particle numbers. These nonlinear couplings give rise to all sorts of feedbacks, especially in the momentum Eq. (42), as the term $\nabla(hv^2)$ is very nonlinear. In the study of Ouyed *et al.* (2018a) we found that these nonlinearities give rise to a richness in behaviours.

In order to simplify our discussion of these processes, we should emphasize that the dynamics of the conversion ultimately rely on the transport of s-quarks into the interface. The s-quarks act as oxidants of the combustion, there must be enough of them available in the interface so that it is energetically favourable that a parcel of hadronic matter converts into quark matter. Let us define the hadronic side of the interface as downstream and the quark matter side as upstream. There will be processes that inject momentum in the interface towards the upstream direction, which transports the s-quarks away from the interface. These processes will be labeled as quenching, as ultimately they prevent s-quarks from arriving downstream and triggering the conversion. On the other hand, there are processes that inject momentum in the downstream direction, accelerating s-quarks towards the interface. These processes will be labeled as enhancing.

We shall use the terminology and analysis adopted in Ouyed *et al.* (2018a) and enumerate both the enhancing and quenching processes (see Fig. 1).

Quenching:

- Flavour equilibration: The weak decays will convert (u,d) into (u,d,s) matter, turning some of the chemical potential energy into heat, elevating the temperature behind the interface. Because the quark EoS has a T^4 dependence, this will lead to an increase of pressure behind the interface, which in turn pushes the s-quarks towards the downstream direction.

- Joule Heating: The weak decays in quark matter lead to an increase of electron neutrino number density behind the interface. Since higher neutrino density leads to a higher chemical potential, a neutrino chemical potential is imposed in the interface, with the quark side having the higher potential. As neutrinos diffuse from higher to lower chemical potential, they deposit the difference in chemical potential as heat. This increases the temperature behind the front and thus the pressure, which accelerates the interface towards the downstream direction.
- Electron capture: Since the lowest binding energy of (u,d,s) matter is when roughly there are the same number of u,d,s quarks, which is equivalent to a zero electron fraction, the equilibrium seeking weak decays will eat up the electrons and convert them into neutrinos in order to make the quantities of u,d, and s quarks more symmetric. Since usually electrons have a higher chemical potential than neutrinos, these reactions release heat and raise the temperature behind the front.
- Neutrino Pressure: Neutrinos deposit momentum into the reaction zone, in the downstream direction. This accelerates the interface and the transport of s-quarks.

Now we shall enumerate the quenching processes:

- Electron Pressure: As mentioned above, electrons are eaten up by the weak decay reactions in order to equilibrate (u,d,s) matter. Since the most stable form of (u,d,s) matter is symmetric in the number of u,d, and s quarks, charge neutrality will lead to the weak decays to transform electrons into neutrinos. These reactions will create a an electron number density dip behind the interface. This will lead to the electron EoS triggering a pressure gradient that pushes the front interface upstream backwards.
- Neutrino Cooling: Since the neutrinos carry energy density away, as the neutrinos leak, there will be reduction of the temperature behind the front. Such a reduction of temperature behind the interface will push the front backwards, into the upstream direction.

All these processes are related to the nonlinear term in the momentum Eq. (42), that is $\nabla(hv^2)$, since these processes lead to nonlinear feedbacks that affect the direction and magnitudes of the fluid velocities, as the change in chemical composition and temperature affect the Eq. (42)'s source term, ∇P.

Let us be more precise about the meaning of a feedback. A feedback process is one where the output of a system is fed into its input as well. A positive feedback leads to the amplification of certain output variable that is in turn plugged into the input, exponentially amplifying it. A negative feedback in contrast, negates the change of an output.

Given the nonlinear couplings across the the reaction diffusion advection equations, we should expect all sorts of feedbacks. However, recently Ouyed *et al.* (2018a) explored a particular source of positive feedbacks, which are leptons. One of the positive feedbacks that seemed to dominate is related to the electron pressure process noted above.

2.4. *Numerical simulations*

The quark-nova group has built a multi-physics simulation code called Burn-UD. This code is specifically tailored for solving the reaction-diffusion-advection equations enumerated in the previous subsection. The code specifically simulates a time-dependent reaction zone between hadrons and quarks, which includes various hydrodynamic, radiative and chemical processes. Weak decay terms, neutrino transport, and the EoSs of hadronic, quark, neutrinos, and electrons, are coupled to the hydrodynamic equations that regulate mass conservation, momentum conservation, and entropy (or energy) conservation.

The timescales and length-scales used in Burn-UD are microphysical, in other words, representative of the reaction zone. Degenerate matter in the cores of neutron stars often can have fluid velocities in the order of $v = 10^{-3}$c due to convective motion caused by degeneracy pressures and thermal gradients. From here we can begin to start calculating the length-scales and timescales to set up the simulation. If the weak-decay timescale, is 10^{-8} s, then a d quark will be advected by fluid velocities for $\tau = 10^{-8}$ s until it decays into an s quark. This will define the length of the reaction zone. The length of the reaction zone therefore is $l = v \times \tau \sim 0.1$ cm.

From these timescales one can estimate the size of the timestep and the length of the computational zone Δx. Assuming we need at least ten zones to resolve the length of the reaction zone, then $\Delta x \sim 0.01$ cm. From here we can calculate the timestep size necessary for the stability of the hydrodynamic scheme. If the sound speed of degenerate matter is of the order of $v_s \sim c$ then we can calculate the necessary timestep through the Courant–Friedrichs–Lewy condition:

$$\frac{v_s \Delta t}{\Delta x} \leq 1. \tag{44}$$

The Courant–Friedrichs–Lewy condition simply states that information must not propagate beyond the length of a computational zone during a timestep for stability reasons. This condition is necessary but not sufficient for the stability of explicit schemes of hydrodynamics. Respecting this condition, we get a timestep of the order of 10^{-13} s. The whole timescale of the simulation corresponds to the timescale of weak decay, as it is slower than hydrodynamics — therefore the timescale of 10^{-8} s corresponds to about 10^5 timesteps.

In order to couple all these different physical processes, specialized numerical techniques had be to be applied to Burn-UD. The advection terms in the equations were evolved with a fourth order Runge Kutta time integration scheme, and spatially discretized with a third order, flux limited scheme. The flux limiter was applied because the higher order spatial scheme can lead to numerical instabilities in sharp gradients of the conserved quantities, so the flux limiter automatically imposes a first order scheme at sharp interfaces.

For the neutrino transport terms, we had to use an implicit time integration scheme. This is because the timescales of neutrino transport are shorter than for hydrodynamics. This can be proved by looking at the Courant–Friedrichs–Lewy conditions, which roughly represent how long does it take for information to propagate through a computational zone. The sound speed of degenerate matter is of the order of c, so the hydrodynamic flux is equal to $F_H \sim u \times c$, where u is a conserved quantity. Meanwhile, for neutrinos following Fick's law (which applies for the high densities of a compact star), in other words, neutrinos that "diffuse" through a random walk, the diffusion flux if we assume a mean free path λ_C is equal to $F_D \sim \lambda c/\Delta x$, where Δx is the length of the computational zone. Calculating the ratio we find,

$$\frac{F_D}{F_H} = \frac{\lambda}{\Delta x}. \tag{45}$$

Using the fiducial values of $\lambda = 1\,\text{cm}$ and $\Delta x = 0.01\,\text{cm}$, we find a ratio of 100, which means neutrinos transport information roughly one hundred times faster than fluid motions! In such a case, using an explicit temporal integration scheme would be extremely limiting, as the fast neutrinos would need a timestep that is 100 times smaller than the hydrodynamic scheme, which would make the simulation computationally intensive. Implicit schemes, do not have the same problem of information propagating too fast through a computational zone, as implicit schemes, rather than using the information of the past timestep and propagating it to the next timestep, which is the case for explicit schemes, implicit schemes solve a system of linear equations that include both past and future states simultaneously. The disadvantage of implicit schemes is that they are more numerically complicated to implement. Furthermore, implicit schemes tend to be generally of lower order than explicit integration schemes, which "washes out" sharp gradients in an artificial sense. Still in our case, using an implicit scheme was necessary for the neutrino case, as otherwise the simulation would have been computationally prohibitive.

Niebergal *et al.* (2010) ran the first simulations with the Burn-UD code. Their code was one dimensional, and assumed that quark matter had already deconfined, and therefore approximated the fuel side as two flavour quark matter without making use of a hadronic EoS. They also assumed that the combustion was instantiated in cold, $T = 0.1\,\text{MeV}$ neutron matter at a low $Y_e \sim 0.1$ proton fraction. The code used equilibrium seeking reaction terms (Anand *et al.*, 1997) — which led to a combustion front powered by the entropy release of two flavoured quark matter equilibrating into three flavoured quark matter. Neutrino physics were implemented as a leakage scheme, which involved modelling neutrino emission as an entropy sink.

One of the most important result of these initial studies was that the burning speeds were $v = 0.001c$–$0.01c$ which was orders of magnitude faster than the results calculated in previous literature. The authors argued that the reasons of the higher burning speeds were related to certain erroneous simplifications previous studies did, such as equating burning speed with fluid velocities, and not considering entropy release from the chemical reactions.

Finally, another mayor result was the discovery of a new mechanism for slowing down the burning speed, which was neutrino physics. Niebergal *et al.* (2010) discovered that when they enhanced the neutrino cooling, the simulation would compute lower burning speeds than when the neutrinos were turned off. This finding motivated a semi-analytical study were the authors solved the jump conditions for combustion from two flavoured to three flavoured quark matter, parameterizing neutrino cooling as a reduction of temperature of the three flavoured quark ash. They found that for temperature reductions as low as 0.1 MeV, the burning front could halt.

This halting effect is related to two counteracting tendencies. First, there is a fermi momenta term in the EoS that reduces the pressure as the three flavoured ash gets equilibrated. This is because three flavoured quark matter is softer than (u,d) matter, since the inclusion of a third degree of freedom reduces the fermi momenta of the quarks. However, there is a counteracting term in pressure that is actually increased through chemical equilibration, since the release of heat stiffens the EoS through a temperature term proportional to T^4. When there is no neutrino cooling, the T^4 term dominates over the fermi momenta term that is reduced through equilibration, giving the burning front a net positive speed. However, if temperature is reduced beyond a certain critical point, the T^4 term becomes smaller than the Fermi momenta term, causing the burning speed to become zero. Since the temperature dependence is to the fourth power, even a small reduction of 0.1 MeV can cause the burning front to halt. The authors argued that this halting process could give rise to a new type of hydrodynamic-leptonic instability — an instability that couples hydrodynamic processes with weak interaction process. They referred to this instability as a deleptonization instability, since the halting is triggered by the deleptonization through neutrino leakage.

Although, Niebergal *et al.* (2010) diagnosticated the existence of a deleptonization instability through semi-analytic arguments, the existence of this instability was confirmed numerically by Ouyed *et al.* (2018a, 2018b). Ouyed *et al.* (2018a, 2018b)'s treatment differed from the previous treatment of Niebergal *et al.* (2010) through the introduction of more sophisticated neutrino physics — with a flux-limited, diffusion scheme for neutrino transport that replaced the more basic leakage scheme in the previous treatment of Niebergal *et al.* Furthermore, Ouyed *et al.* (2018a, 2018b) also added the EoSs of electrons and neutrinos to the simulation, which effectively tracked change in pressure gradients due to the evolution of neutrino and electron number densities. Another difference is that Ouyed *et al.* (2018a, 2018b)'s treatment considered a protoneutron star as initial conditions, i.e., high proton fraction ($Y_e \geq 0.2$) and high temperature ($T \geq 10\,\mathrm{MeV}$). In contrast, Niebergal *et al.* (2010)'s initial treatment considered a cold neutron star (0.1) and low proton fraction ($Y_e = 0.1$). The assumption of proto-neutron star justified the treatment of neutrino transport as a Fick's law of diffusion.

The inclusion of neutrino transport and EoS effects of neutrinos and electrons gave rise to the deleptonization instability originally hinted by Niebergal *et al.* (2010). The equilibration of (u,d) to (u,d,s) matter behind the interface converts much of the initial electron content into neutrinos that get diffused out of the quark matter. Neutrinos diffusing out of the interface carry away entropy that in turn leads to a reduction of temperature behind the front. Furthermore, the depletion of electrons behind the front creates an electron dip that leads to a degeneracy pressure gradient that pushes against the interface, slowing it down. Under certain choices of parameters, the combined effect of the entropy leakage caused by neutrino emission and the degeneracy pressure gradient caused by the electron dip can quench almost completely the front. In fact, it turned out that in some cases the burning front only revived when the neutrinos are trapped, since the trapped neutrinos would deposit entropy and momentum in the reaction zone, triggering a positive burning speed.

To return to our discussion of nonlinearities, we can parse the halting processes as a positive feedback. This positive feedback ends up halting the burning front in the following way: electron capture behind the interface causes a dip in the electron number density, triggering pressure gradients that push the interface backwards. As the interface begins to slow down because of the electron pressure, this in turn makes the electron dip sharper, as the interface is moving too slow forward in order to "smoothen" the electron gradient. A more acute electron gradient leads to a more acute pressure gradient, slowing the interface further. As the interface slows down, the reaction processes that enhance the burning, such as heat released through flavour equilibration, or neutrino momentum injected due to the flow of neutrinos, are also weakened, as the interface is absorbing less fuel to "burn".

This circuit of cause and effect that amplifies the pressure gradient that slows down the burning can halt it completely for certain cases, such as when the neutrinos are free streaming (Fig. 2).

This very important effect halting induced by lepton physics implies a couple of important things. First, that lepton physics are at least as important as the details of the high density EoS. Much literature focuses on details of the EoS of quark matter and hadronic matter as the most uncertain part of the study, where lepton physics is often not included. Since leptons are crucial to whether the burning front halts or revives, then their details are at least as equally important as the details of the EoS. Another important lesson derived from lepton physics is that multidimensional simulations are necessary to calculate the true timescale of the combustion process. Much of existing studies (e.g., Furusawa *et al.* (2016)) simply assume a steady state approach to the reaction zone and extrapolate the burning speed to the scales of the whole compact star. Yet, since leptons can halt the burning front, anisotropies in the lepton number density could cause uneven halting across the interface, leading to instabilities and the loss of a well defined interface. Such an effect may render the "laminar" approximation of the

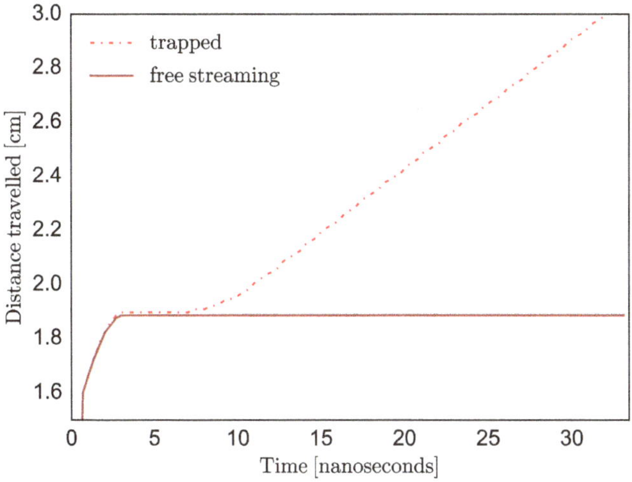

Fig. 2. Distance travelled by the combustion front as a function of time from a numerical simulation. The line labeled as "free streaming" represents the burning front with neutrinos free streaming, while the line labeled as "trapped" plots the burning front with trapped neutrinos. Notice how the front halts for the remainder of the simulation for the free streaming case. The thermodynamic parameters for the simulation were an initial temperature of $T = 20\,\mathrm{MeV}$, an initial lepton fraction of $Y_{\mathrm{L}} = 0.2$, and an initial baryonic density of $n_{\mathrm{B}} = 0.35\,\mathrm{fm}^{-3}$. Caption and figure taken from Ouyed *et al.* (2018b).

flame ill-suited and requires a multidimensional simulation to calculate the true timescale of the combustion process of the whole star.

2.5. *Deleptonization instability*

As mentioned in the above sections, the coupling of hydrodynamics, chemical production of entropy, and the chemical evolution and transport of leptons, lead to the creation of instabilities in the burning interface. This is because of the existence of feedbacks that may amplify initial perturbations in the front.

Niebergal *et al.* (2010) outlined how the deleptonization instability would look in a multidimensional situation for the cold neutron star case. Since neutrinos carry away energy density, they decrease the entropy behind the burning front, decreasing the temperature. Since chemical equilibration behind the front softens the non-thermal part of the EoS through an introduction of the s-quark degree of freedom, the decrease of thermal pressure can lead to a situation where the ash exerts less pressure than the fuel, making the interface halt.

Now, in the context of a multidimensional situation, the interface will not be homogenous, and instead there will be inhomogenities across the interface in density, lepton number density, and temperature. In such a situation, some parts of the interface will halt before others, since the feedbacks that halt the interface are sensitive to lepton density and temperature, as mentioned in the above sections. Uneven halting will create shearing and wrinkling in the interface.

Ouyed *et al.* (2018a) reinforced the possibility of this instability by noticing that the implementation of more sophisticated lepton physics reinforced the instability. However, not only the leakage of neutrinos reduce the temperature, and therefore the thermal pressure sustaining the front, but at the same time, the loss of leptons reduces the degeneracy pressure behind the front of electrons and neutrinos, intensifying the instability. In fact, Ouyed *et al.* (2018a) proved that the interface can halt with numerical simulations, which was previously only suggested with semi-analytical toy models.

This instability complicates the analysis, since the existence of this process would nullify the approximation of a well defined front. At this point, the 1D analysis that we have previously done breaks down. So the question of the final fate of the compact star remains unknown. However we can explore qualitatively some of the possible routes of this instability. Here we sketch three possible routes:

- Quenching: It could be that the deleptonization instability leads to almost complete quenching (as in the 1D case). Quenching in this case does not mean that the burning speed is zero, rather that, there isn't a ram pressure behind the front that powers and sustains the combustion. Instead, combustion is driven through the slower process of fuel accretion — the under-pressurizing behind the front causes negative, fluid velocities that point inward, accreting the fuel into the interface. Such a process would be really slow with negligible mass ejection.
- Detonation: Since the wrinkling of the interface through the instability would lead to a larger surface area of the front, the instability could accelerate the interface since more fuel is absorbed per unit of time. If acceleration leads to supersonic burning speed, this could generate a detonation wave. This detonation wave would disrupt the outer layers of the compact star, causing mass ejection. Mass ejection is only possible in processes that are faster than the speed of sound, since otherwise the matter will remain bound by gravity and simply expand.

- Core collapse: Another interesting outcome is the one of core collapse. If the under-pressurizing of the ash due to energy loss of neutrinos is fast enough, the compact star could suffer a quark core collapse, and lead to an explosion similar to the supernova. A similar mechanism has been known throughout the literature in the context of quark deconfinement in supernova, where a phase transition from hadronic to quark matter if it appears abruptly could lead to a sudden softening of the EoS that triggers core collapse. In the case of the combustion explored in the context of the quark-nova, a sudden softening of the EoS can happen through the loss of energy through neutrinos.

3. Conclusion

According to the BTWH, the reason why hadronic matter does not spontaneously decay into (u,d,s) matter is that there is an intermediate higher energy state of (u,d) matter. In order to diminish this energy barrier, there needs to be enough s-quarks available to trigger the combustion process. Another way of stating this, is that s-quarks act as a catalyst that lowers the free energy barrier, allowing hadronic matter to decay into a lower state of (u,d,s) matter. This energy barrier could explain why (u,d,s) matter is much more scarce than hadronic matter to the point that we have not detected the former. Assuming s-quark seeding, the next challenge is to understand the details of the conversion proper which we explored in details in the quark-nova model. Simulations of the micro-physics of the quark-nova show that neutrino physics couple nonlinearly to the system, and can lead to a deleptonization instability; previously hinted at through semi-analytic, parameterized arguments (Niebergal *et al.*, 2010). When explored in time-dependent, numerical simulations (Ouyed *et al.*, 2018a, 2018b), it was discovered that the inclusion of a hadronic EoS can generate nonlinear thermodynamic effects where the coupling of neutrino transport and the free energy of the hadronic EoS can lead to quenching of the hadronic-to-quark-matter burning. This quenching appears since the hadronic fuel can absorb neutrinos emitted by the hot 3QM, which can lead to the erection of a free energy barrier that makes combustion thermodynamically unfavourable.

The largest consequence of the deleptonization instability is that the combustion front will not be laminar. Instead, it will wrinkle. This wrinkling would most likely lead to the acceleration of the burning speed, since wrinkling leads to a larger surface area of the flame, and therefore a higher rate of burning. Ultimately, the final fate of a neutron star combusting into a quark star would depend on this deleptonization instability. However, such instability can only be proved through a multidimensional code that takes into account lepton physics, so the question of the final fate of the compact star undergoing combustion is still an open question.

References

Anand, J., Goyal, A., Gupta, V., and Singh, S. (1997). Burning of two-flavour quark matter into strange matter in neutron stars and in super-nova cores, *The Astrophysical Journal*, **481**(2), p. 954.

Benvenuto, O. and Horvath, J. (1989). Evidence for strange matter in supernovae?, *Physical Review Letters*, **63**(7), p. 716.

Bodmer, A. (1971). Collapsed nuclei, *Physical Review D*, **4**(6), p. 1601.

Coll, B. (1976). Combustion fronts in relativistic magnetohydrodynamics. In *Annales de L'Institut Henri Poincare Section (A) Physique Theorique*, **25**, p. 363.

Demorest, P., Pennucci, T., Ransom, S., Roberts, M., and Hessels, J. (2010). A two-solar-mass neutron star measured using shapiro delay, *Nature*, **467**(7319), p. 1081.

Drago, A., Lavagno, A., and Parenti, I. (2007). Burning of a hadronic star into a quark or a hybrid star, *The Astrophysical Journal*, **659**(2), p. 1519.

Drago, A. and Pagliara, G. (2016). The scenario of two families of compact stars, *The European Physical Journal A*, **52**(2), p. 41.

Furusawa, S., Sanada, T., and Yamada, S. (2016). Hydrodynamical study on the conversion of hadronic matter to quark matter: I. shock-induced conversion, *Physical Review D*, **93**(4), p. 043018.

Gao, Z.-F., Shan, H., Wang, W., and Wang, N. (2017). Reinvestigation of the electron fraction and electron fermi energy of neutron star, *Astronomische Nachrichten*, **338**(9–10), p. 1066.

Iwamoto, N. (1982). Neutrino emissivities and mean free paths of degenerate quark matter, *Annals of Physics*, **141**(1), p. 1.

Jaikumar, P., Reddy, S., and Steiner, A. W. (2006). Strange star surface: A crust with nuggets, *Physical Review Letters*, **96**(4), p. 041101.

Jaikumar, P., Meyer, B. S., Otsuki, K., and Ouyed, R. (2007). Nucleo-synthesis in Neutron-rich Ejecta from Quark-novae, *Astronomy & Astrophysics*, **471**(1), p. 227.

Johnson, K. (1975). The MIT bag model, *Acta Phys. Polon. B*, **6**, p. 865.

Keränen, P., Ouyed, R., and Jaikumar, P. (2005). Neutrino emission and mass ejection in Quark-novae, *The Astrophysical Journal*, **618**(1), p. 485.

Lugones, G. (2016). From quark drops to quark stars, *The European Physical Journal A*, **52**(3), p. 53.

Niebergal, B., Ouyed, R., and Jaikumar, P. (2010). Numerical simulation of the hydrodynamical combustion to strange quark matter, *Physical Review C*, **82**(6), p. 062801.

Olinto, A. V. (1987). On the conversion of neutron stars into strange stars, *Physics Letters B*, **192**(1–2), p. 71.
Oppenheimer, J. R. and Volkoff, G. M. (1939). On massive neutron cores, *Physical Review*, **55**, p. 374.
Ouyed, R., Dey, J., and Dey, M. (2002). Quark-nova, *Astronomy & Astrophysics*, **390**(3), p. L39.
Ouyed, A., Ouyed, R., and Jaikumar, P. (2018a). Numerical simulation of the hydrodynamical combustion to strange quark matter in the trapped neutrino regime, *Phys. Lett. B*, **77**(10), p. 184.
Ouyed, A., Ouyed, R., and Jaikumar, P. (2018b). Hadron-quark combustion as a nonlinear, dynamical system, *Universe*, **4**(3), p. 51.
Pérez-García, M.Á., Silk, J., and Stone, J. R. (2010). Dark matter, neutron stars, and strange quark matter, *Physical Review Letters*, **105**(14), p. 141101.
Pons, J. A., Steiner, A. W., Prakash, M., and Lattimer, J. M. (2001). Evolution of protoneutron stars with quarks, *Physical Review Letters*, **86**(23), p. 5223.
Sasaki, M., Suyama, T., Tanaka, T., and Yokoyama, S. (2018). Primordial black holes perspectives in gravitational wave astronomy, *Classical and Quantum Gravity*, **35**(6), p. 063001.
Sukhbold, T. and Woosley, S. (2016). The most luminous supernovae, *The Astrophysical Journal Letters*, **820**(2), p. L38.
Terazawa, H. (1979). Super-hypernuclei in the quark-shell model, *Tokyo University Report*, **INS336**.
Tolman, R. C. (1934). *Relativity, Thermodynamics, and Cosmology*, Clarendon, Oxford, UK.
Weber, F. (2005). Strange quark matter and compact stars, *Progress in Particle and Nuclear Physics*, **54**(1), p. 193.
Witten, E. (1984). Cosmic separation of phases, *Physical Review D*, **30**(2), p. 272.

https://doi.org/10.1142/9789811220944_0003

Chapter 3

Astrophysics of Super-dense Matter: A Strangeon Conjecture

Chengjun Xia[*,§], Xiaoyu Lai[†,¶] and Renxin Xu[‡,||]

[*] *Center for Gravitation and Cosmology, College of Physical Science and Technology, Yangzhou University, Yangzhou 225009, China*

[†] *Department of Physics and Astronomy Hubei University of Education Wuhan 430205, P.R. China*

[‡] *School of Physics and KIAA Peking University Beijing 100871, P. R. China*

[§] *cjxia@yzu.edu.cn*

[¶] *laixy@hue.edu.cn*

[||] *r.x.xu@pku.edu.cn*

Abstract: What if normal baryonic matter is squeezed so that atomic nuclei come in close contact? This is a problem implying a broad interest in astrophysics, but is not solved because of non-perturbative issues of the strong interaction. The weak interaction also plays an important role there: although it is in the mainstream that neutronization occurs to make a state with asymmetric isospin, we are addressing that quark flavour-symmetry from 2 (u and d quarks) to 3 (u, d and s quarks) could be restored during squeezing. The latter is called strangeonization. Strangeon is conjectured to be the constituent of bulk strong matter, as an analogy of nucleon for an atomic nucleus. We will explain strangeon and strangeon matter in this chapter, as well as discuss their observational manifestations in astrophysics.

Keywords: dense matter; pulsars; elementary particles; cosmic ray; dark matter.

1. Introduction

The baryonic matter, taking on various forms in our Universe, is well described by the Standard Model of particle physics. However, the state of matter at extremely high densities created by the gravitational collapse of massive stars is still far from being understood, which is essential for us to unveil the nature of pulsar-like compact stars. The problem lies in the fact that so far no one is able to solve quantum chromodynamics (QCD) in the regions that are inaccessible by lattice QCD or perturbation theories, which involve dense matter at vanishing temperatures.[1] Due to the significant non-perturbative QCD effects, the inner structure of pulsar-like compact stars as well as the equation-of-state (EoS) of supra-nuclear dense matter are challenging in both physics and astronomy.

To understand the properties and phase states inside compact stars, significant efforts were made in the past decades, where many possibilities were proposed. In principle, there are six flavours of quarks as fundamental particles that could constitute baryonic matter, among which the up (u), down (d) and strange (s) quarks are called light quarks due to their small current masses. Atomic nucleus, the most familiar baryonic matter to us, is composed of nucleons that are made of only u and d valence quarks. In other words, the nucleon is the constituent of a two-flavour nucleus. For pulsar-like compact stars, however, it is still under debate if their main constitution is two-flavoured or three-flavoured matter, where the latter case is the focus of this chapter.

Three-flavoured bulk matter composed of nearly equal numbers of light (u, d and s) free quarks is speculated to be the absolutely stable ground state of strongly-interacting matter, which is known as Witten's conjecture (Witten, 1984). It is worth noting that, three-flavoured matter may not necessarily be composed of deconfined quarks. Instead, due to the non-perturbative effects of strong interaction, quarks inside pulsar-like compact stars may be grouped into clusters, similar to the cases of two-flavoured matter where u and d quarks are actually grouped into nucleons. Although Witten's conjecture was proposed based on "strange quark matter" that composed of deconfined and almost free quarks, it could still holds even if the quarks are clustered, i.e., the generalized Witten's conjecture. In this chapter, we will explain the notion of "strangeon" made of a cluster

[1] It is possible to solve SU(2) QCD with two colors (Bornyakov *et al.*, 2018), but not for the realistic situations with three colours.

of several light quarks, whose name is simply an abbreviation of "strange nucleon" (with strangeness degree of freedom).

This chapter is organized as follows. Beginning with an introduction about the strangeons and strangeon matter in Section 2, we propose that the pulsar-like compact stars born in supernova explosions would be actually strangeon stars, whose basic properties are given in Section 3. Astrophysics of strangeon stars are discussed in Section 4, including observations that could be understood from the surface and global properties of strangeon stars, as well as the consequences of merging double strangeon stars. The summary is given in Section 5.

2. Strangeons and Strangeon Matter

For colorsinglet multiquark states comprised of approximately equal numbers of u, d, and s quarks, we refer them as strangeons, which are quark clusters with baryon number $A > 1$, i.e., H-dibaryons, strange tribaryons, etc. The properties of strangeons in vacuum were studied extensively with lattice QCD (Beane *et al.*, 2011; Inoue *et al.*, 2011; Sasaki *et al.*, 2020) and QCD-inspired models such as the MIT bag model (Aerts *et al.*, 1978; Jaffe, 1977a,1977b; Liu and Wong, 1982; Maezawa *et al.*, 2005; Maltman, 1992; Mulders *et al.*, 1980), non-relativistic quark cluster model (Oka *et al.*, 1983; Oka and Takeuchi, 1991; Shen *et al.*, 1999; Straub *et al.*, 1988), Skyrme model (Balachandran *et al.*, 1984; Jaffe and Korpa, 1985; Kopeliovich *et al.*, 1992; Yost and Nappi, 1985), and diquark model (Lee and Yasui, 2009). It was shown that strangeons are likely unstable or weakly bound in vacuum. Nevertheless, the interaction among strangeons may be strongly attractive (Glendenning and Schaffner–Bielich, 1998; Lai *et al.*, 2013; Sakai *et al.*, 1997), then strangeons are expected to bind together and form a strangeon condensed matter, which may become more stable than nuclear matter. In such cases, the super-dense matter in bulk at realistic densities of pulsars, i.e., $n \approx 2$–$10n_0$ with n_0 being the saturated nuclear matter density, could be strangeon matter (Lai *et al.*, 2013; Lai and Xu, 2009). The corresponding compact stars are thus called strangeon stars (Xu, 2018).

2.1. *Neutronization or strangeonization?*

In 1932, Landau argued that the process of neutronization will take place in a "gigantic nucleus" when "the density of matter becomes so great that atomic nuclei come in close contact" (Landau, 1932). More specifically, protons are converted into neutrons via the weak reaction $p + e^- \rightarrow n + \nu_e$, so that the gigantic nucleus is formed mostly by neutrons, i.e., a neutron star. After that, Baade and Zwicky (1934) proposed that neutron stars can be formed in supernovae. It was not until 1967 that Hewish *et al.* (1968) detected the radio pulsars with neutron stars being the possible candidates. More pulsar-like objects were observed later on, e.g., millisecond pulsars, magnetars, dim isolated neutron stars, central compact objects (CCO), etc.

Nevertheless, despite the elegant physical picture proposed by Landau *et al.*, the nucleons are not elementary particles as initially considered in the 1930s. In fact, it was later shown that nucleons are comprised of u and d quarks, while there are in total six flavours of quarks (u, d, s, c, t, and b) forming various types of hadrons (Gell-Mann, 1964). Combined with their spin and color degrees of freedom, we expect extremely rich phase structure for cold dense matter (Alford *et al.*, 2008; McLerran, 2009). In particular, one should at least include the possible emergence of strangeness, with s quarks being just ~100 MeV more massive than u and d quarks, which is easily accessible inside compact stars.

In the final stage of core collapse supernova, baryonic matter in the core region of a massive star is being compressed drastically by gravity, where the density is expected to reach up to ten times of the nuclear saturation density. Consequently, the Fermi energy of electrons can reach ~500 MeV if the baryonic matter was comprised of same amounts of neutrons and protons. Then it is energetically favourable to eliminate electrons via weak reactions. Particularly, as illustrated in Fig. 1, there are two types of reaction processes that may take place in the core region of a massive star, i.e., the neutronization process $e^- + p \to n + \nu_e$ with nucleons being the fundamental degrees of freedom and the strangenization process $e^- + u \to s + \nu_e$ and $u + d \to s + u$ with quarks being the degrees

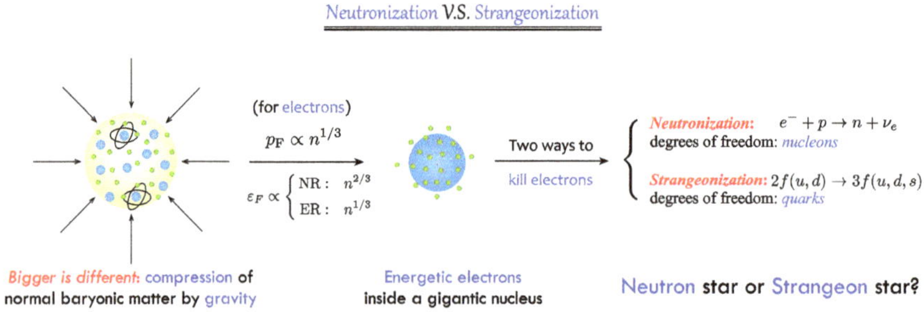

Fig. 1. Neutronization and Strangenization as two competing processes to eliminate energetic electrons (Xu and Guo, 2017).

of freedom. While neutronization eliminates electrons, strangenization can further reduce the energy and leads to the flavour symmetry for strongly interacting matter. Both processes are relevant to the nature of pulsar-like objects, as summarized in Fig. 1, to be neutron stars or strange stars.

Due to the complexity of non-perturbative QCD, so far it is still unclear what is the state of matter inside compact stars. In the past decades, many speculations about the nature of pulsars were proposed. In Fig. 2, we demonstrate different models for the inner structures of pulsars. Among them, hadron stars and hybrid stars are considered as conventional neutron stars, while quark stars and strangeon stars are strange stars with light flavour symmetry. In hadron stars, quarks are still confined within nucleons, hyperons, and/or mesons, while quark stars contain mostly deconfined quarks. A hybrid star includes ingredients of both the hadron stars and quark stars, where quark matter forms the core and hadronic matter constitute the outer regions with a possible quark-hadron mixed phase located in between (Xia *et al.*, 2020a). Unlike hadron stars or quark stars, the strangeon stars consist of individual clusters of quarks called strangeons.

Meanwhile, as indicated in Fig. 2, conventional neutron stars (hadron stars and hybrid stars) are gravity-bound, while strange stars (quark stars and strangeon stars) are self-bound by strong force with distinctive quark-vacuum interfaces on their surfaces. Such a feature difference can be very useful for us to identify them in astrophysical observations. In particular, even though the cores of conventional neutron stars may be made of various combinations of hadrons and deconfined quarks in chemical equilibrium, the astromaterials in the outer regions are well known and extensively studied (Caplan and Horowitz, 2017). Starting from the core-crust transition

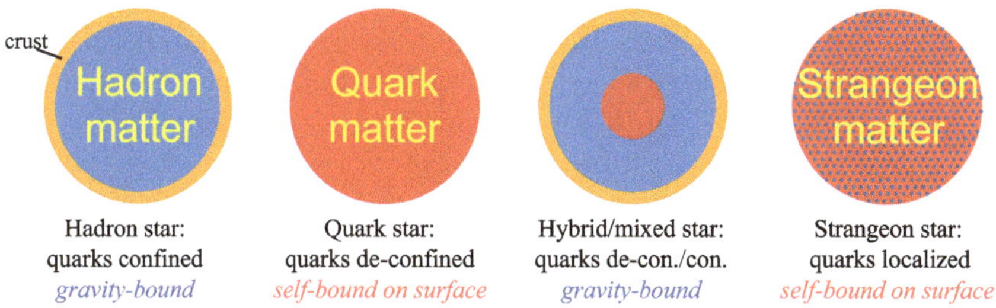

Fig. 2. Different models for the inner structures of pulsars: Hadron stars and hybrid stars are considered as conventional neutron stars, while strangeness plays an important role for strange stars (quark star and strangeon star) with three-light-flavour symmetry restored (Lai and Xu, 2017).

region to the surface of a neutron star, one expects an inner crust made of nuclear pastas, an outer crust made of Coulomb crystals of nuclei and highly degenerate electron gas, an ocean of ions and degenerate electrons, and a thin layer of atmosphere, where the density drops to zero as these materials are bound by gravity alone. Nevertheless, there is no clear observational evidence for any ordinary matter to exist on the surfaces of neutron stars, given the absent observation of any atomic emission or absorption lines in pulsars (Burwitz *et al.*, 2003; Drake *et al.*, 2002). Such a situation has a natural explanation for self-bound strange stars, where their surfaces are comprised of electrons in continuum states. Similar to traditional quark stars, the composition of strangeon stars is almost the same from centre to surface. The residual interaction between strangeons bind them together and form a condensed matter, leading to non-zero densities on the surface with vanishing pressure, which is similar to strange quark stars as well.

2.2. *Strange quark matter or strangeon matter?*

The existence of three-light-flavour symmetry might be understood qualitatively by analyzing the energy scale. Quarks are often described by the indirectly measured mass parameters since they are confined within hadrons, which are the key ingredients to estimate the active quark-flavour degrees of freedom at certain energy scales. The current masses of both u and d quarks are just a few MeV while that of s quark is about 100 MeV (Particle Data Group, 2016). For nuclei or nuclear matter with density $n_0 \approx 0.16\ \mathrm{fm}^{-3}$, the average distance between quarks is $\Delta x \sim 0.5$ fm, corresponding to the energy scale $E_{\mathrm{nucl}} \sim 400$ MeV according to Heisenberg's relation $\Delta x \cdot pc \sim \hbar c \simeq 200\ \mathrm{MeV} \cdot \mathrm{fm}$. Then the existence of three-flavour nuclei in dense matter with density $n \gtrsim n_0$ is possible if we neglect the effects of chiral condensates, because in such cases u and d quarks are the lightest and E_{nucl} is larger than the mass difference between s quarks and u/d quarks (Bodmer, 1971). In stable nuclei with sizes $\sim$ fm, however, the valence strangeness degree of freedom is missing because the mass of s quark is much larger than that of u/d quarks, where any newly created s quarks would quickly decay into u and d quarks via weak reactions. This may not be the case for strongly interacting matter in bulk, where the light-flavour (u, d and s) symmetry restoration may occur.

To demonstrate this explicitly, let us first take a look at the structure of an atomic nucleus. Ordinary nuclei, composed of nucleons, have two flavours with u and d quarks. It is then interesting to think about the fact that our baryonic matter is two-flavour (isospin) symmetric. A philosophical explanation could be: micro-nuclei are too small to have three-flavour symmetry, but bigger is different. The electric charges of u and d quarks are respectively $+2/3$ and $-1/3$, so strongly interacting matter with isospin symmetry should be positively charged, and electrons are required to maintain electric neutrality. The possibility of electrons inside a micro-nucleus is negligible because the nucleus radius ~ 1 fm is much smaller than the Compton wavelength of electrons $\lambda_e \sim 10^3$ fm. Therefore, electrons contribute negligible energy for micro-nuclei (Xu and Guo, 2017). For a gigantic nucleus with its size much larger than λ_e (corresponding to number of nucleons $A > 10^9$), however, it might be three-flavour symmetric. If the two-flavour symmetry stays in macro-nucleus, electron-positron pair creation inevitably takes place due to the large Coulomb potential (Xia *et al.*, 2020b), then it becomes a huge Thomson atom with electrons inside it. As illustrated in Fig. 3, electrons will then contribute a high Fermi energy to the total energy of the system, indicating a large chemical potential

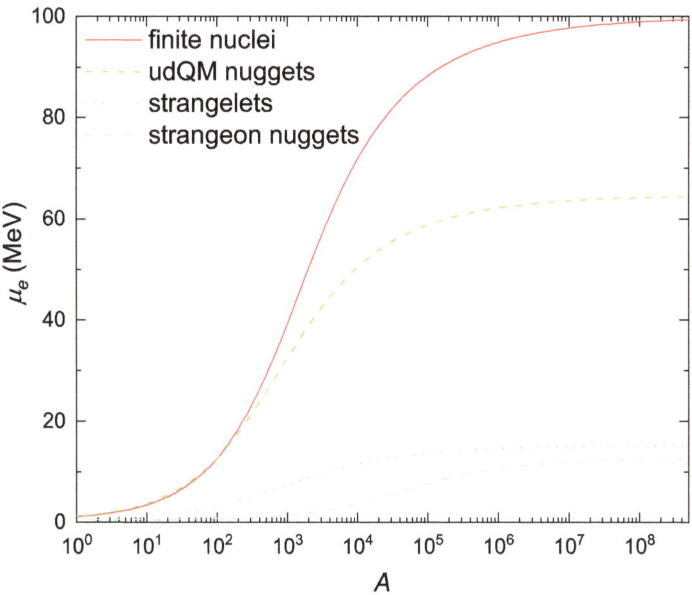

Fig. 3. The electron chemical potentials on the surfaces of finite nuclei, udQM nuggets, strangelets, and strangeon nuggets, where the β-stability condition is adopted. For numerical details, please refer to Xia *et al.* (2020b).

$\mu_e \sim 100$ MeV. Similarly, $\mu_e \sim 65$ MeV even for an object made of deconfined u and d quarks, i.e., the udQM nuggets (Holdom *et al.*, 2018). The situation becomes different if strangeness is included: no electrons exist if the matter is composed of equal numbers of u, d and s quarks. In this case, the three-flavour symmetry, an analogy of two-flavour symmetry in micro-nucleus, may result in a ground state of gigantic nuclei. Strangeon, the constituent of an gigantic nucleus, is also an analogy of nucleon which is the constituent of a micro-nucleus. In practice, as indicated in Fig. 3, the mass difference between s and u/d quarks slightly breaks the three-flavour symmetry, where the Fermi energy of electrons is much smaller than that of two-flavour cases with $\mu_e \sim 10$ MeV.

It is worth noting that, rational thinking about stable strangeness dates back to 1970s (Bodmer, 1971) and bulk strange objects called strangelets are speculated to be the absolutely stable ground state of strong-interacting matter, which is known as Witten's conjecture (Witten, 1984). As illustrated in Fig. 3, the corresponding chemical potential of electrons is reduced as well with $\mu_e \sim 15$ MeV. Although Witten's conjecture was proposed based on the matter composed of almost free quarks, we can make an extension that it still holds for localized quarks, i.e., strangeons. In some of our previous papers, we have used the word "quark-cluster", which have actually the same meaning as strangeons. In this chapter, we will demonstrate that the three-flavour matter could be possible in a "gigantic nucleus", which is then a three-flavour "nucleus", and "strangeons" are its constituent.

3. Strangeon Star

After the formation of a strangeon star, similar to ions in white dwarfs, we expect strangeons with three-flavour symmetry freeze and form a solid state as the star cools down. The reason for this is two-fold: (1) due to its large mass m_q, the quantum wave packet of a free strangeon scales $\lambda_{\rm q} \sim h/(m_q c)$, which is smaller than the typical inter strangeon distance a and can be further reduced if it is trapped in the potential well with depth of $> \hbar c/a \simeq 100$ MeV (2 fm/a); and (2) a large ratio of the strong interaction energy $U(a)$ to the thermal energy at a relatively low temperature $U(a)/kT \gtrsim 10^{1\sim 2}$. A star made of strangeon matter would then be a solid star, where strangeons form typically a simple lattice.

According to Lai and Xu (2009), the interaction between strangeons can be treated with a Lennard-Jones potential, i.e.,

$$U(r) = 4\epsilon \left[\left(\frac{\sigma}{r}\right)^{12} - \left(\frac{\sigma}{r}\right)^{6} \right], \tag{1}$$

where ϵ is the depth of the potential and σ characterize the range of interaction. Based on various considerations (Lai and Xu, 2009), the typical values for the parameters are taken as $\epsilon \approx 50$–$100\,\mathrm{MeV}$ and $\sigma \approx 1$–3 fm. More recently, instead of adopting the phenomenological potential of Eq. (1), we have proposed a linked bag model for strangeon matter (Miao *et al.*, 2020). The corresponding energy per unit cell is obtained with

$$E = \sum_{i=u,d,s,e} (\Omega_i + N_i \mu_i) + B a^3 - \frac{z_0}{r_{\mathrm{bag}}} \frac{\omega}{4\pi}, \tag{2}$$

where Ω_i, N_i and μ_i represent the thermodynamic potential, total particle number, and chemical potential of particle type i obtained in the Fermi-gas approximation. The bag constant B and parameter z_0 vary with density and are fixed by reproducing the saturation properties of nuclear matter (Miao *et al.*, 2020). The variable ω represents the solid angle of the remaining bag and decreases from 4π once the bags are linked. If the bag takes up the entire volume of the lattice cell with $\omega = 0$, a deconfinement phase transition takes place and Eq. (2) is restored to its original MIT bag model description of quark matter.

As an example, in Fig. 4 we present the energy per baryon of nuclear matter, hyperonic matter, and strangeon matter in compact stars. Comparing the dash-dotted and dashed curves, we notice the energy per baryon of nuclear matter is decreased once hyperons emerge at $n \approx 2n_0$. For strangeons with larger number of quarks (9 in this case), the energy is further reduced and can be smaller than 930 MeV, i.e., fulfilling the generalized Witten's conjecture. Consequently, the EoSs become stiffer for strangeon matter with large zero-pressure densities at $2 \sim 3n_0$.

Based on the EoSs indicated in Fig. 4(a), the internal structures of neutron stars and hyperon stars can be obtained by solving the TOV equation (Oppenheimer and Volkoff, 1939), which is the solution for static and spherically symmetric gravitational sources in general relativity. Nevertheless, the TOV solution assumes perfect fluid and may be not suitable for solid strangeon stars, where the local pressure can be anisotropic.

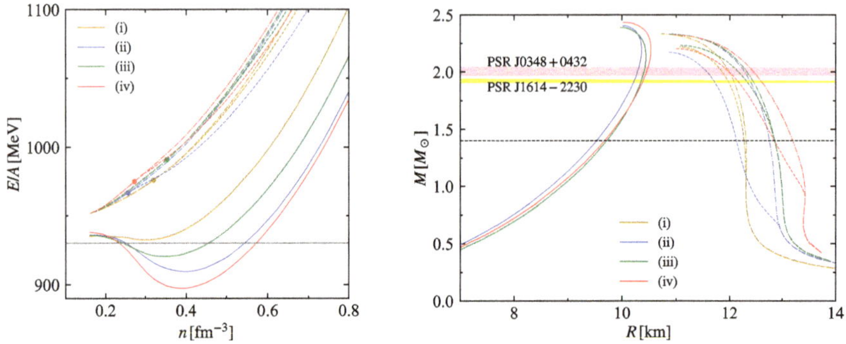

(a) (b)

Fig. 4. (a) The energy per baryon of nuclear matter (dash-dotted), hyperonic matter (dashed), and strangeon matter (solid) inside compact stars, where four different parameter sets are adopted. (b) The corresponding mass–radius relations. The horizontal line in the left panel represents the energy per baryon of ^{56}Fe, while the solid dots on the dashed curves correspond to the onset densities of hyperons. For more details, please refer to Miao *et al.* (2020).

The radial pressure gradient can then be partially balanced by the tangential shear force in strangeon stars (Karlovini and Samuelsson, 2004). The local anisotropic force in solid strangeon stars may be built up from the elastic energy as they spin down and/or cool down, where both the ellipticity and size are altered. As will be discussed later, the release of the elastic and gravitational energies may have significant astrophysical implications. To numerically obtain the structure of a solid strangeon star, for simplicity, we consider only the spherically symmetric sources. In Newtonian gravity, the equilibrium configuration of static anisotropic matter can be obtained with (Herrera and Santos, 1997):

$$\mathrm{d}P/\mathrm{d}r = -Gm(r)\rho/r^2 + 2(P_\perp - P)/r, \tag{3}$$

where P and $P_\perp$ correspond to the radial and tangential pressures, ρ the mass density, G the gravitational constant, and $m(r) = \int_0^r 4\pi r^2 \rho(x)\mathrm{d}x$ the enclosed mass within a sphere of radius r. In Einstein's gravity, the equilibrium equation becomes (Xu *et al.*, 2006)

$$\frac{\mathrm{d}P}{\mathrm{d}r} = -\frac{Gm(r)\rho}{r^2}\frac{(1+\frac{P}{\rho c^2})(1+\frac{4\pi r^3 P}{m(r)c^2})}{1-\frac{2Gm(r)}{rc^2}} + \frac{2\epsilon}{r}P, \tag{4}$$

which assumes $P_\perp = (1+\epsilon)P$. Note that Eq. (4) is restored into the TOV equation in case of isotropic pressure with $\epsilon = 0$. It is evident that the radial pressure gradient $|dP/dr|$ decreases at $\epsilon > 0$, so that the mass of a strangeon star may become larger.

In Fig. 4(b), we present the typical mass–radius relations of neutron stars, hyperon stars, and strangeon stars corresponding to the EoSs indicated in Fig. 4(a), which are obtained assuming $\epsilon = 0$. The observational masses of PSR J1614-2230 ($1.928 \pm 0.017\, M_\odot$) (Demorest *et al.*, 2010; Fonseca *et al.*, 2016) and PSR J0348+0432 ($2.01 \pm 0.04\, M_\odot$) (Antoniadis *et al.*, 2013) are indicated with horizontal bands. The horizontal dashed line corresponds to the mass ($M = 1.4\, M_\odot$) of the canonical-mass star. Due to the stiffer EoSs, the maximum masses $M_{\rm tov}$ of strangeon stars are higher than those of neutron stars and hyperon stars, which can be increased slightly if $\epsilon > 0$. A further investigation on all possible combinations of parameters suggests that $M_{\rm tov}$ of strangeon stars can even exceed $2.5\, M_\odot$, which is compatible with the recent measured massive object (2.50–$2.67\, M_\odot$) in a compact binary coalescence of GW190814 (LVC, 2020). The radius of the canonical-mass star ranges from 9.5 km to 13 km, where strangeon stars typically have smaller radii and larger maximum mass than those of conventional neutron stars.

For non-zero ϵ with $P_\perp > P$, a sudden decrease of ϵ can lead to a substantial release in gravitational energy as the star's radius decreases (10^{48} erg if $\epsilon \approx 10^{-4}$ and $M \approx 1 M_\odot$), while an additional release in the tangential strain energy is expected in starquakes. Generally speaking, the differences of radius, gravitational energy, and moment of inertia in strangeon stars due to non-zero ϵ increase proportionally to M and ϵ, which suggests that a bigger change in ϵ with higher mass will lead to a more evident event. In the superflares of SGRs, the typical energy release is of $10^{44\sim47}$ erg, which can be produced in a giant starquake with $\epsilon \lesssim 10^{-4}$ (Xu *et al.*, 2006). A sudden change of ϵ may also lead to a jump of the spin frequency with $\Delta\Omega/\Omega = -\Delta I/I$. For example, for strangeon stars with $M = (0.1 \sim 1.4) M_\odot$ and $\epsilon = 10^{-9} \sim 10^{-4}$, glitch activities with $\Delta\Omega/\Omega \approx 10^{-10} \sim 10^{-4}$ are expected to occur. In such cases, we suspect that a giantflare is usually accompanied by a high-amplitude glitch.

4. Astrophysics of Strangeon Star

Despite the lack of conclusive evidence for pulsar-like compact objects being strangeon stars, there are various observational hints supporting this proposal, as will be discussed in the following based on their surface

(Section 4.1) and global (Section 4.2) properties, as well as the implications of merging strangeon stars (Section 4.3).

4.1. *Surface properties*

Dirfting subpulses. Even though pulsar-like compact objects have various manifestations, most of them are populated by radio pulsars. A number of magnetospheric emission models were developed to describe the pulsar radio radiative processes. Among them, the Ruderman-Sutherland model explains clear drifting sub-pulses for the first time (Ruderman and Sutherland, 1975). In their seminal paper, a vacuum gap above the polar cap of a pulsar was proposed, where the sparks produced by the inner-gap breakdown result in the subpulses with the observed drifting feature caused by $\mathbf{E} \times \mathbf{B}$. Despite their success, the Ruderman-Sutherland model can hardly work for conventional neutron stars, where strong magnetic field and low temperature on surfaces of pulsars with $\mathbf{\Omega} \cdot \mathbf{B} < 0$ are required. Unfortunately, this strict condition usually cannot be satisfied according to various calculations, where the binding energy of Fe at the neutron star surface is < 1 keV (Flowers *et al.*, 1977; Lai, 2001) and is not sufficient to reproduce the vacuum gap. Such a problem is often referred to as "binding energy problem", which might be alleviated within a partially screened inner gap model (Gil *et al.*, 2006) for neutron stars with $\mathbf{\Omega} \cdot \mathbf{B} < 0$.

Alternatively, such a problem does not exist for strangeon stars since there is sufficient binding energy for any $\mathbf{\Omega} \cdot \mathbf{B}$. In Yu and Xu (2011), a quantitative investigation on the magnetospheric activity of bare strange quark star was carried out. Since quarks are confined by strong interaction on the surface, the binding energy can be viewed as infinite comparing with electromagnetic interaction. The electrons on strange quark star's surface are prevented from streaming into the magnetosphere by the potential barrier of the vacuum gap, while their total energy on the Fermi surface is none-zero. The binding energy of electrons is then the difference between the height of the potential barrier and total energy. In the bare strangeon star model, both positively and negatively charged particles are strongly bound to the surface to form a vacuum gap above its polar cap, which naturally explains the drifting (even bi-drifting) subpulses observed in radio pulsars (Qiao *et al.*, 2004; Xu *et al.*, 1999).

In addition, a recent hot topic of fast radio bursts could also be interesting events to reveal the magnetospheric activity of strangeon stars

(Du *et al.*, 2020; Luo *et al.*, 2020; Wang *et al.*, 2018a, 2020b), where more efforts are required on this aspect in our future.

X-ray spectral lines. On the surfaces of neutron stars or strange stars with crusts, there exist atmospheres comprised of normal materiel, which are expected to leave atomic features in their thermal X-ray emissions (Romani, 1987). The construction of advanced facilities such as *Chandra* and *XMM-Newton* were then proposed to detect those lines, where the spectrum determined by radiative transfer in atmosphere should differ substantially from Planck-like one. With those observations, one expects to find out the chemical composition and magnetic field of the atmosphere, and eventually constrain the stellar mass and radius according to the redshift and pressure broadening of spectral lines. However, unfortunately, so far none of the spectral features has been identified, which may hint a fundamental weakness of the neutron star models. Although conventional neutron star models cannot be ruled out completely by the absence of atomic thermal spectra since very strong surface magnetic fields could lead to a featureless spectrum as well (Ho and Lai, 2003; Turolla *et al.*, 2004), bare strange (quark or strangeon) stars can naturally explain the general observation (Xu, 2002), being almost without atoms there on the surfaces. It worth mentioning that due to the strangeness barrier, however, a thin atmosphere that is optically thin in X-ray bands can form on the surface of a strangeon star if accretion takes place.

Nevertheless, further observations have revealed the absorption lines in pulsar-like objects. For example, the best absorption features were detected for the CCO 1E 1207.4-5209 in the supernova remnant PKS 1209-51/52, at $\sim 0.7\,\text{keV}$ and $\sim 1.4\,\text{keV}$ (Bignami *et al.*, 2003; Mereghetti *et al.*, 2002; Sanwal *et al.*, 2002). Those absorption features, however, turn out to be of electron–cyclotron origin instead of the atomic transitions of ionized helium in a strong magnetic field, so that 1E 1207 could be a bare strange star with a surface magnetic field of $\sim 10^{11}$ G (Xu *et al.*, 2003). In fact, more observations on both spectra feature (Bignami *et al.*, 2003) and precise timing (Gotthelf and Halpern, 2007) supported the electron–cyclotron model of 1E 1207.

Due to the large electron density in strange stars, the simple single particle approximation might not apply. To resolve this, Xu *et al.* investigated the global motion of the electron seas on strange stars' surfaces with a strong magnetic field (Xu *et al.*, 2012). It was shown that the magnetic field have significant impacts on the hydrodynamic surface fluctuations of the electron sea, and the seas may undergo hydrocyclotron oscillations with the eigenfrequencies determined by $\omega(l) = \omega_c/[l(l+1)]$, where $l = 1, 2, 3, ...$ and $\omega_c = eB/mc$ is the cyclotron frequency. The similar absorption strengths at 0.7 keV and 1.4 keV of 1E 1207.4-5209 can be understood in this hydrocyclotron oscillations model, where these two lines with l and $l+1$ have nearly equal intensity. The strength of the first harmonic, however, is much smaller than that of the fundamental in the electron–cyclotron model. Besides these absorption features, the detected lines around $(17.5, 11.2, 7.5, 5.0)$ keV in the burst spectrum of SGR 1806-20 and those in other dead pulsars (e.g., radio quiet compact objects) may be of hydrocyclotron origin as well (Xu *et al.*, 2012).

Supernova and gamma-ray bursts. The radiation fireballs of gamma-ray bursts (GRBs) and supernovae are known to move towards the observer with a high Lorentz factor as a whole (Paczynski, 1986), where the bulk Lorentz factor of GRBs is on the order of $\Gamma \sim 10^2 - 10^3$ (Meszaros *et al.*, 1998). In such cases, the total mass of baryons in an ultra-relativistic fireball should remain small to avoid carrying out too much energy from the central engine, i.e., the so-called "baryon contamination". Such a requirement is challenging if a conventional neutron star is the central engine, where the number of baryons loaded with the fireball is unlikely to be small since the star is gravity-bound and the luminosity of fireball is extremely high. Nevertheless, the tension can be eased if the CCO are strangeon stars. The bare and self-bound surface of a strangeon star naturally separates baryonic matter from the photons and leptons that dominate the fireball, where baryons are confined by strong force but $e^{\pm}$-pairs, photons and neutrino pairs can escape from the surface. In such cases, the surface of strangeon stars automatically generates a low baryon condition for GRBs and supernovae (Cheng and Dai, 1996; Ouyed *et al.*, 2005; Paczynski and Haensel, 2005).

It is still difficult to successfully simulate supernovae in the neutrino-driven explosion models of neutron stars. In the strangeon star scenario, however, the bare surfaces may be essential for the explosions of both core and accretion-induced collapses (Xu, 2005). Due to its ultrahigh surface temperature, a newly born strangeon star would radiate thermal emission that is not restricted by the Eddington limit (Haensel *et al.*, 1991). In such a photon-driven scenario, the strong radiation pressure caused by the enormous thermal emissions might play an important role in core-collapse supernovae (Chen *et al.*, 2007). Calculations have revealed that the radiation pressure is able to push the overlying mantle away through photon–electron scattering with energy as large as $\sim 10^{51}$ ergs. Such a photon-driven mechanism is thus promising to alleviate the current difficulty in core-collapse supernovae, where a strangeon star is formed inside the collapsing core. The recently observed highly super-luminous supernova ASASSN-15lh, with a total observed energy $(1.1 \pm 0.2) \times 10^{52}$ ergs, can also be understood if a very massive strangeon star with its mass approaching $M_{\rm max}$ forms (Dong *et al.*, 2016).

The neutrino burst detected during supernova SN 1987A can be interpreted in the framework of strangeon star model as well (Yuan *et al.*, 2017). Within a few tens of seconds after its formation, a liquid-solid phase transition takes place in a strangeon star at temperature ~1–2 MeV, where huge amount of energy is released via photons and neutrinos. As illustrated in Yuan *et al.* (2017), the temperature of a proto-strangeon star drops rapidly on a timescale of millisecond, and then the star becomes homothermal during a liquid-solid phase transition. The solid strangeon stars cool down more drastically after the phase transition. The conventional neutron stars, on the contrary, cool down more smoothly and have no interceptive indication during the supernova neutrino burst (Pons *et al.*, 1999). It is found that a strangeon star with melting temperature ~1 MeV well reproduce the burst events of SN1987A.

Strangeness barrier. A strangeon is made of approximately equal numbers of u, d, and s quarks (three-flavour) instead of just u and d quarks (two-flavour). The transition from normal two-flavoured nuclear matter into three-flavoured strangeon matter requires the weak reaction $u + d \to s + u$, which is not easy and occurs only once in a large number of collisions ($\gg$1). To account for this kind of difficulty, we have introduced a strangeness barrier (Xu, 2014). Consequently, most of the ordinary nuclei falling onto a strangeon star's surface might bounce back, so that a strangeon star may be surrounded by a hot corona or an atmosphere, or even a crust depending on the accretion rate. In such cases, the strangeness barrier could help us to understand the redshifted ($z = 0.009$) O VIII Ly-α emission line, the blackbody radiation area of 4U 1700+24 (Xu, 2014), the low mass function measured for the puzzling X-ray binary, as well as the Type-I X-ray bursters (Li *et al.*, 2015).

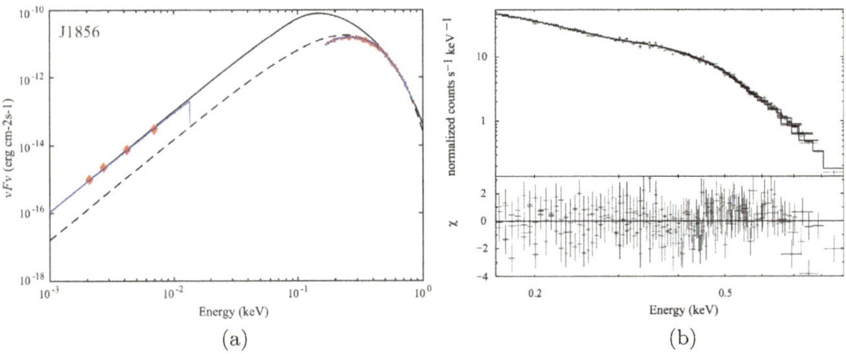

Fig. 5. Bremsstrahlung emission ((a), solid line) and X-ray data fitting (b) of RX J1856.5-3754. (a) The dashed line represents the blackbody spectrum extrapolated from the X-ray data, while the red diamonds indicate the optical observations from the Hubble Space Telescope photometry (Kaplan *et al.*, 2011). For numerical details, please refer to Wang *et al.* (2018b).

Additionally, the strangeness barrier may be responsible for the optical/UV excess puzzle of X-ray dim isolated neutron stars (XDINSs), which are characterized by Planckian spectra in X-ray bands but show optical/ultraviolet (UV) excesses exptrapolated from X-ray spectra. On the surface of a strangeon star, the accretion of interstellar medium (ISM) could form a plasma atmosphere due to strangeness barrier, where a radiative model of bremsstrahlung emission well reproduce the spectra of the seven XDINSs from optical/UV to X-ray bands (Wang *et al.*, 2017). As an example, in Fig. 5 we present the bremsstrahlung emission and X-ray data fitting of a strangeon star atmosphere for RX J1856.5-3754 (Wang *et al.*, 2018b). The solid curve in Fig. 5(a), is calculated in the bremsstrahlung radiation model of a plasma atmosphere above a strangeon star, which is larger than the pure blackbody extrapolation from the X-ray data and well reproduce the optical/UV data from the Hubble Space Telescope photometry (Kaplan *et al.*, 2011). Note that the strangeon star's atmosphere is similar to the upper layer of a normal neutron star's atmosphere, but with almost homogeneous electron (or ion) temperature. Therefore, thermal X-rays from lower layer of conventional neutron star atmosphere are prohibited, resulting in an excess of the optical/UV photons. Hard X-ray cut-off is also natural in our model, while the optical/UV spectral deviations from a Rayleigh–Jeans distribution may be due to the non-uniformity of the atmosphere (Wang *et al.*, 2018b).

4.2. *Global properties*

Rigidity. According to Yuan *et al.* (2017), the temperature of a proto-strangeon star drops rapidly on a timescale of millisecond after creation, then the strangeon matter solidifies and forms crystalline structures as the temperature drops below a melting temperature after few tens of seconds. A spinning rigid body precesses naturally either by itself or by torque, but it is difficult for a fluid one to precess. In such cases, the observation of possible precession in B1821-11 (Stairs *et al.*, 2000) and others could suggest a global solid structure for pulsar-like objects, e.g., a solid strangeon star. Low-mass strangeon stars ($\lesssim 10^{-2} M_\odot$) with radii of a few kilometers are gravitationally force-free, so that their surfaces may be irregular and asteroid-like. Free or torque-induced precession can then be excited easily, with larger amplitude for lower-mass strangeon stars. The masses of anomalous X-ray pulsars/soft gamma-ray repeaters (AXPs/SGRs) may approach the mass-limit ($>1.5 M_\odot$) in the accretion-induced quake (AIQ) model (Xu, 2007). In such cases, these objects manifest no or weak precession as observed, even though they are more likely to be surrounded by dust disks due to their larger masses comparing with CCOs/DTNs.

Additionally, the elastic stresses of cold strangeon matter could support mountains on strangeon stars, which are responsible for the asymmetric matter distributions about its rotation axis with a non-zero ellipticity. A fast rotating strangeon star will then emit gravitational waves continuously, which are expected to be observed by gravitational wave detectors in the kilo-Hz band. It is worth mentioning that recent observations have put strong constraints on the ellipticities ($<10^{-8}$) of PSR J0437-4715 and PSR J0711-6830 (Abbott *et al.*, 2020), which rules out the existence of large mountains on those stars. At the same time, the ellipticities of young pulsars such as Vela (J0835-4510) and Crab (J0534+2200) pulsars may be much larger.

Starquakes. It is difficult to explain the glitch phenomenons of pulsars in the framework of strange quark stars, with the internal strange quark matter being in fluid states (Alpar, 1987; Benvenuto *et al.*, 1990). Such a problem can be resolved in a solid strangeon star model by considering the glitches as starquakes, which occurs once the strain energy develops to a critical value. There are huge amounts of gravitational and elastic energy being released after the quake of a solid strangeon star, which is then accompanied by a large spin-change. The released energy can then power some extreme events detected in AXPs/SGRs and GRBs. Two types of starquakes are expected in the solid strangeon star model, i.e., bulk-invariable (Vela-like, type I) and bulk-variable (AXP/SGR-like, type II) starquakes, with energy release negligible for the former but significant for the latter.

Based on the strangeon star model of pulsar glitch (Lai *et al.*, 2018b; Wang *et al.*, 2020a), the shear modulus of strangeon matter is constrained to be order of 10^{34}erg/cm^3 to account for the observed glitch activities. Further theoretical estimation on the shear modulus of strangeon matter at various density and temperature is essential for us to understand various behaviours of pulsar timing irregularities, which will be carried out in our future works (Miao *et al.*, 2020). By extending the model for normal glitches (Zhou *et al.*, 2004), one can model pulsar's slow glitches (Peng and Xu, 2008) that is not well understood in neutron star models. A typical energy $10^{44\sim47}$ erg is released in the superflares of SGRs, which can be produced in a giant type II starquake (for slow rotators) (Xu *et al.*, 2006). Both types of glitches with (type II) and without (type I, for fast rotators) X-ray enhancement can be understood as starquakes of solid strangeon stars (Zhou *et al.*, 2014). In addition, many efforts were devoted to connect the GRB central engines to SGRs' flares, in order to understand different GRB light-curves observed, especially the internal-plateau X-ray emission (Dai *et al.*, 2011; Xu and Liang, 2009).

Significant stellar oscillations are usually accompanied by starquakes (especially for type II) in pulsars, which will excite the magnetospheric activity in the polar cap region. As the toroidal oscillation of the star propagates into the magnetosphere, additional voltage due to unipolar induction alters the Goldreich-Julian charge density (Goldreich and Julian, 1969) and consequently influences particle acceleration (Lin *et al.*, 2015). In such cases, the onset of radio emission after glitches/flares in AXPs/SGRs could be initiated by the oscillation-driven magnetic activities in solid strangeon stars, where the transient radio signals are caused by the activation of a pulsar's inner gap due to an oscillation-induced voltage enhancement.

Mass and radius of strangeon stars. As indicated in Fig. 4(a), the EoSs of strangeon matter are usually stiffer than that of nuclear/hyperonic matter. The reason is two-fold: (1) strangeons are non-relativistic particles due to their large mass; and (2) there exists a strong short-range repulsion between strangeons in analogy to the repulsive core of nuclear force. Based on the EoSs of strangeon matter, the structures of strangeon stars in perfect spherical shapes ($\epsilon = 0$) can be obtained by solving the TOV equation (Oppenheimer and Volkoff, 1939). A stiffer EoS for strangeon matter implies a larger maximum mass, while the existence of low mass strangeon stars is a direct consequence of self-bound surfaces. This can be observed in Fig. 4(b), where strangeon stars with high maximum masses ($>2M_\odot$) and very low masses ($<10^{-2}M_\odot$) are obtained (Lai and Xu, 2009; Miao *et al.*, 2020).

The radio observations of pulsars in binary systems with strong Shapiro delay signatures imply that the mass of pulsars can reach $2M_\odot$ (Antoniadis *et al.*, 2013; Demorest *et al.*, 2010; Fonseca *et al.*, 2016), with the most massive one (PSR J0740+6620) being measured as $2.14^{+0.10}_{-0.09}M_\odot$ (Cromartie *et al.*, 2020). In addition, there possibly exist even more massive compact stars in the universe, e.g., the recent measured massive object (2.50–2.67 $M_\odot$) in a compact binary coalescence of GW190814 (LVC, 2020). The massive pulsars indicate a stiff EoS for their internal material, which has ruled out various EoS models for neutron stars and strange quark stars. The strangeon star model, however, could not be ruled out by massive pulsars. The observations of more massive pulsars (e.g., $>2.5M_\odot$) may even be a strong evidence to support strangeon star model, and give further constraints on model parameters.

By assuming that the touchdown flux corresponds to Eddington luminosity and including the obscure effect, the mass and radius of 4U 1746-37 can by constrained by the photospheric radius expansion (PRE) bursts (Li *et al.*, 2015), which turns out to be small. It is thus difficult to consider 4U 1746-37 as a neutron star or a strange star with crust, while it may be a strangeon star with small radius. Beside this, there may be more observational hints for low-mass strangeon stars. For example, by fitting the spectral data of the thermal radiation components in some pulsar-like objects according to the Planck spectrum, their radii can be estimated in blackbody models, which are usually much smaller than 10 km (Pavlov *et al.*, 2004). No pulsation with periods longer than ∼0.68 s were found in the CCO of Cas A, and the stellar radius and mass are constrained to be $R = (4 \sim 5.5)\,\mathrm{km}$ and $M \lesssim 0.8M_\odot$ in hydrogen neutron star atmosphere models (Pavlov and Luna, 2009).

Another low-mass strangeon star could be 4U 1700+24. Because of the strangeness barrier on the surface, a strangeon star may be surrounded by a hot corona, an atmosphere, or even a crust depending on its accretion rate. Both the redshifted O VIII Ly-α emission line (only $z = 0.009$) and the variation in the blackbody radiation area (with an inferred scale of $\sim$10–100 m) can be understood naturally if 4U 1700+24 is a low-mass strangeon star with weak wind accretion (Xu, 2014). The mass function $f_o = (1.8 \pm 0.9) \times 10^{-5} M_\odot$ is obtained by observing the G-type red giant company (Galloway *et al.*, 2002), from which the derived mass of compact star should be much lower than $1M_\odot$ unless there is geometrically fine-tuning (inclination angle $i < 2^\circ$ (Xu and Guo, 2017)). The three independent observations of redshift, hot spot and mass function suggest that 4U 1700+24 may be a low mass strangeon star.

Future observations with more advanced facilities (e.g., FAST, NICER, aLIGO, aVirgo, KAGRA, and SKA) could provide more observational hints for the nature of dense matter. The pulsar mass measurements provide opportunities in finding more massive pulsars, while the radii can be obtained with the measurements of the momentum of inertia, spectroscopic, pulse profile, and gravitational waveform. The possible detection of sub-millisecond pulsars could provide a clear evidence for (low-mass) strange stars, since the spin period of conventional neutron stars is limited by the Kepler frequency and should be less than $\sim 0.5 M_1^{1/2} R_6^{-3/2}$ ms ($R_6 = R/10^6\,\mathrm{cm}$).

4.3. *Merging strangeon stars*

The direct observation of gravitational waves has started a new era of multi-messenger astrophysics. On August 18, 2017 the gravitational waves emitted by the merger of binary neutron stars were observed by the Advanced LIGO and Advanced Virgo gravitational-wave detectors (LVC, 2017). About 1.7 s later, a short γ-ray burst GRB 170817A lasted $\sim$2 s was detected (Goldstein *et al.*, 2017; LIGO Scientific Collaboration *et al.*, 2017), which is associated with the gravitational-wave signal GW170817. In the following few weeks, the ejecta from the binary system quickly evolves into a standard neutron-rich environment for r-process nucleosynthesis, where significant amount of unstable nuclei close to neutron drip line are produced (Kasen *et al.*, 2017; Kasliwal *et al.*, 2017; Watson *et al.*, 2019). The radioactive decay of these unstable nuclei then powers a rapidly evolving and supernova-like transient counterpart AT2017gfo, which was often referred to as "kilonova" (Li and Paczyński, 1998), "macronova" (Kulkarni, 2005), or "mergernova" (Gao *et al.*, 2015; Yu *et al.*, 2013).

The bolometric light curve at various wavelength can then be interpreted by such radioactivity-powered transient (Cowperthwaite *et al.*, 2017; Smartt *et al.*, 2017; Villar *et al.*, 2017), containing a low-opacity ($\kappa \sim 10^{-1}\,\mathrm{cm}^2\,\mathrm{g}^{-1}$) component ("blue" component, light r-process) whose luminosity peaks at $\sim 10^{42}$ erg s^{-1} 1 day after the merger, and a high-opacity ($\kappa \sim 10 - 100\,\mathrm{cm}^2\,\mathrm{g}^{-1}$) component ("red" component, heavy r-process) whose luminosity peaks at about 1 week later.

A combined analysis of the binary neutron star merger event GRB 170817A-GW170817-AT 2017gfo have put strong constraints on the radius ($11.9^{+1.4}_{-1.4}$ km) and tidal deformability $70 \leq \Lambda_{1.4} \leq 580$ for $1.4M_\odot$ neutron stars (LVC, 2018). Due to the lack of information on the post-merger remnant, however, the current observation can not exclude other possible origins for GW170817, e.g., the merger of binary strange stars. The observational tidal deformability was adopted to constrain parameters in MIT bag model, which imply the maximum mass of strange quark stars to be $\leq 2.18M_\odot$ (Zhou *et al.*, 2018).

The dimensionless tidal deformability of a compact star is determined by

$$\Lambda = \frac{2k_2}{3}\left(\frac{R}{GM}\right)^5, \tag{5}$$

where the gravity constant is $G = 6.707 \times 10^{-45}\ \mathrm{MeV}^{-2}$. The second Love number k_2 measures how easily a star is deformed by an external tidal field, which can be fixed by perturbation theory (Damour and Nagar, 2009; Hinderer *et al.*, 2010; Postnikov *et al.*, 2010), i.e.,

$$\begin{aligned} k_2 = {} & \frac{8c^5}{5}(1-2c)^2[2+2c(y_R-1)-y_R]/\{2c[6-3y_R+3c(5y_R-8)] \\ & + 4c^3[13-11y_R+c(3y_R-2)+2c^2(1+y_R)] \\ & + 3(1-2c)^2[2-y_R+2c(y_R-1)]\ln(1-2c)\}. \end{aligned} \tag{6}$$

Here $c = M/R$ is the compactness of the star and $y_R = y(R)$ is fixed by the variable $y(r)$ at the surface $r = R$, where $y \equiv rH'/H$ with H being the perturbation to the metric and satisfies the following differential equation:

$$r\frac{dy(r)}{dr} + y(r)^2 + y(r)F(r) + r^2Q(r) = 0, \tag{7}$$

with

$$F(r) = \{1 - 4\pi r^2[\varepsilon(r) - P(r)]\}\left[1 - \frac{2m(r)}{r}\right]^{-1}, \tag{8}$$

$$Q(r) = 4\pi\left[5\varepsilon(r) + 9P(r) + \frac{\varepsilon(r) + P(r)}{\partial P(r)/\partial\varepsilon(r)} - \frac{6}{4\pi r^2}\right]\left[1 - \frac{2m(r)}{r}\right]^{-1}$$
$$- \frac{4m(r)^2}{r^4}\left[1 + \frac{4\pi r^3 P(r)}{m(r)}\right]^2\left[1 - \frac{2m(r)}{r}\right]^{-2}. \tag{9}$$

Equation (7) is solved simultaneously with the TOV equation, where

$$m(r) = \int_0^r 4\pi r^2\rho(x)\mathrm{d}x, \tag{10}$$

as in Eq. (4). Note that for self-bound strange stars (quark stars and strangeon stars), the continuity condition on the surface requires

$$y(R_+) = y(R_-) - \frac{\varepsilon(R_-)}{4\pi M/R^3}. \tag{11}$$

If pulsar-like compact stars are in fact strangeon stars, a much larger parameter range is permited by the constraints of both the tidal deformability from GW170817 and mass of PSR J0740+6620. As an example, the tidal deformabilities of various compact stars predicted by the linked bag model (Miao *et al.*, 2020) are presented in Fig. 6, where the corresponding EoSs are indicated in Fig. 4(a). In general, Λ decreases monotonously with stellar mass. The tidal deformability of a strangeon star is found to be much smaller than that of conventional neutron stars, and fulfills the GW170817 constraint of $\Lambda_{1.4} < 580$ at various combinations of parameters (LVC, 2018). Combined with the mass measurements of pulsars such as PSR J0740+6620 with $2.14^{+0.10}_{-0.09} M_\odot$ (Cromartie *et al.*, 2020), many EoSs for the conventional neutron star failed to meet the constraints. A possibly more massive compact star (2.50–2.67 $M_\odot$) measured in a compact binary coalescence of GW190814 (LVC, 2020) could further shrink the parameter window for the conventional neutron star, while strangeon stars can naturally reach a large maximum mass of 2.8 $M_\odot$ (Lai *et al.*, 2019). As indicated in Fig. 7, if the Lennard-Jones potential in Eq. (1) is adopted for the interactions among strangeons, both the constraints of $\Lambda_{1.4}$ and M_{tov} can be easily fulfilled in a large parameter space for strangeon stars. Note that n_{s} is the surface density of strangeon stars and is linked to σ.

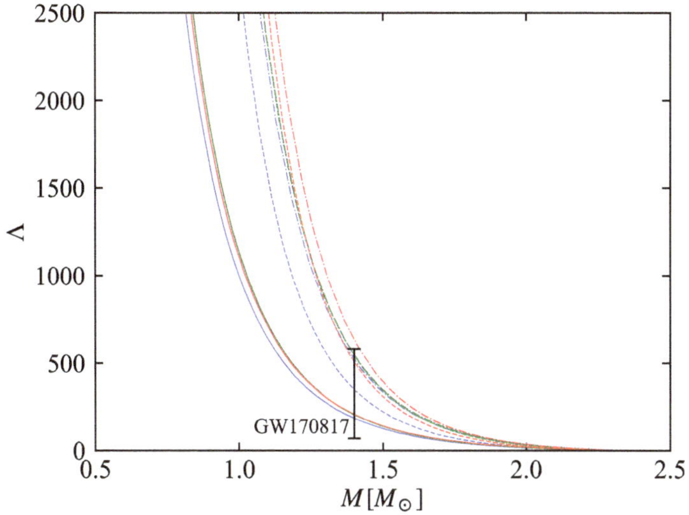

Fig. 6. The tidal deformability as a function of stellar mass, in correspondence to Fig. 4. The GW170817 constraint ($\Lambda_{1.4} = 190^{+390}_{-120}$) is indicated (LVC, 2018).

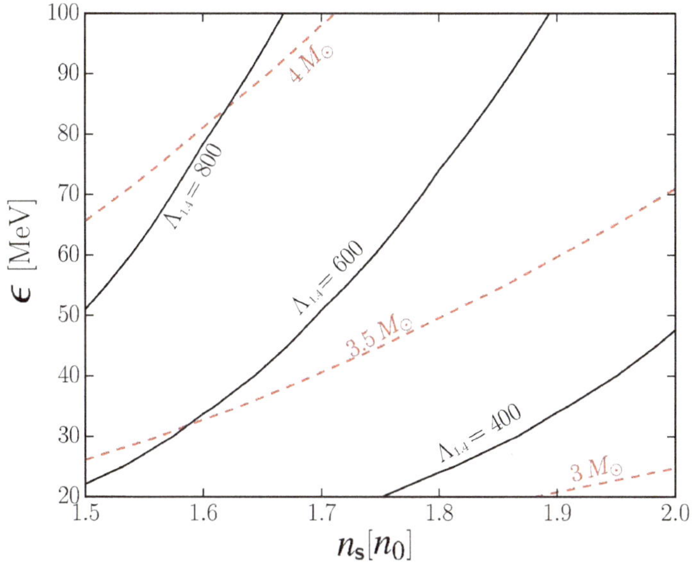

Fig. 7. Constraints on the equation-of-state parameters for strangeon stars: ϵ and $n_{\rm s}$, where the corresponding tidal deformability of a 1.4 $M_\odot$ star ($\Lambda_{1.4}$) and TOV maximum mass are presented (Lai *et al.*, 2019).

Even for the cases with $\Lambda_{1.4} \lesssim 300$, we note that the maximum mass can still reach 2.8 $M_\odot$ (Lai *et al.*, 2019).

Similar to the merger of binary neutron stars, strangeon nuggets are formed and ejected via tidal disruption and squeezing. The decay of strangeon nuggets provides an important energy source for the bolometric light curve of the following strangeon kilonova (Lai *et al.*, 2018a). The temperature of the ejected strangeon nuggets could be up to tens of MeV according to the numerical simulations of binary strange quark stars (Bauswein *et al.*, 2010), so they evaporate various particles to cool down. In Fig. 8(a), we present the emission rates per surface area for various evaporation channels (strangeons, neutrons, protons and α) (Lai *et al.*, 2020), which are estimated with statistical model. The emission rates for any strangeon nuggets with $R \gtrsim 10^5$ fm can then be obtained by multiplying it by the surface area of a nugget. It is found that the evaporation channels are dominated by the emission of strangeons at $T > 10$ MeV, and by neutrons at $T < 10$ MeV.

During the merger of double strangeon stars, the temperature is so large that almost all the ejected strangeon nuggets are turned into strangeons within 1 ms. Then strangeons instantly decay into neutrons and proton within 10 ms. The ejecta thus become similar to that ejected in the merger

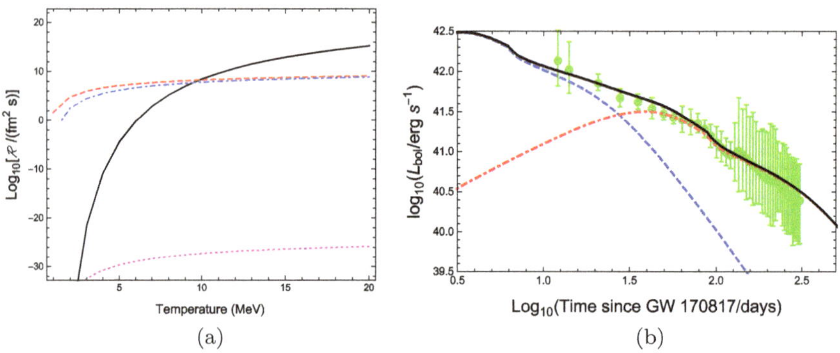

Fig. 8. (a) Emission rates per surface area for evaporating strangeons (solid line), neutrons (dashed line), protons(dash-dotted line) and α (dotted line) as functions of temperature. (b) Bolometric light curve of a strangeon kilonova including two-component ejecta, fitted to the data from Kasliwal *et al.* (2017). The dashed and dash-dotted lines represent the light curves of blue and red components, respectively. The solid line is the result of the combination of the two components. For both blue and red component, the ejected mass $M_{\rm ej} = 10^{-4} M_\odot$, and the opacity $\kappa = 0.1\,{\rm cm}^2\,{\rm g}^{-1}$ and $20\,{\rm cm}^2\,{\rm g}^{-1}$, respectively. For numerical details, please refer to Lai *et al.* (2020).

of double neutron stars, where the neutron-rich matter is ejected from the directions around the equatorial plane and the proton-rich matter is ejected in a broad range of angular directions. The yield of protons due to strangeon decay is expected to be much larger than neutrons, so that the high and low opacity components are naturally created. Adopting the ejecta model introduced in Yu *et al.* (2013), the light curves of blue and red components can be obtained, which is presented in Fig. 8(b). It is found that, despite the large difference in the initial components of the ejecta from merging strangeon stars, the "strangeon kilonova" could have light curves similar to that of neutron kilonova, where supernova-like transient (Kasliwal *et al.*, 2017) can be reproduced.

Finally, we should mention that strangeon star with rigidity may be tested further by kilo-Hz gravitational wave observations in at least two types of events: (1) merger of pulsar-pulsar/pulsar-black hole binaries; and (2) starquake of pulsar-like compact stars. In the meantime, the electromagnetic diagnostics is important as well to remove the degeneracy of physical parameters.

5. Summary and Outlook

It turns out that only a small portion of our universe is made of normal matter ($< 5\%$) that contains mostly atoms with baryonic nuclei in the centre. The micro-nuclei are made of nuclear matter comprised of two-flavoured nucleons (u and d). For a gigantic nucleus formed by the compression of gravity in supernova, we argue it should be in three-flavour symmetry (u, d, and s). The basic component, instead of nucleons, should be strangeons made of almost equal numbers of u, d, and s quarks. The matter inside the pulsar-like compact stars is then strangeon matter made of three-flavour strangeons, and the corresponding stars are called strangeon stars. In this chapter we have introduced the basic idea of strangeon star by extending Landau's initial proposal (Landau, 1932) of gigantic nuclei from two-flavour nucleon to three-flavour strangeon. We demonstrate that the strangeon star model is necessary to understand different manifestations of pulsar-like compact stars, and can be tested by future observations with advanced facilities, including either ground radio telescopes Five-hundred-metre Aperture Spherical radio Telescope (FAST) and Square Kilometre Array radio telescope (SKA), gravitational-wave detectors (aLIGO, aVirgo, KAGRA), or space-based high energy observatories Lightweight Asymmetry and Magnetism Probe (LAMP), enhanced X-ray Timing and Polarimetry (eXTP), and Gravitational wave high energy Electromagnetic Counterpart All-sky Monitor (GECAM).

Acknowledgements

This work was supported by National SKA Program of China (Grant No. 2020SKA0120300), the National Key R&D Program of China (Grant No. 2017YFA0402602), the National Natural Science Foundation of China (Grant Nos. 11705163, 11673002, U1531243, U1831104) and Ningbo Natural Science Foundation (Grant No. 2019A610066).

References

Abbott, R., Abbott, T. D., Abraham, S., Acernese, F., Ackley, K., Adams, A., Adams, C., Adhikari, R. X., Adya, V. B., Affeldt, C., Agathos, M., Agatsuma, K. *et al.* (2020). Gravitational-wave constraints on the equatorial ellipticity of millisecond pulsars, *The Astrophysical. Journal* **902**, 1, p. L21.

Aerts, A. T. M., Mulders, P. J. G., and de Swart, J. J. (1978). Multibaryon states in the bag model, *Physical Review D*, **17**, p. 260.

Alford, M. G., Schmitt, A., Rajagopal, K., and Schäfer, T. (2008). Color superconductivity in dense quark matter, *Reviews of Modern Physics*, **80**, p. 1455.

Alpar, M. A. (1987). Comment on strange stars, *Phys. Rev. Lett.* **58**, p. 2152.

Antoniadis, J., Freire, P. C. C., Wex, N., Tauris, T. M., Lynch, R. S., van Kerkwijk, M. H., Kramer, M., Bassa, C., Dhillon, V. S., Driebe, T., Hessels, J. W. T., Kaspi, V. M., Kondratiev, V. I., Langer, N., Marsh, T. R., McLaughlin, M. A., Pennucci, T. T., Ransom, S. M., Stairs, I. H., van Leeuwen, J., Verbiest, J. P. W., and Whelan, D. G. (2013). A massive pulsar in a compact relativistic binary, *Science*, **340**, p. 1233232.

Baade, W. and Zwicky, F. (1934). On super-novae, *Proceedings of the Nattional Academy of Sciences*, **20**(5), p. 254.

Balachandran, A. P., Barducci, A., Lizzi, F., Rodgers, V. G. J., and Stern, A. (1984). Doubly strange dibaryon in the chiral model, *Physical Review Letter* **52**, p. 887.

Bauswein, A., Oechslin, R., and Janka, H.-T. (2010). Discriminating strange star mergers from neutron star mergers by gravitational-wave measurements, *Physical Review D* **81**, p. 024012.

Beane, S. R., Chang, E., Detmold, W., Joo, B., Lin, H. W., Luu, T. C., Orginos, K., Parreño, A., Savage, M. J., Torok, A., and Walker-Loud, A. (2011). Evidence for a bound *H* dibaryon from lattice qcd, *Physical Review Letter*, **106**, p. 162001.

Benvenuto, O. G., Horvath, J. E., and Vucetich, H. (1990). *Strange-pulsar model, Physical Review Letter*, **64**, p. 713.

Bignami, G. F., Caraveo, P. A., Luca, A. D., and Mereghetti, S. (2003). The magnetic field of an isolated neutron star from X-ray cyclotron absorption lines, *Nature*, **423**, p. 725.

Bodmer, A. R. (1971). Collapsed nuclei, *Physical Review D* **4**, p. 1601.

Bornyakov, V., Braguta, V., Ilgenfritz, E. M., Kotov, A. Y., Molochkov, A., and Nikolaev, A. (2018). Observation of deconfinement in a cold dense quark medium, *JHEP*, **03**, p. 161.

Burwitz, V., Haberl, F., Neuhäuser, R., Predehl, P., Trümper, J., and Zavlin, V. E. (2003). The thermal radiation of the isolated neutron star RX 1856.5-3754 observed with chandra and XMM-Newton, *Astronomy Astrophysies* **399**(3), p. 1109.

Caplan, M. E. and Horowitz, C. J. (2017). Colloquium: Astromaterial science and nuclear pasta, *Review of Modern Physics* **89**, p. 041002.

Chen, A., Yu, T., and Xu, R. (2007). The birth of quark stars: Photon-driven supernovae? *The Astrophysics Journal*, **668**(1), p. L55.

Cheng, K. S. and Dai, Z. G. (1996). Conversion of neutron stars to strange stars as a possible origin of γ-ray bursts, *Physical Review Letter*, **77**, p. 1210.

Cowperthwaite, P. S., Berger, E., Villar, V. A., Metzger, B. D., Nicholl, M., Chornock, R., Blanchard, P. K., Fong, W., Margutti, R., Soares-Santos, M., Alexander, K. D., Allam, S., Annis, J., Brout, D., Brown, D. A., Butler, R. E., Chen, H.-Y., Diehl, H. T., Doctor, Z., Drout, M. R., Eftekhari, T., Farr, B., Finley, D. A., Foley, R. J., Frieman, J. A., Fryer, C. L., García-Bellido, J., Gill, M. S. S., Guillochon, J., Herner, K., Holz, D. E., Kasen, D., Kessler, R., Marriner, J., Matheson, T., Neilsen, E. H., Quataert, E., Palmese, A., Rest, A., Sako, M., Scolnic, D. M., Smith, N., Tucker, D. L., Williams, P. K. G., Balbinot, E., Carlin, J. L., Cook, E. R., Durret, F., Li, T. S., Lopes, P. A. A., Lourenço, A. C. C., Marshall, J. L., Medina, G. E., Muir, J., Muñoz, R. R., Sauseda, M., Schlegel, D. J., Secco, L. F., Vivas, A. K., Wester, W., Zenteno, A., Zhang, Y., Abbott, T. M. C., Banerji, M., Bechtol, K., Benoit-Lévy, A., Bertin, E., Buckley-Geer, E., Burke, D. L., Capozzi, D., Rosell, A. C., Kind, M. C., Castander, F. J., Crocce, M., Cunha, C. E., D'Andrea, C. B., da Costa, L. N., Davis, C., DePoy, D. L., Desai, S., Dietrich, J. P., Drlica-Wagner, A., Eifler, T. F., Evrard, A. E., Fernandez, E., Flaugher, B., Fosalba, P., Gaztanaga, E., Gerdes, D. W., Giannantonio, T., Goldstein, D. A., Gruen, D., Gruendl, R. A., Gutierrez, G., Honscheid, K., Jain, B., James, D. J., Jeltema, T., Johnson, M. W. G., Johnson, M. D., Kent, S., Krause, E., Kron, R., Kuehn, K., Nuropatkin, N., Lahav, O., Lima, M., Lin, H., Maia, M. A. G., March, M., Martini, P., McMahon, R. G., Menanteau, F., Miller, C. J., Miquel, R., Mohr, J. J., Neilsen, E., Nichol, R. C., Ogando, R. L. C., Plazas, A. A., Roe, N., Romer, A. K., Roodman, A., Rykoff, E. S., Sanchez, E., Scarpine, V., Schindler, R., Schubnell, M., Sevilla-Noarbe, I., Smith, M., Smith, R. C., Sobreira, F., Suchyta, E., Swanson, M. E. C., Tarle, G., Thomas, D., Thomas, R. C., Troxel, M. A., Vikram, V., Walker, A. R., Wechsler, R. H., Weller, J.,

Yanny, B., and Zuntz, J. (2017). *The electromagnetic counterpart of the binary neutron star merger LIGO/virgo GW170817. II. UV, optical, and near-infrared light curves and comparison to kilonova models,* The Astrophysics Journal, **848**(2), p. L17.

Cromartie, H. T., Fonseca, E., Ransom, S. M., Demorest, P. B., Arzoumanian, Z., Blumer, H., Brook, P. R., DeCesàr, M. E., Dolch, T., Ellis, J. A., Ferdman, R. D., Ferrara, E. C., Garver-Daniels, N., Gentile, P. A., Jones, M. L., Lam, M. T., Lorimer, D. R., Lynch, R. S., McLaughlin, M. A., Ng, C., Nice, D. J., Pennucci, T. T., Spiewak, R., Stairs, I. H., Stovall, K., Swiggum, J. K., and Zhu, W. (2020). Relativistic shapiro delay measurements of an extremely massive millisecond pulsar, *Nature Astronomy*, **4**, p. 72.

Dai, S., Li, L., and Xu, R. (2011). The plateau of gamma-ray burst: hint for the solidification of quark matter? *Science China Physics, Mechanics and Astronomy*, **54**, p. 1541.

Damour, T. and Nagar, A. (2009). Relativistic tidal properties of neutron stars, *Physical Review D* **80**, p. 084035.

Demorest, P. B., Pennucci, T., Ransom, S. M., Roberts, M. S. E., and Hessels, J. W. T. (2010). A two-solar-mass neutron star measured using shapiro delay, *Nature*, **467**, p. 1081.

Dong, S., Shappee, B. J., Prieto, J. L., Jha, S. W., Stanek, K. Z., Holoien, T. W.-S., Kochanek, C. S., Thompson, T. A., Morrell, N., Thompson, I. B., Basu, U., Beacom, J. F., Bersier, D., Brimacombe, J., Brown, J. S., Bufano, F., Chen, P., Conseil, E., Danilet, A. B., Falco, E., Grupe, D., Kiyota, S., Masi, G., Nicholls, B., Olivares E., F., Pignata, G., Pojmanski, G., Simonian, G. V., Szczygiel, D. M., and Woźniak, P. R. (2016). Asassn-15lh: A highly super-luminous supernova, *Science*, **351** (6270), p. 257.

Drake, J. J., Marshall, H. L., Dreizler, S., Freeman, P. E., Fruscione, A., Juda, M., Kashyap, V., Nicastro, F., Pease, D. O., Wargelin, B. J., and Werner, K. (2002). Is rx j1856.5-3754 a quark star? *The Astrophysics Journal*, **572**(2), p. 996.

Du, S., Wang, W., Wu, X., and Xu, R. (2020). On the geometry and environment of repeating FRBs, *Monthly Notices of the Royal Astronomical Society*, **500**(4), p. 4678.

Flowers, E. G., Ruderman, M. A., Lee, J. F., Sutherland, P. G., Hillebrandt, W., and Mueller, E. (1977). Variational calculation of ground-state energy of iron atoms and condensed matter in strong magnetic fields. *The Astrophysics Journal* **215**, p. 291.

Fonseca, E., Pennucci, T. T., Ellis, J. A., Stairs, I. H., Nice, D. J., Ransom, S. M., Demorest, P. B., Arzoumanian, Z., Crowter, K., Dolch, T., Ferdman, R. D., Gonzalez, M. E., Jones, G., Jones, M. L., Lam, M. T., Levin, L., McLaughlin, M. A., Stovall, K., Swiggum, J. K., and Zhu, W. (2016). The nanograv nine-year data set: Mass and geometric measurements of binary millisecond pulsars, *The Astrophysics Journal*, **832**(2), p. 167.

Galloway, D. K., Sokoloski, J. L., and Kenyon, S. J. (2002). Correlated radial velocity and x-ray variations in HD 154791/4u 1700+24, *The Astrophysics Journal*, **580**(2), p. 1065.

Gao, H., Ding, X., Wu, X.-F., Dai, Z.-G., and Zhang, B. (2015). GRB 080503 late afterglow re-brightening: Signature of a magnetar-powered merger-nova, *The Astrophysics Journal*, **807**(2), p. 163.

Gell-Mann, M. (1964). A schematic model of baryons and mesons, *Phys. Lett.* **8**(3), p. 214.

Gil, J., Melikidze, G., and Zhang, B. (2006). Formation of a partially screened inner acceleration region in radio pulsars: Drifting subpulses and thermal x-ray emission from polar cap surface, *The Astrophysics Journal*, **650**(2), p. 1048.

Glendenning, N. K. and Schaffner–Bielich, J. (1998). Neutron star constraints on the H dibaryon, *Physical Review C* **58**, p. 1298.

Goldreich, P. and Julian, W. H. (1969). Pulsar electrodynamics, *The Astrophysical Journal*, **157**, p. 869.

Goldstein, A., Veres, P., Burns, E., Briggs, M. S., Hamburg, R., Kocevski, D., Wilson-Hodge, C. A., Preece, R. D., Poolakkil, S., Roberts, O. J., Hui, C. M., Connaughton, V., Racusin, J., von Kienlin, A., Canton, T. D., Christensen, N., Littenberg, T., Siellez, K., Blackburn, L., Broida, J., Bissaldi, E., Cleveland, W. H., Gibby, M. H., Giles, M. M., Kippen, R. M., McBreen, S., McEnery, J., Meegan, C. A., Paciesas, W. S., and Stanbro, M. (2017). An ordinary short gamma-ray burst with extraordinary implications: FermiGBM detection of GRB 170817a, *The Astrophysics Journal* **848**(2), p. L14.

Gotthelf, E. V. and Halpern, J. P. (2007). Precise timing of the x-ray pulsar 1e 1207.4-5209: A steady neutron star weakly magnetized at birth, *The Astrophysics Journal*, **664**(1), p. L35.

Haensel, P., Paczynski, B., and Amsterdamski, P. (1991). Gamma-ray bursts from colliding strange stars, *The Astrophysics Journal*, **375**, p. 209.

Herrera, L. and Santos, N. (1997). Local anisotropy in self-gravitating systems, *Physic Reports*, **286**(2), p. 53.

Hewish, A., Bell, S. J., Pilkington, J. D. H., Scott, P. F., and Collins, R. A. (1968). Observation of a rapidly pulsating radio source, *Nature*, **217**, p. 709.

Hinderer, T., Lackey, B. D., Lang, R. N., and Read, J. S. (2010). Tidal deformability of neutron stars with realistic equations of state and their gravitational wave signatures in binary inspiral, *Physical Review D*, **81**, p. 123016.

Ho, W. C. G. and Lai, D. (2003). Atmospheres and spectra of strongly magnetized neutron stars–II. The effect of vacuum polarization, *Monthly Notices of the Royal Astronomical Society*, **338**(1), p. 233.

Holdom, B., Ren, J., and Zhang, C. (2018). Quark matter may not be strange, *Physical Review Letter*, **120**, p. 222001.

Inoue, T., Ishii, N., Aoki, S., Doi, T., Hatsuda, T., Ikeda, Y., Murano, K., Nemura, H., and Sasaki, K. (2011). Bound *H* dibaryon in flavour su(3) limit of lattice qcd, *Physical Review Letter*, **106**, p. 162002.

Jaffe, R. and Korpa, C. (1985). Semiclassical quantization of the dibaryon skyrmion, *Nuclear Physics B*, **258**, p. 468.

Jaffe, R. L. (1977a). Perhaps a stable dihyperon, *Physical Review Letter*, **38**, p. 195.

Jaffe, R. L. (1977b). Perhaps a stable dihyperon, *Physical Review Letter*, **38**, p. 617.

Kaplan, D. L., Kamble, A., van Kerkwijk, M. H., and Ho, W. C. G. (2011). New optical/ultraviolet counterparts and the spectral energy distributions of nearby, thermally emiting, isolated neutron stars, *The Astrophysics Journal*, **736**(2), p. 117.

Karlovini, M. and Samuelsson, L. (2004). Elastic stars in general relativity: III. stiff ultrarigid exact solutions, *Classical and Quantum Gravity*, **21**(19), p. 4531.

Kasen, D., Metzger, B., Barnes, J., Quataert, E., and Ramirez-Ruiz, E. (2017). Origin of the heavy elements in binary neutron-star mergers from a gravitational wave event, *Nature*, **551**, p. 80.

Kasliwal, M. M., Nakar, E., Singer, L. P., Kaplan, D. L., Cook, D. O., Van Sistine, A., Lau, R. M., Fremling, C., Gottlieb, O., Jencson, J. E., Adams, S. M., Feindt, U., Hotokezaka, K., Ghosh, S., Perley, D. A., Yu, P.-C., Piran, T., Allison, J. R., Anupama, G. C., Balasubramanian, A., Bannister, K. W., Bally, J., Barnes, J., Barway, S., Bellm, E., Bhalerao, V., Bhattacharya, D., Blagorodnova, N., Bloom, J. S., Brady, P. R., Cannella, C., Chatterjee, D., Cenko, S. B., Cobb, B. E., Copperwheat, C., Corsi, A., De, K., Dobie, D., Emery, S. W. K., Evans, P. A., Fox, O. D., Frail, D. A., Frohmaier, C., Goobar, A., Hallinan, G., Harrison, F., Helou, G., Hinderer, T., Ho, A. Y. Q., Horesh, A., Ip, W.-H., Itoh, R., Kasen, D., Kim, H., Kuin, N. P. M., Kupfer, T., Lynch, C., Madsen, K., Mazzali, P. A., Miller, A. A., Mooley, K., Murphy, T., Ngeow, C.-C., Nichols, D., Nissanke, S., Nugent, P., Ofek, E. O., Qi, H., Quimby, R. M., Rosswog, S., Rusu, F., Sadler, E. M., Schmidt, P., Sollerman, J., Steele, I., Williamson, A. R., Xu, Y., Yan, L., Yatsu, Y., Zhang, C., and Zhao, W. (2017). Illuminating gravitational waves: A concordant picture of photons from a neutron star merger, *Science*, **358**, 6370, p. 1559.

Kopeliovich, V., Schwesinger, B., and Stern, B. (1992). SU(3) symmetry breaking in strange dibaryon configurations, *Nuclear Physics A*, **549**, p. 485.

Kulkarni, S. (2005). *Modeling supernova-like explosions associated with gamma-ray bursts with short durations,* ArXiv preprint ArXiv:astro-ph/0510256.

Lai, D. (2001). Matter in strong magnetic fields, *Reviews of Modern Physics*, **73**, p. 629.

Lai, X., Xia, C., Yu, Y., and Xu, R. (2020). Merging strangeon stars: The ejecta and light curves, ArXiv preprint ArXiv:2009.06165.

Lai, X. and Xu, R. (2017). Strangeon and strangeon star, *Journal of Physics: Conference Series*, **861**, p. 012027.

Lai, X., Zhou, E., and Xu, R. (2019). Strangeons constitute bulk strong matter: Test using gw 170817, *European Physics Journal A*, **55**(4), p. 60.

Lai, X. Y., Gao, C. Y., and Xu, R. X. (2013). H-cluster stars, *Monthly Notices of the Royal Astronomical Society*, **431**(4), p. 3282.

Lai, X. Y. and Xu, R. X. (2009). Lennard-Jones quark matter and massive quark stars, *Monthly Notices of the Royal Astronomical Society*, **398**(1), p. L31.

Lai, X.-Y., Yu, Y.-W., Zhou, E.-P., Li, Y.-Y., and Xu, R.-X. (2018a). Merging strangeon stars, *Research in Astronomy and Astrophysics*, **18**(2), p. 024.

Lai, X. Y., Yun, C. A., Lu, J. G., L., G. L., Wang, Z. J., and Xu, R. X. (2018b). Pulsar glitches in a strangeon star model, *Monthly Notices of the Royal Astronomical Society*, **476**(3), p. 3303.

Landau, L. (1932). On the theory of stars, *Physikalische Zeitschriff Sowjetunion*, **1**, p. 285.

Lee, S. H. and Yasui, S. (2009). Stable multiquark states with heavy quarks in a diquark model, *European Physics Journal C*, **64**, p. 283.

Li, L.-X. and Paczyński, B. (1998). Transient events from neutron star mergers, *The Astrophysics Journal*, **507**(1), p. L59.

Li, Z.-S., Qu, Z.-J., Chen, L., Guo, Y.-J., Qu, J.-L., and Xu, R.-X. (2015). An ultra-low-mass and small-radius compact object in 4u 1746-37? *The Astrophysics Journal* **798**(1), p. 56.

LIGO Scientific and Virgo Collaborations (2017). Gw170817: Observation of gravitational waves from a binary neutron star inspiral, *Physical Review Letter*, **119**, p. 161101.

LIGO Scientific and Virgo Collaborations (2018). Gw170817: Measurements of neutron star radii and equation of state, *Physical Review Letter*, **121**(16), p. 161101.

LIGO Scientific and Virgo Collaborations (2020). GW190814: Gravitational waves from the coalescence of a 23 solar mass black hole with a 2.6 solar mass compact object, *The Astrophysics Journal*, **896**(2), p. L44.

LIGO Scientific Collaboration, Virgo Collaboration, Fermi gamma-ray burst monitor, and integral (2017). Gravitational waves and gamma-rays from a binary neutron star merger: GW170817 and GRB 170817a, *The Astrophysics Journal*, **848**(2), p. L13.

Lin, M.-X., Xu, R.-X., and Zhang, B. (2015). Oscillation-driven magnetospheric activity in pulsars, *The Astrophysics Journal*, **799**(2), p. 152.

Liu, K. and Wong, C. (1982). MIT bag model with center-of-mass correction, *Physical Letter B* **113**(1), p. 1.

Luo, R., Wang, B. J., Men, Y. P., Zhang, C. F., Jiang, J. C., Xu, W. Y. L. K. J. H. J. L. Z. B. C. R. N. C. M. Z., H. and Wang, Chen, X. L., Gan, H. Q., Guo, Y. J., Hao, L. F., Huang, Y. X., Jiang, P., Li, H., Li, J., Li, Z. X., Luo, J. T., Pan, J., Pei, X., Qian, L., Sun, J. H., Wang, M., Wang, N., Wen, Z. G., Xu, R. X., Xu, Y. H., Yan, J., Yan, W. M., Yu, D. J., Yuan, J. P., Zhang, S. B., and Zhu, Y. (2020). Diverse polarization angle swings from a repeating fast radio burst source, Nature, **586**, p. 693.

Maezawa, Y., Hatsuda, T., and Sasaki, S. (2005). Strange tribaryons as nonaquark states, *Progress of Theoretical and Experimental Physics*, **114**, p. 317.

Maltman, K. (1992). How secure are predictions for dibaryons in the MIT bag model? *Physics Letter B*, **291**(4), p. 371.

McLerran, L. (2009). The phase diagram of qcd and some issues of large n_c, *Nuclear Physics B (Proc. Suppl.)* **195**, p. 275.

Mereghetti, S., Luca, A. D., Caraveo, P. A., Becker, W., Mignani, R., and Bignami, G. F. (2002). Pulse phase variations of the x-ray spectral features in the radio-quiet neutron star 1e 1207-5209, *The Astrophysics Journal* **581**(2), p. 1280.

Meszaros, P., Rees, M. J., and Wijers, R. A. M. J. (1998). Viewing angle and environment effects in gamma-ray bursts: Sources of afterglow diversity, *The Astrophysics Journal* **499**(1), p. 301.

Miao, Z.-Q., Xia, C.-J., Lai, X.-Y., Maruyama, T., Xu, R.-X., and Zhou, E.-P. (2022). A bag model of matter condensed by the strong interaction, *International Journal of Modern Physics E* **31**(04), p. 2250037

Mulders, P. J., Aerts, A. T., and de Swart, J. J. (1980). Multiquark states. iii. Q^6 dibaryon resonances, *Physical Review D*, **21**, p. 2653.

Oka, M., Shimizu, K., and Yazaki, K. (1983). The dihyperon state in the quark cluster model, *Physics Letter B*, **130**, p. 365.

Oka, M. and Takeuchi, S. (1991). Instanton Induced Interaction and the Strange Dibaryons, *Nuclear Physics A*, **524**, p. 649.

Oppenheimer, J. R. and Volkoff, G. M. (1939). On massive neutron cores, *Physical Review*, **55**, p. 374.

Ouyed, R., Rapp, R., and Vogt, C. (2005). Fireballs from quark stars in the color-flavour locked phase: Application to gamma-ray bursts, *The Astrophysics Journal*, **632**(2), p. 1001.

Paczynski, B. (1986). Gamma-ray bursters at cosmological distances, *The Astrophysics Journal*, **308**, p. L43.

Paczynski, B. and Haensel, P. (2005). Gamma-ray bursts from quark stars, *Monthly Notices of the Royal Astronomical Society*, **362**(1), p. L4.

Particle Data Group (2016). Review of particle physics, *Chinese Physics C*, **40**(10), 100001.

Pavlov, G. G. and Luna, G. J. M. (2009). A dedicated chandra acis observation of the central compact object in the Cassiopeia A supernova remnant, *The Astrophysics Journal*, **703**(1) p. 910.

Pavlov, G. G., Sanwal, D., and Teter, M. A. (2004). Central compact objects in supernova remnants, in F. Camilo and B. M. Gaensler (eds.), *Young Neutron Stars and Their Environments*, Vol. 218, p. 239.

Peng, C. and Xu, R. X. (2008). Pulsar slow glitches in a solid quark star model, *Monthly Notices of the Royal Astronomical Society*, **384**(3), p. 1034.

Pons, J. A., Reddy, S., Prakash, M., Lattimer, J. M., and Miralles, J. A. (1999). Evolution of proto-neutron stars, *The Astrophysics Journal* **513**(2), p. 780.

Postnikov, S., Prakash, M., and Lattimer, J. M. (2010). Tidal love numbers of neutron and self-bound quark stars, *Physical Review D*, **82**, p. 024016.

Qiao, G. J., Lee, K. J., Zhang, B., Xu, R. X., and Wang, H. G. (2004). A model for the challenging "bi-drifting" phenomenon in PSR j0815+09, *The Astrophysics Journal* **616**(2), p. L127.

Romani, R. W. (1987). Model Atmospheres for Cooling Neutron Stars, *The Astrophysics Journal*, **313**, p. 718.

Ruderman, M. A. and Sutherland, P. G. (1975). Theory of pulsars: polar gaps, sparks, and coherent microwave radiation. *The Astrophysics Journal*, **196**, p. 51.

Sakai, T., Mori, J., Buchmann, A., Shimizu, K., and Yazaki, K. (1997). The interaction between H-dibaryons, *Nuclear Physics A*, **625**(1), p. 192.

Sanwal, D., Pavlov, G. G., Zavlin, V. E., and Teter, M. A. (2002). Discovery of absorption features in the x-ray spectrum of an isolated neutron star, *The Astrophysics Journal* **574**(1), p. L61.

Sasaki, K., Aoki, S., Doi, T., Gongyo, S., Hatsuda, T., Ikeda, Y., Inoue, T., Iritani, T., Ishii, N., Murano, K., and Miyamoto, T. (2020). $\lambda\lambda$ and nξ interactions from lattice qcd near the physical point, *Nuclear Physics A*, **998**, p. 121737.

Shen, P., Zhang, Z., Yu, Y., Yuan, X., and Yang, S. (1999). H dihyperon in quark cluster model, *J. Phys. G*, **25**, p. 1807.

Smartt, S., *et al.* (2017). A kilonova as the electromagnetic counterpart to a gravitational-wave source, *Nature*, **551**, 7678, p. 75.

Stairs, I., Lyne, A., and Shemar, S. (2000). Evidence for free precession in a pulsar, *Nature*, **406**, p. 484.

Straub, U., Zhang, Z.-Y., Brauer, K., Faessler, A., and Khadkikar, S. (1988). Binding energy of the dihyperon in the quark cluster model, *Physics Letter B*, **200**, p. 241.

Turolla, R., Zane, S., and Drake, J. J. (2004). Bare quark stars or naked neutron stars? the case of RX j1856.5-3754, *The Astrophysics Journal*, **603**(1), p. 265.

Villar, V. A., Guillochon, J., Berger, E., Metzger, B. D., Cowperthwaite, P. S., Nicholl, M., Alexander, K. D., Blanchard, P. K., Chornock, R., Eftekhari, T., Fong, W., Margutti, R., and Williams, P. K. G. (2017). The combined ultraviolet, optical, and near-infrared light curves of the kilonova associated with the binary neutron star merger GW170817: Unified data set, analytic models, and physical implications, *The Astrophysics Journal*, **851**(1), p. L21.

Wang, W., Lu, J., Tong, H., Ge, M., Li, Z., Men, Y., and Xu, R. (2017). The optical/UV excess of x-ray-dim isolated neutron stars. i. bremsstrahlung emission from a strangeon star atmosphere, *The Astrophysics Journal*, **837**(1), p. 81.

Wang, W., Luo, R., Yue, H., Chen, X., Lee, K., and Xu, R. (2018a). FRB 121102: A starquake-induced repeater? *The Astrophysics Journal*, **852**(2), p. 140.

Wang, W. H., Lai, X. Y., Zhou, E. P., Lu, J. G., Zheng, X. P., and Xu, R. X. (2020a). Pulsar glitches in a strangeon star model. II. The activity, *Monthly Notices of the Royal Astronomical Society*, **500**(4), p. 5336.

Wang, W.-Y., Feng, Y., Lai, X.-Y., Li, Y.-Y., Lu, J.-G., Chen, X., and Xu, R.-X. (2018b). The optical/UV excess of x-ray-dim isolated neutron star II. nonuniformity of plasma on a strangeon star surface, *Research in Astronomy and Astrophysics* **18**(7), p. 082.

Wang, W.-Y., Xu, R., and Chen, X. (2020b). On the magnetospheric origin of repeating fast radio bursts, *The Astrophysics Journal*, **899**(2), p. 109.

Watson, D., Hansen, C. J., Selsing, J., Koch, A., Malesani, D. B., Andersen, A. C., Fynbo, J. P. U., Arcones, A., Bauswein, A., Covino, S., Grado, A., Heintz, K. E., Hunt, L., Kouveliotou, C., Leloudas, G., Levan, A., Mazzali, P., and Pian, E. (2019). Identification of strontium in the merger of two neutron stars, *Nature (London)*, **574**(7779), p. 497.

Witten, E. (1984). Cosmic separation of phases, *Physical Review D*, **30**, p. 272.

Xia, C.-J., Maruyama, T., Yasutake, N., Tatsumi, T., Shen, H., and Togashi, H. (2020a). Systematic study on the quark-hadron mixed phase in compact stars, *Physical Review D*, **102**, p. 023031.

Xia, C.-J., Xue, S.-S., Xu, R.-X., and Zhou, S.-G. (2020b). Supercritically charged objects and electron-positron pair creation, *Physical Review D*, **101**(10), p. 103031.

Xu, R. (2007). AXPs/SGRs: Magnetars or quark-stars? *Advances in Space Research*, **40**(10), p. 1453.

Xu, R. (2018). Strong matter: Rethinking philosophically, *Science China Physics, Mechanics and Astronomy*, **61**(10), p. 109531.

Xu, R. and Guo, Y. (2017). Strange matter: A state before black hole, in Centennial of General Relativity: A celebration, World Scientific, Singapoure, C.A.Z. Vasconcellos (ed.), p. 119.

Xu, R. and Liang, E. (2009). X-ray flares of gamma-ray bursts: Quakes of solid quark stars? *Science china Physics, Mechanics and Astronomy,* **52**, p. 315.

Xu, R. X. (2002). A thermal featureless spectrum: Evidence for bare strange stars? *The Astrophysics Journal*, **570**(2), p. L65.

Xu, R. X. (2005). 1E 1207.4-5209: A low-mass bare strange star? *Monthly Notices of the Royal Astronomical Society*, **356**(1), p. 359.

Xu, R.-X. (2014). A solution to the puzzling symbiotic x-ray system 4U 1700+24, *Research in Astronomy and Astrophysics* **14**(6), p. 617.

Xu, R. X., Bastrukov, S. I., Weber, F., Yu, J. W., and Molodtsova, I. V. (2012). Absorption features caused by oscillations of electrons on the surface of a quark star, *Physical Review D*, **85**, p. 023008.

Xu, R. X., Qiao, G. J., and Zhang, B. (1999). PSR 0943+10: A bare strange star? *The Astrophysics Journal*, **522**(2), p. L109.

Xu, R. X., Tao, D. J., and Yang, Y. (2006). The superflares of soft γ-ray repeaters: giant quakes in solid quark stars? *Monthly Notices of the Royal Astronomical Society*, **373**(1), p. L85.

Xu, R.-X., Wang, H.-G., and Qiao, G.-J. (2003). A note on the discovery of absorption features in 1e 1207.4-5209, *Chinese Physics Letter*, **20**(2), p. 314.

Yost, S. A. and Nappi, C. R. (1985). The mass of the H dibaryon in a chiral model, *Physical Review D*, **32**, p. 816.

Yu, J. W. and Xu, R. X. (2011). Magnetospheric activity of bare strange quark stars, *Monthly Notices of the Royal Astronomical Society*, **414**(1), p. 489.

Yu, Y.-W., Zhang, B., and Gao, H. (2013). Bright merger-nova from the remnant of a neutron star binary merger: A signature of a newly born, massive, millisecond magnetar, *The Astrophysics Journal*, **776**(2), p. L40.

Yuan, M., Lu, J.-G., Yang, Z.-L., Lai, X.-Y., and Xu, R.-X. (2017). Supernova neutrinos in a strangeon star model, in *Research Astronomy and Astrophysics*, **17**(9), p. 092.

Zhou, A., Xu, R., Wu, X., and Wang, N. (2004). Quakes in solid quark stars, *Astroparticle Physics*, **22**(1), p. 73.

Zhou, E. P., Lu, J. G., Tong, H., and Xu, R. X. (2014). Two types of glitches in a solid quark star model, *Monthly Notices of the Royal Astronomical Society*, **443**(3), p. 2705.

Zhou, E.-P., Zhou, X., and Li, A. (2018). Constraints on interquark interaction parameters with gw170817 in a binary strange star scenario, *Physical Review D*, **97**, p. 083015.

https://doi.org/10.1142/9789811220944_0004

Chapter 4

Fast-spinning and Highly Magnetized White Dwarfs

Edson Otoniel[*,†,**], Jaziel G. Coelho[†,‡,††], Manuel Malheiro[§,‡‡] and Fridolin Weber[¶,‖,§§]

[*] *Instituto de Formação de Educadores, Universidade Federal do Cariri Av. Tenente Raimundo Rocha N 1639 Bairro Cidade Universitária Juazeiro do Norte — Ceará CEP 63048-080, Brazil*

[†] *Núcleo de Astrofísica e Cosmologia (Cosmo-Ufes) & Departamento de Física, Universidade Federal do Espírito Santo, 29075–910, Vitória, ES, Brazil*

[‡] *Divisão de Astrofísica, Instituto Nacional de Pesquisas Espaciais Av. dos Astronautas, 1.758 — Jardim da Granja, São José dos Campos — SP, 12227-010, Brazil*

[§] *Departamento de Física, Instituto Tecnológico de Aeronáutica Praa Marechal Eduardo Gomes, 50 — Vila das Acacias, São José dos Campos — SP, 12228-900, Brazil*

[¶] *Department of Physics, San Diego State University San Diego, CA 92182, USA*

[‖] *Center for Astrophysics and Space Sciences University of California at San Diego La Jolla, CA 92093, USA*

[**] *edson.otoniel@gmail.com*

[††] *jaziel.coelho@ufes.br*

[‡‡] *manuelmalheiro@gmail.com*

[§§] *fweber@ucsd.edu*

Abstract: As is well known, the existence of fast and ultra-massive white dwarfs (WDs) has been discovered in recent years. We estimate the mass density thresholds for the onset of electron capture reactions and pycnonuclear reactions in the cores of fast, massive and highly magnetized white dwarfs (HMWDs) and discuss the impact of microscopic stability and rapid rotation on the structure and stability of different

WD constitutions. We confirm that fast rotation increases the mass of a WD by up to 10%, while the central density may drop by one to two orders of magnitude, depending on stellar mass and rate of rotation. We also note that the central densities of the rotating WDs are smaller than those of the non-rotating stars, since less pressure is to be provided by the equation of state (EoS) in the rotating case, and that the maximum-mass limit slightly decreases when lattice contributions are taken into account, which soften the EoS mildly. The softening leads to WDs with somewhat smaller radii and therefore smaller Kepler periods. Overall, one sees that very massive and magnetic $^{12}C + ^{16}O$ WDs have rotational Kepler periods on the order of 0.5 s. Pycnonuclear reactions are triggered in these WDs at masses that are markedly smaller than the maximum WD masses. The corresponding rotational periods turn out to be in the 5 s (~0.2 Hz) range.

Keywords: white dwarfs; rotation; magnetic fields.

1. Introduction

The increasing data from observational white dwarf (WD) surveys has confirmed the existence of highly magnetized white dwarfs (HMWDs) with surface magnetic fields between 10^6 an 10^9 G (see Kepler *et al.*, 2010, 2013, 2015; Külebi *et al.*, 2009, 2010a,b, for recent results of the Sloan Digital Sky Survey (SDSS) for HMWDs). Indeed, polar magnetic fields vary from 7 MG (V2301Oph) up to 240 MG (AR UMa) whereas intermediate polar fields have magnitudes of ~4–30 MG, with the highest value of ~32 MG for V405Aur (Ferrario *et al.*, 2015). For isolated WDs, the magnetic field is in the range of $\sim 10^3$–10^9 G. Close to 10^7 G, there are several examples of isolated WDs such as WD 0806+376 with a field of 3.97×10^7 G and WD 1017-367 with $B = 6.5 \times 10^7$ G (see Ferrario *et al.*, 2015, for another examples of HMWDs). Most of them have been shown to be massive and responsible for the high-mass peak at 1 $M_\odot$ of the white dwarf mass distribution. Examples are the WD REJ 0317–853 which has a mass of $M \approx 1.35\ M_\odot$ and a magnetic field of $B \approx (1.7\text{–}6.6) \times 10^8$ G (Barstow *et al.*, 1995; Külebi *et al.*, 2010a,b), PG 1658+441 with $M \approx 1.31\ M_\odot$ and $B \approx 2.3 \times 10^6$ G (Liebert *et al.*, 1983; Schmidt *et al.*, 1992), and PG 1031+234 with the highest magnetic field of $B \approx 10^9$ G (Külebi *et al.*, 2009; Schmidt *et al.*, 1986). The existence of very massive WDs has been revealed in several studies (Althaus *et al.*, 2005, 2007; Camisassa *et al.*, 2019; Castanheira *et al.*, 2013; Curd *et al.*, 2017; Gentile *et al.*, 2018; Hermes *et al.*, 2013; Jiménez-Esteban *et al.*, 2018; Sousa *et al.*, 2020a).

In addition, WDs rotate with periods of days or even years. Recently a pulsating WD named AR Scorpii has been discovered (see Marsh *et al.*, 2016), which emits electromagnetic radiation ranging from X-rays to radio waves, pulsing in brightness with a period of 1.97 min. Furthermore, the X-ray emission has been proposed to be caused by a pulsar-like emission from a fast-spinning WD, suggesting that AR Sco is primarily a rotation-powered pulsar (RPP). Besides that, other sources have been proposed as candidates for WD pulsars. In particular, AE Aquarii, with a short spin period of $P = 33.08$ s (Terada *et al.*, 2008). Furthermore, the X-ray Multimirror Mission XMM-Newton observatory has observed a WD rotating faster than AE Aquarii. It has been shown by Mereghetti *et al.* (2009) that the X-ray pulsator RX J0648.0-4418, which belongs to the binary systems HD 49798/RX J0648.0-4418, is a massive $M = 1.28$ $M_\odot$ white dwarf with a radius of around $R = 3000\,\text{km}$ and a very small spin period of $P = 13.2\,\text{s}$. Also, Ashley *et al.* (2020) reported the discovery of strong pulsations with a period of 38.9 s in the *HST* ultraviolet data of V1460 (2020). Very recently, Lopes *et al.* (2020) reported on XMM-*Newton* observations of CTCV J2056-3014 revealing that this object is an X-ray-faint intermediate polar harboring an extremely fast-spinning WD with a spin period of 29.6 s. Shortly thereafter, we used the equation of state (EoS) of a carbon WD accounting for electron–ion, electron–electron, and ion–ion interactions. For this EoS, we determined the mass density thresholds for the onset of pycnonuclear fusion reactions and investigated the impact of microscopic stability and of fast-rotation on the structure and stability of WDs. Our analysis has lead to a minimum mass for CTCV J2056-3014 of 0.56 $M_\odot$, and a maximum mass of around 1.38 $M_\odot$ (see Otoniel *et al.*, 2020, for details).

It is worth mentioning that several models of highly magnetized and fast-spinning WDs have been proposed (see Cáceres *et al.*, 2016; Coelho and Malheiro, 2014; Gentile *et al.*, 2018; Lobato *et al.*, 2016; Malheiro *et al.*, 2012; Mukhopadhyay *et al.*, 2020; Sousa *et al.*, 2020a,b; Vishal and Mukhopadhyay, 2014, and references therein). More recently, Kalita nd Mukhopadhyay (2019) show that continuous gravitational waves (GWs) can be emitted from rotating magnetized WDs and will possibly be detected by the upcoming GW detector (see also Sousa *et al.*, 2020a,b). Furthermore, as pointed out in Sousa *et al.* (2020b), for the first time in the literature, it is estimated the GWs counterpart for soft gamma repeaters (SGRs) and anomalous X-ray pulsars (AXPs) described as WD pulsars. They have found that some SGRs/AXPs can be observed by the space detectors BBO and DECIGO. In addition, the same authors have estimated the GW amplitude and luminosity for three binary systems that have a fast-spinning magnetized WD, namely, AE Aquarii, AR Scorpii and RX J0648.0-4418 (see Sousa *et al.*, 2020a, for details).

Also, Rueda *et al.* (2013) have proposed a scenario that the optical/IR data are explained by the WD photosphere and by assuming the existence of an accretion disk (Rueda *et al.*, 2013). Recently, a new scenario has been proposed to explain the spectral energy distribution of 4U0142+61, from mid-infrared up to hard X-rays (see Borges *et al.*, 2020, and references therein). In this model, the persistent emission comes from an accreting isolated massive and HMWD surrounded by a debris disk, having gas and dusty regions. This scenario is inspired by the periodic flux modulation and by the presence of mid-IR emission, which is rare for neutron stars (NSs) (appears only in about 0.3% of all isolated NSs). On the other hand, the presence of mid-IR in WDs is quite common. Debes *et al.* (2011) found that about 7% of all isolated WDs exhibit an excess of mid-IR emission.

It has also been proposed by Subramanian and Mukhopadhyay (2015) that the light curves of some peculiar superluminous Ia supernovae could be explained by super-Chandrasekhar white dwarfs (SChWDs) whose masses are greater than the Chandrasekhar mass limit, $M_{\rm Ch} = 1.44\ M_{\odot}$. From a theoretical point of view, WDs can rotate at periods as short as $P \sim 0.5\,$s (Boshkayev *et al.*, 2013a,b) and they can be formed in double WD mergers (Becerra *et al.*, 2018; Garcia-Berro *et al.*, 2012; Rueda *et al.*, 2013). Livio and Pringle (1998) and Livio (1999) have investigated the role of rotation for the maximum mass of a uniformly rotating WD.

General relativistic effects on uniformly rotating WDs have been studied more recently in Boshkayev *et al.* (2013a,b). In particular, fast-spinning WDs with extreme magnetic fields have been investigated in order to see how huge magnetic fields would change the mass and the radius of these stars, affecting their spherical symmetry and stabilities (see, e.g., Angel, 1978; Bera and Bhattacharya, 2014, 2016; Das *et al.*, 2013; Das and Mukhopadhyay, 2015; Franzon and Schramm, 2015). It has been shown by Franzon and Schramm (2015) that massive WDs can support magnetic fields at their center of up to $\sim 10^{15}$ G if they are non-rotating, and $\sim 10^{14}$ G if they are rapidly rotating. According to Das and Mukhopadhyay (2012), the effects of an extremely large and uniform magnetic field on the EoS of a WD could increase its critical mass up to 2.58 $M\odot$. This mass limit is reached for extremely large magnetic fields of $\sim 10^{18}$ G.

Also, a general relativistic magneto-hydrodynamic (GRMHD) framework that describes rotating and magnetized axisymmetric WDs has been explored sequentially by Bera and Bhattacharya (2016) and Subramanian and Mukhopadhyay (2015). Nevertheless, as already discussed in the literature (Chamel *et al.*, 2014; Chamel and Fantina, 2015; Coelho *et al.*, 2014), the breaking of spherical symmetry due to magnetic fields and microphysical effects, such as electron capture reactions and pycnonuclear reactions, can severely limit the magnetic fields inside WDs. In addition, calculations employing magneto-hydrodynamics (MHDs) and a polytropic EoS with an adiabatic index $\Gamma = 4/3$ were performed in order to investigate the stability of SChWDs, considering poloidal and toroidal magnetic field configurations (see Das and Mukhopadhyay, 2015; Subramanian and Mukhopadhyay, 2015, and references therein). It was found that huge magnetic fields, up to 5×10^{15} G, may exist inside such high-mass WDs, and that their masses can be in the range of 2–3 $M_{\odot}$. The internal magnetic fields can be much stronger than at the surface (see Franzon and Schramm, 2015; Otoniel *et al.*, 2019, and references therein). Theoretical studies suggest that the magnetic field in the interior can be up to three orders of magnitude larger, i.e., $\sim 10^{12}$ G, than the largest observed surface magnetic fields, $\sim 10^8$–10^9 G (Franzon and Schramm, 2015). Recently, we have also shown that maximum-mass WDs could have masses of around 2.14 $M_{\odot}$ for a central magnetic field of $\sim 3.85 \times 10^{14}$ G, which indicates that highly magnetized white dwarfs may play a crucial role for the interpretation of superluminous type Ia supernovae. Furthermore, we note that pycnonuclear fusion reactions limit the central WD density to $\lesssim 9.35 \times 10^9\,\mathrm{g/cm^3}$, which implies that the equatorial radii of WDs cannot be much smaller than around 1100 km (see Otoniel *et al.*, 2019). It is worth noting that general relativistic effects do not change the maximum mass of a WD but have a significant impact on the stellar radius (see, for instance, Carvalho *et al.*, 2018).

Nuclear reactions have a determining role in the existence and evolution of several astrophysical environment (see Liccardo *et al.*, 2018). These nuclear reactions have been found to occur over a significant range of stellar densities (Gasques *et al.*, 2005), including the density range found in the interiors of WDs (Boshkayev *et al.*, 2013a). Recently, *static* WD calculations also showed that central energy densities are limited by nuclear fusion reactions and inverse β-decay (see Chamel *et al.*, 2013, 2014; Chamel and Fantina, 2015). We have also investigated the properties of stationary and strongly magnetized WDs, taking into account the electron capture and pycnonuclear fusion reactions instabilities (Malheiro *et al.*, 2021; Otoniel *et al.*, 2019).

Here, we explore the properties of rotating WDs endowed with strong magnetic fields ($\sim 10^{12}$ G) using a modern EoS which accounts for electron-ion interactions among lattice nuclei, and which makes use of the latest experimental atomic mass data (Audi *et al.*, 2012; Wang *et al.*, 2012). We estimate the mass density thresholds for the onset of electron capture reactions and pycnonuclear reactions in the cores of massive and HMWDs, and explore the impact of microscopic stability and fast rotation on the structure and stability of different WD constitutions. We also note that the presence of $^{12}C + ^{16}O$ lattices in WDs reduces the Kepler periods (increases the Kepler frequencies) of WDs mildly, which has its origin in the softening of the EoS caused by the lattice. The softening leads to WDs with somewhat smaller radii and therefore smaller Kepler periods.

This chapter is organized as follows. In Section 2, we revisit the physics of an electron gas subjected to high magnetic fields. In Section 3, we introduce the model used for the EoS of HMWDs, which accounts for electron–ion interactions among lattice nuclei. In Section 4, we discuss the role of the electron capture and pycnonuclear fusion reactions for the structure and stability of HMWDs. In Section 5, we present the structure equations of WDs in hydrostatic equilibrium and present limits of the rotational frequencies of strongly magnetized WDs. The Kepler (mass shedding) frequency is used to set an absolute limit on stable stellar rotation. Features on non-spherical stars are discussed in Section 6. Finally, in Section 7, we present a summary of the main findings and concluding remarks.

2. Electron Gas in Strong Magnetic Fields

The increase of more sensitive techniques of observation and detection of WDs (Eisenstein *et al.*, 2006), coupled with advanced calculations of the properties of fermionic matter under extreme physical condition, has led to considerable interest in theoretical studies of the structure and evolution of WDs (Chamel *et al.*, 2014; Marsh *et al.*, 2016; Mereghetti *et al.*, 2009; Terada *et al.*, 2008). Strong magnetic fields are known to modify the EoS used to describe the global structure of WDs (see, for instance, Coelho *et al.*, 2014; Das and Mukhopadhyay, 2013, 2015; Otoniel *et al.*, 2015, and references therein). Below we briefly summarize the equations which describe a relativistic electron gas (Canuto and Chiu, 1968; Potekhin and Chabrier, 2000, 2013; Potekhin and Yakovlev, 2012) in the presence of a constant magnetic field. The Lagrangian of such a system is given by (we work here with Gaussian units) (Canuto and Chiu, 1968)

$$\mathcal{L} = \mathcal{L}_e + \mathcal{L}_{\text{int}} + \mathcal{L}_\gamma, \tag{1}$$

where the first term on the right-hand-side describes free relativistic electrons, and is given by

$$\mathcal{L}_e = -\frac{1}{2}\hbar c(\bar{\psi}\gamma_\mu \partial_\mu \psi - \partial_\mu \bar{\psi}\gamma_\mu \psi) - m_e c^2 \bar{\psi}\psi, \tag{2}$$

where c denotes the speed of light, m_e the electron mass, and $\hbar$ Planck's constant. The quantity ψ is the electron wave function, and $\bar{\psi} = \psi^\dagger \gamma_0$, where $\psi^\dagger$ is the Hermitian conjugate of ψ.

Interactions among the electrons are described by

$$\mathcal{L}_{\text{int}} = ie\bar{\psi}\gamma_\mu A_\mu \psi, \tag{3}$$

and photon contributions are taken into account through

$$\mathcal{L}_\gamma = -\frac{1}{2}\frac{\partial A_\mu}{\partial x_\nu}\frac{\partial A_\mu}{\partial x_\nu}, \tag{4}$$

where $\partial_\mu \equiv \frac{\partial}{\partial x_\mu}$ with Greek indices running from 1 to 4. The γ matrices are defined in Pauli–Dirac representation. The components of the four-vector potential $A_\mu = (i\phi, \mathbf{A})$ are the scalar electric potential ϕ and the three-vector potential $\mathbf{A}$, which is parallel to the z-axis, according to the gauge (Canuto and Chiu, 1968)

$$A_\mu(x) = \delta_{\mu 2}\; x_1\; \mathbf{B}, \tag{5}$$

where $\mathbf{B}$ denotes a constant external magnetic field.

To determine the EoS we consider the grand-canonical partition function,

$$Z = Tr[e^{-\beta(H_e - \mu_e N_e)}], \tag{6}$$

where $\beta \equiv 1/k_B T$ with k_B the Boltzmann constant and T the temperature. The quantities H_e, μ_e, and N_e denote the Hamiltonian, chemical potential, and average number density of electrons, respectively. The latter is given by $N_e = \frac{1}{\beta}\frac{\partial \ln Z}{\partial \mu_e}$. The grand-canonical potential is given by

$$\Omega = -\frac{1}{\beta}\ln Z, \tag{7}$$

which leads for the electron gas to

$$\Omega_e = -\sum_n g_n \frac{2eB}{\hbar c} \int_{-\infty}^{\infty} \frac{dp_z}{(2\pi)^2} \sqrt{p_z^2 c^2 + s_e(n)^2}$$
$$-\frac{1}{\beta} \sum_n g_n \frac{2eB}{\hbar c} \int_{-\infty}^{\infty} \frac{dp_z}{(2\pi)^2}$$
$$\times \left[\ln(1 e^{-\beta(E_e-\mu_e)}) + \ln(1 + e^{-\beta(E_e+\mu_e)})\right]. \tag{8}$$

The quantity $g_n = 2 - \delta_{n0}$ describes the degeneracy of the Landau levels. For an occupied ground state, $n = 0$, one has $g_n = 1$. For $n \geq 1$ one has $g_n = 2$, which corresponds to doubly degenerate states. The quantity s_e is given by

$$s_e(n) = \sqrt{m_e^2 c^4 \left(1 + 2\frac{B}{B_c} n\right)}, \tag{9}$$

and the electron energies E_e are defined as

$$E_e = \sqrt{p_z^2 c^2 + m_e^2 c^4 \left(1 + 2\frac{B}{B_c} n\right)}, \tag{10}$$

where $B_c = \frac{m_e^2 c^3}{e\hbar} \approx 4.14 \times 10^{13}$ G is the so-called critical magnetic field. The second integral on the right-hand-side of Eq. (8) is the finite-temperature contribution of electrons to Ω_e. For the limiting case of zero temperature, one obtains

$$\lim_{T\to 0} \frac{1}{\beta} \ln(1 + e^{-\beta(E_e-\mu_e)}) = (\mu_e - E_e)\theta(\mu_e - E_e), \tag{11}$$

$$\lim_{T\to 0} \frac{1}{\beta} \ln(1 + e^{-\beta(E_e+\mu_e)}) = 0. \tag{12}$$

The energy density of the system in the parallel and perpendicular directions to the magnetic field B in terms of the grand-canonical potential can be written as

$$\mathcal{E}_e = \Omega_e + \mu_e N_e + \frac{B^2}{8\pi}. \tag{13}$$

Therefore, it follows immediately that

$$\mathcal{E}_e = \sum_{n=0}^{n_{\max}} \frac{eB}{(2\pi)^2 \hbar c} \left[\mu_e \sqrt{\mu_e^2 - s_e(n)^2} + s_e(n)^2 \ln \frac{\mu_e + \sqrt{\mu_e^2 - s_e(n)^2}}{s_e(n)}\right] + \frac{B^2}{8\pi}. \tag{14}$$

Similarly, the pressure parallel to the magnetic field follows from the grand-canonical potential as

$$P_{\parallel} = -\Omega_e - \frac{B^2}{8\pi}, \tag{15}$$

so that

$$P_{\parallel} = \sum_{n=0}^{n_{\max}} g_n \frac{eB}{(2\pi)^2\hbar c}\left[\mu_e\sqrt{\mu_e^2 - s_e(n)^2} - s_e(n)^2 \ln\frac{\mu_e + \sqrt{\mu_e^2 - s_e(n)^2}}{s_e(n)}\right] - \frac{B^2}{8\pi}. \tag{16}$$

The perpendicular pressure is given by

$$P_{\perp} = -\Omega_e - B\mathcal{M}_e + \frac{B^2}{8\pi}, \tag{17}$$

where $\mathcal{M}_e = -\left(\frac{\partial\Omega_e}{\partial B}\right)$ is the magnetization. This leads for the perpendicular pressure to

$$\begin{aligned} P_{\perp} = {} & \sum_{n=0}^{n_{\max}} g_n \frac{eB}{(2\pi)^2\hbar c}\left[\mu_e\sqrt{\mu_e^2 - s_e(n)^2} - s_e(n)^2 \right. \\ & \left. \times \ln\frac{\mu_e + \sqrt{\mu_e^2 - s_e(n)^2}}{s_e(n)}\right] - \frac{em_e}{4\pi^2}\left(\sum_{n=0}^{n_{\max}} g_n \left[\mu_e\sqrt{\mu_e^2 - s_e(n)^2}\right.\right. \\ & \left.\left. - \left[s_e(n)^2 + 2s_e(n)\mathcal{C}_e\right] \ln\frac{\mu_e + \sqrt{\mu_e^2 - s_e(n)^2}}{s_e(n)}\right]\right) \\ & + \frac{B^2}{8\pi}, \end{aligned} \tag{18}$$

where $\mathcal{C}_e$ is given by $\mathcal{C}_e = \frac{B}{B_c} n m_e c^2 / \sqrt{2\frac{B}{B_c}n + 1}$. It is possible to obtain a relationship between $P_{\parallel}$ and $P_{\perp}$ of the following form (see, e.g., Paret *et al.*, 2015a,b; Strickland and Dexheimer, 2012)

$$P_{\perp} = P_{\parallel} - \mathcal{M}_e B + B^2/(4\pi). \tag{19}$$

In Fig. 1, we show the parallel and longitudinal pressures as a function of mass density for different magnetic field values. One sees that magnetic fields up to 10^{12} G have essentially no impact on the EoS. This begins to change for magnetic fields on the order of 10^{13} G, in which case the pressure becomes anisotropic, that is, the pressure in the direction parallel to B differs from the pressure in the orthogonal direction to B.

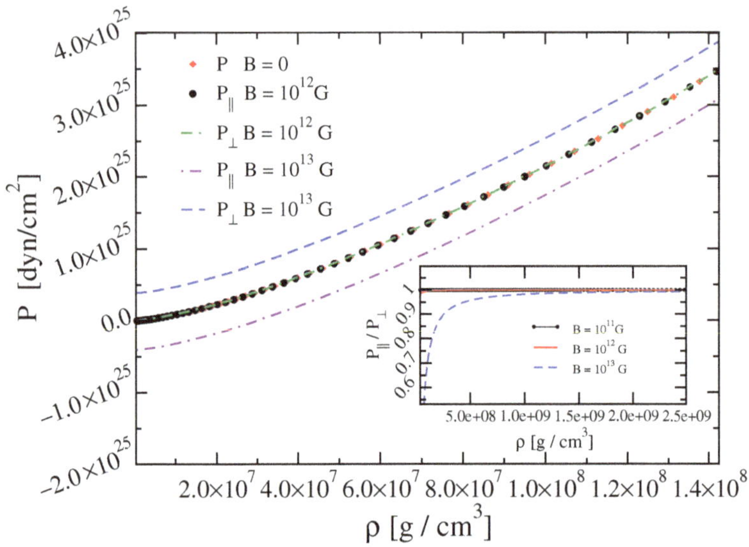

Fig. 1. Parallel and perpendicular pressures as a function of mass density, ρ. The orange diamond represents the $B = 0$ case. The inset shows $P_\parallel / P_\perp$ for magnetic fields ranging from 10^{11} to 10^{13} G.

3. Crystal Lattice

We assume that the core of a HMWD is composed of fully ionized ions forming a regular body-centered-cubic (bcc) crystal lattice (see Chamel *et al.*, 2014, and references therein), referred to as regular crystal lattice (RCL) for short. Numerical simulations suggest that it is essential to take into account the effects of the Coulomb interaction in the EoS (Chamel and Haensel, 2008; Pearson *et al.*, 2011), particularly when determining the maximum stable mass of a stationary WD (i.e., non-rotating) (Rotondo *et al.*, 2011; Salpeter, 1961). This path has been followed for the investigations presented here. Originally, this model was used to describe the outer crust of a NS and later on to model the cores of super-dense WDs (Pearson *et al.*, 2011).

The pressure at the center of a WD is essentially given by the pressure of degenerate electrons. However, due to the regular crystal lattice, the total pressure of WD matter is given by the sum of the degenerate electron and ionic lattice pressures,

$$P = P_e + P_L(Z, Z'), \tag{20}$$

Table 1. Lattice constants C, α, γ and parameters $(1-\alpha-\gamma)$ and ξ for a bcc crystal lattice structure of the model of Coldwell–Horsfall and Maradudin.

Lattice	C	α	γ	$(1-\alpha-\gamma)$	ξ
bcc	−1.444231	0.389821	0.389821	0.220358	0.5

Note: See Chamel *et al.* (2014) for details.

where the pressure of the lattice, P_L, is given in terms of the energy density, $\mathcal{E}_L$, of the ionic lattice according to

$$P_L(Z, Z') = \frac{1}{3}\mathcal{E}_L, \tag{21}$$

with the lattice energy density given by (Chamel *et al.*, 2014)

$$\mathcal{E}_L = Ce^2 n_e^{4/3} f(Z, Z'). \tag{22}$$

The function $f(Z, Z')$ in Eq. (22) is given by

$$f(Z, Z') = \frac{\alpha Z^2 + \gamma Z'^2 + (1-\alpha-\gamma)ZZ'}{(\xi Z + (1-\xi)Z')^{4/3}}, \tag{23}$$

with Z and Z' being the proton numbers of two different ions. The quantities C, α, γ are lattice constants. The values of the lattice constants together with ξ are listed in Table 1.

The total energy density is composed by the appropriate combination of degenerate electron gas, ions, ionic lattice, and the rest energy density of electrons as follows:

$$\mathcal{E} = n_x M(Z, A)c^2 + n_{x'} M(Z', A')c^2 + \mathcal{E}_e + \mathcal{E}_L - n_e m_e c^2. \tag{24}$$

The quantities n_x and $n_{x'}$ are the number densities of nuclei with masses $M(Z, A)$ and $M(Z', A')$ (see, e.g., Audi *et al.*, 2012; Wang *et al.*, 2012, for the recent experimental nuclear masses of these nuclei).

4. Electron Capture and Pycnonuclear Reactions in HMWDs

Highly magnetized WDs with high central densities can became unstable to pycnonuclear and/or electron capture reactions (see Chamel *et al.*, 2013, 2014; Chamel and Fantina, 2015; Coelho *et al.*, 2014, and references therein). Electron capture by atomic nuclei, $A(N, Z) + e^- \to A(N+1, Z-1) + \nu_e$, leads to more neutron rich atomic nuclei in the cores of WDs. The loss of electrons in these reactions leads to an overall softening of the EoS of WD matter. The rate for this type of reaction is given by (Chamel *et al.*, 2013)

$$\rho_\beta > \rho_\beta^{\min} = \frac{B^\beta(A,Z)^{3/2}}{\sqrt{2}\pi^2\lambda^3}\frac{A}{Z}m, \tag{25}$$

where λ is the electron Compton wavelength, m is the neutron mass, and $B^\beta(A,Z)$ stands for the following relation:

$$B^\beta(A,Z) \equiv \frac{1}{2}\left(\frac{\mu_e^\beta(A,Z)}{m_e c^2}\right)^2. \tag{26}$$

The electron chemical potential is given by (Audi *et al.*, 2012; Wang *et al.*, 2012)

$$\mu_e \geq \mu_e^\beta(A,Z) \equiv \Delta(A,Z-1) - \Delta(A,Z) + mc^2, \tag{27}$$

where $\Delta(A,Z)$ denotes the excess of nuclei mass, which for magnetic field strengths less than 10^{17} G is independent of the magnetic field (Pearson *et al.*, 2011).

As discussed by Pearson *et al.* (2011) and Chamel *et al.* (2013), pycnonuclear fusion reactions might establish a more stringent limit on the microscopic stability of matter than electron capture reactions in ultra-magnetized WDs (Gasques *et al.*, 2005; Yakovlev *et al.*, 2006). The rates of these reactions are however rather uncertain. The threshold density of pycnonuclear reactions is given by (Chamel *et al.*, 2013; Pearson *et al.*, 2011)

$$\rho_{\text{pyc}} > \rho_{\text{pyc}}^{\min} = \frac{B^\beta(2A,2Z)^{3/2}}{\sqrt{2}\pi^2\lambda^3}\frac{A}{Z}m. \tag{28}$$

Tables 2 and 3 show the threshold densities of inverse electron capture reactions and pycnonuclear fusion reactions in dense WD matter. Figure 2 illustrates the EoS of both dense non-magnetic as well as magnetic WD matter, with and without lattice contributions. The vertical lines show the densities at which pycnonuclear and electron capture (inverse-β) reactions set in (see the numerical values listed in Tables 2 and 3). It is worth mentioning that the threshold mass density above which matter becomes unstable against electron captures for different compositions are consistent with the results discussed by Chamel and Fantina (2015). In addition, the inclusion of the electron–ion interactions increases ρ_β, and the heavier the elements the larger the effect (see Chamel and Fantina, 2015). Lattice contributions

Table 2. Mass densities at which electron capture reactions in dense WD matter made of a bcc crystal lattice set in.

${}^A_Z$X	ρ_β (g/cm^3)	$\rho^\star_\beta$ (g/cm^3)	ρ^0_β (g/cm^3)	$\rho^{\rm BOSK}_\beta$ (g/cm^3)
${}^4_2$He	1.32×10^{11}	1.28×10^{11}	1.72×10^{11}	1.39×10^{11}
${}^{12}_6$C	3.12×10^{10}	2.93×10^{10}	3.94×10^{10}	3.97×10^{10}
${}^{16}_8$O	1.55×10^{10}	1.43×10^{10}	1.92×10^{10}	1.94×10^{10}
${}^{30}_{26}$Fe	1.05×10^9	8.76×10^8	1.18×10^{19}	1.18×10^9

Notes: The values associated with $\rho^\star_\beta$ ignore contributions from electron–ion interactions. The densities listed in column ρ^0_β are for zero magnetic fields, those in column $\rho^{\rm BOSK}_\beta$ are for the BOSK exponent of Boshkayev *et al.* (2013b).

Table 3. Mass densities at which pycnonuclear reactions in dense WD matter made of a bcc crystal lattice set in.

${}^A_Z$X	$\rho_{\rm pyc}$ (g/cm^2)	$\rho^\star_{\rm pyc}$ (g/cm^2)	$\rho^0_{\rm pyc}$ (g/cm^2)
${}^4_2{\rm He} + {}^4_2{\rm He} \to {}^8_4{\rm Be}$	5.17×10^9	4.93×10^9	6.63×10^{10}
${}^{12}_6{\rm C} + {}^{12}_6{\rm C} \to {}^{24}_{12}{\rm Mg}$	2.67×10^9	2.40×10^9	3.22×10^9
${}^{16}_8{\rm O} + {}^{16}_8{\rm O} \to {}^{32}_{16}{\rm S}$	1.37×10^8	1.20×10^8	1.61×10^8

Notes: The threshold densities are computed from Eq. (28). Electron–ion interactions have been neglected in the values listed for $\rho^\star$. The data shown in column $\rho^0_{\rm pyc}$ are for zero magnetic field.

soften the EoS, as can be seen in Fig. 2(b). The EoS of magnetized matter is slightly softer than the EoS of non-magnetized matter.

5. Stellar Structure Equations

The structure of WDs is governed by hydrostatic equilibrium, where gravity is balanced by the outward pressure generated by a relativistic electron gas. Recently, it has been noted that contributions from general relativity (GR) ought to be taken into account when modeling the structure of WDs (Boshkayev *et al.*, 2013a,b; Carvalho *et al.*, 2015; Rotondo *et al.*, 2011). This is accomplished by making use of the Tolman–Oppenheimer–Volkoff (TOV) equations for the calculation of the properties of non-rotating WDs (Oppenheimer and Volkoff, 1939; Tolman, 1939). Adopting the geometric unit system ($G = c = 1$), the TOV equation is given by

$$\frac{dP(r)}{dr} = -\frac{(\mathcal{E}(r) + P(r))(M(r) + 4\pi P(r) r^3)}{r^2(1 - 2M(r)/r)}, \tag{29}$$

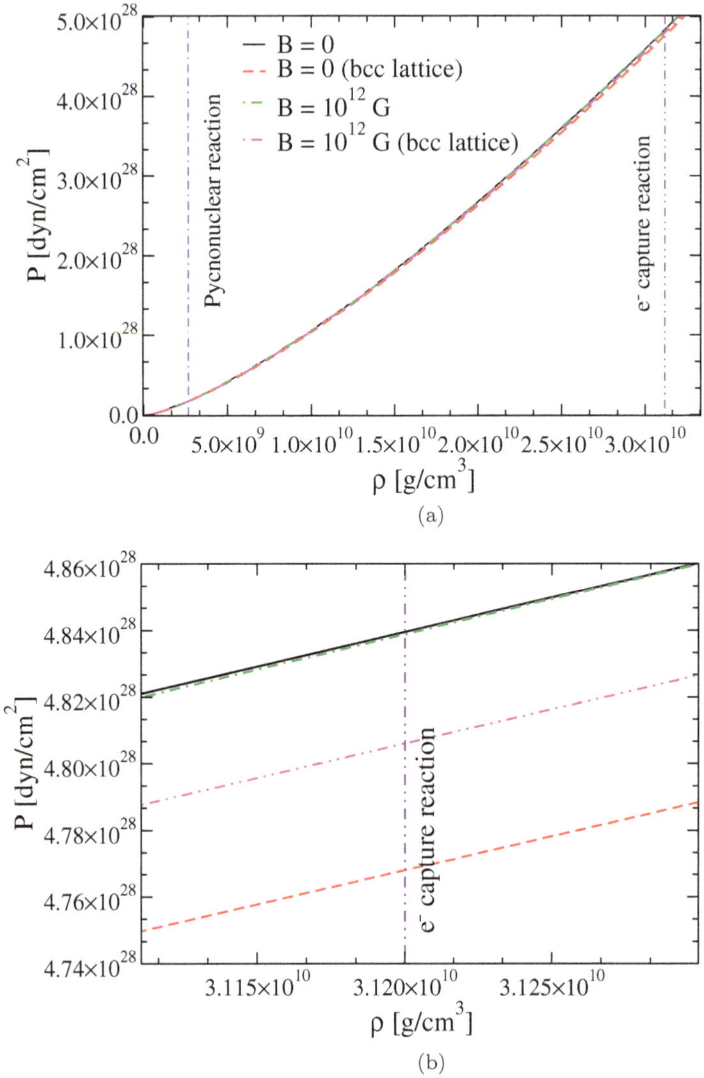

Fig. 2. Pressure-mass density relations for $B = 0$ and $B = 10^{12}$ G, with and without lattice contributions. An enlargement of the high-density part of panel (a) is shown in panel (b). The black (red) dashed lines are for $B = 0$ without (with) lattice contributions. The green dot-dashed (brown double dot-dashed) line are for $B = 10^{12}$ G without (with) lattice contributions. The vertical lines mark the onsets of pycnonuclear and electron capture (inverse-β) reactions.

where $M(r)$ denotes the gravitational mass contained inside a sphere of radius r. The amount of this mass is given by

$$\frac{dM(r)}{dr} = 4\pi\mathcal{E}(r)r^2. \tag{30}$$

As mentioned just above, the TOV equations describe the structure of non-rotating stellar objects. In order to account for rotational effects, Einstein's field equations need to be solved for a metric which accounts for rotational deformation and the dragging of local inertial frames (Friedman *et al.*, 1986; Hartle and Thorne, 1967, 1968). Such a metric is given by (Friedman *et al.*, 1986; Weber, 2005a,b)

$$ds^2 = -e^{2\nu}dt^2 + e^{2\psi}(d\phi - \omega dt)^2 + e^{2\mu}d\theta^2 + e^{2\lambda}dr^2, \tag{31}$$

where the metric functions ν, ψ, μ, λ and the frame dragging frequency, ω, depend on the radial coordinate r, the polar angle θ, and implicitly on the star's rotational frequency Ω (see Weber, 2005b, for details). The latter is taken to be in the range of $0 \leq \Omega \leq \Omega_K$, with Ω_K denoting the Kepler (mass shedding) frequency. Ω_K sets an absolute upper limit on rapid rotation (Glendenning and Weber, 1994; Weber and Glendenning, 2012, 1991) and is given by (Friedman *et al.*, 1986)

$$\Omega_{\mathrm{K}} = \omega + \frac{\omega'}{2\psi'} + e^{\nu-\psi}\sqrt{\frac{\nu'}{\psi'} + \left(\frac{\omega'}{2\psi'}e^{\psi-\nu}\right)^2}, \tag{32}$$

where the primes refer to partial derivatives with respect to the radial coordinate (Weber, 2005b). The classical Newtonian limit of Eq. (32) is given by

$$\Omega_{\mathrm{K}} = \sqrt{M/R^3}, \tag{33}$$

where M denotes the star's gravitational mass and R its equatorial radius. The Kepler period, P_K, follows from Ω_{K} as

$$P_{\mathrm{K}} \equiv \frac{2\pi}{\Omega_{\mathrm{K}}}. \tag{34}$$

The Kepler frequencies of carbon (^{12}C) white dwarfs with different magnetic fields and compositions are shown in Table 4. The masses and radii of the ^{12}C WDs of this table are shown in columns 3 and 4 of Table 5. As can be seen, the general relativistic Kepler frequencies are smaller than the Newtonian frequencies, which is due to the existence of an axisymmetric instability region in the general relativistic case, which is absent in the

Table 4. Kepler frequencies of ^{12}C white dwarfs for different magnetic fields ($B = 0$ and $B = 10^{12}$ G), with and without bcc lattice contributions.

Frequency	Method	$B = 0$ (no bcc lattice) Hz	$B = 10^{12}$ G (bcc lattice) Hz	$B = 10^{12}$ G (no bcc lattice) Hz
Ω_K^β	GR	2.058	1.952	1.617
Ω_K^β	Newton	2.116	2.014	1.654
Ω_K^{GR}	GR	1.689	1.720	1.707
Ω_K^{GR}	Newton	1.742	1.776	1.762
$\Omega_K^{\rm pyc}$	GR	0.727	0.698	0.667
$\Omega_K^{\rm pyc}$	Newton	0.757	0.729	0.696

Notes: The frequencies have been computed classically (Newton) from Eq. (33) as well as GR from Eq. (32) using the formalism of Hartle and Thorne (Hartle and Thorne, 1967, 1968; Weber and Glendenning, 2012). The masses and radii of these stars are shown in Table 5.

Newtonian calculation (see Boshkayev *et al.*, 2013a). We also note that the bcc WDs shown in Table 4 have somewhat higher mass shedding frequencies than the WDs made of matter where ion–lattice contributions are ignored. This has its origin in the smaller equatorial radii of the bcc WDs caused by the attractive nature of the ion–lattice interactions.

6. Spherical Symmetry Breaking

It was shown by Chandrasekhar and Fermi (1953) that the figure of equilibrium of an incompressible spherical fluid with an internal uniform magnetic field that matches an external dipole field is no longer represented by a sphere (see also Coelho *et al.*, 2014, and references therein) but by an oblate spheroid instead. For a spherical fluid, the lowest-order surface correction can be expressed as (Chandrasekhar and Fermi, 1953)

$$r(\mu) = R + \epsilon P_l(\mu), \tag{35}$$

where $P_l(\mu)$ is the Legendre polynomial of order l and $\mu = \cos\theta$ (θ denoting the polar angle). The quantity ϵ in Eq. (35) satisfies $\epsilon \ll R$ and measures the deviation from a spherical configuration. In lowest order, only terms up to second order in l contribute (Chandrasekhar and Fermi, 1953). The polar, $R_{\rm p}$, and equatorial, $R_{\rm eq}$, radii are then given by

$$R_{\rm p} = R + \epsilon P_l(1), \tag{36}$$

$$R_{\rm eq} = R + \epsilon P_l(0), \tag{37}$$

One can readily show that

$$\frac{\epsilon}{R} = \frac{R_{\mathrm{eq}} - R_{\mathrm{p}}}{R}, \tag{38}$$

for a non-spherical deformation with $l = 2$.

A magnetized star becomes unstable when a constant internal magnetic field deforms such a star to an oblate spheroid ($\epsilon < 0$) (see Coelho *et al.*, 2014, for details). The stellar deformation proceeds until a limiting value for ϵ/R, given by

$$\frac{\epsilon}{R} = -\frac{15}{8}\frac{B^2 R^4}{GM^2}, \tag{39}$$

is reached. Using this condition, it is possible to determine which maximum magnetic field can be supported by a magnetized star, until the object becomes unstable due to the breaking of spherical symmetry. Using the expression for the gravitational potential energy, $-3GM^2/5R$, one has

$$\frac{1}{4}B_c^2 R^3 = -\frac{3}{5}\frac{GM^2}{R}, \tag{40}$$

and the deformation parameter given by Eq. (39) can be written as

$$\frac{\epsilon}{R} = -4.5\, B_\star^2, \tag{41}$$

where $B_\star = B/B_c$. It is worth mentioning that when the (constant) internal magnetic field of a magnetized star is close to the limiting value established by the virial theorem, the star tends to take on a highly oblate form, and thus becoming unstable (see Chandrasekhar and Fermi, 1953; Coelho *et al.*, 2014, and references therein).

7. Discussion and Conclusions

The main objective of this chapter is to discuss lattice instabilities and the mass-radius relationship of HMWDs. Emphasis is also put on the role of stellar rotation on the properties of such objects.

Figure 3 shows the mass–radius relationships of HMWDs containing ^{12}C(50%) + ^{16}O(50%) lattices in their interiors, for magnetic field strengths ranging from zero to 4.4×10^{13} G. As can be seen, magnetic fields increase both the mass and the radius of a WD, and the effects become more pronounced as the strength of B increases, which is consistent with the results shown by Ostriker and Hartwick (1968); Suh *et al.* (2000). It is also known from the literature that the radii of white dwarfs with a given gravitational

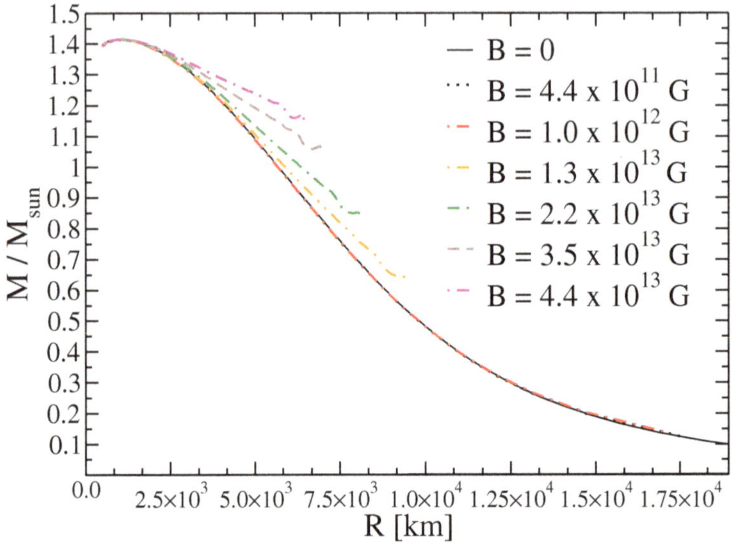

(a)

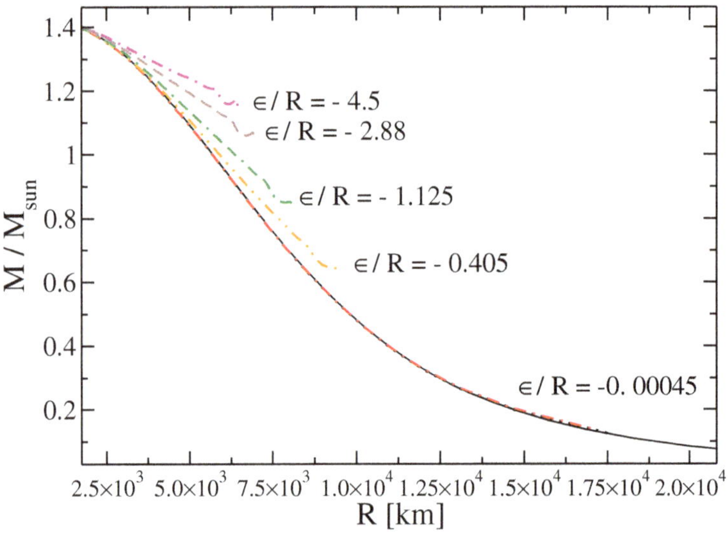

(b)

Fig. 3. (a) Mass-radius relationship of ^{12}C (50%) + ^{16}O (50%) white dwarfs with magnetic fields up to 4.4×10^{13} G. (b) Radial stellar deformation parameter, ϵ/R (see text), of the WDs shown in (a). The fact that $\epsilon/R < 0$ indicates that the stars possess oblate shapes.

mass increase with B. We confirm this feature for the EoS of our study. For a heavy WD, with a gravitational mass of around 1.2 $M_\odot$, for instance, we find that strong magnetic fields may lead to a radius increase of a few thousand kilometers, depending on the strength of the magnetic field. In Fig. 3(b), we show the deformation parameter, ϵ/R, of the magnetized WD shown in Fig. 3(a). One sees that the deformation of the WDs increases (i.e., ϵ/R becomes more negative) with increasing B values. The fact that the deformation parameter is negative reveals that the WDs become more and more oblate if the magnetic field strength increases to larger values. This shows that spherical symmetry, and the concept of isotropic pressures in HMWDs, is no longer valid for large B values, a phenomenon also noted in Fig. 1.

Figure 4 shows the gravitational mass of non-rotating an well as rotating ^{12}C white dwarfs, as a function of central mass density. The latter rotate at the general relativistic Kepler frequency Ω_K, given by Eq. (32), at which mass shedding sets an absolute limit on rapid rotation. The masses of white dwarfs rotating at Ω_K are up to 0.15 $M_\odot$ more massive than non-rotating WD. The magnetic fields carried by the WDs of this figure is $B = 10^{12}$ G. Several additional important characteristic features are

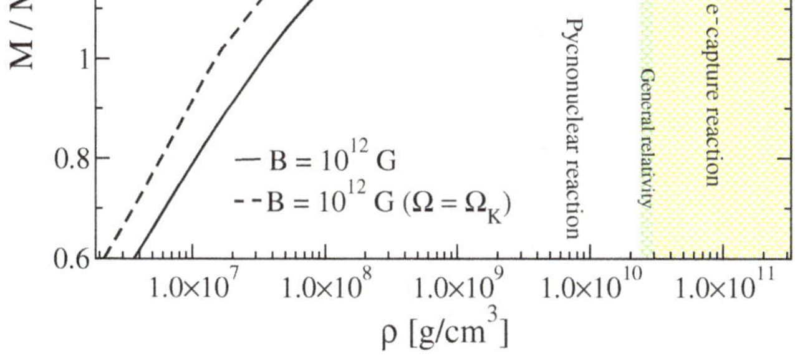

Fig. 4. Gravitational mass as a function of central mass density for a ^{12}C WD with a magnetic field of $B = 1.0 \times 10^{12}$ G. The colored regions represent areas where WD matter is subject to pycnonuclear reactions, an axisymmetric general relativistic instability, and electron capture reactions.

highlighted next. First, the light gray area shows the density at which pycnonuclear reactions are active inside of these WDs. These reactions set in at around 2×10^9 g/cm^3. The black square shows the location of the WD at the pycnonuclear reaction boundary. The mass (in units of $M_\odot$) and radius (in km) of this object are presented in parenthesis. Second, the narrow green area marks the density range over which an axisymmertric, general relativistic instability exists. Third, the yellow area shows the density regime for which the inverse-β reaction is present. The black circle and black triangle show the mass and radius values of white dwarfs located at the threshold densities at which these instabilities set in. Sumarizing, we find that the general relativistic instability sets in before inverse-β reaction occur in WD matter. The lowest threshold of all instabilities, however, is set by pycnonuclear fusion reactions. The rates of these reactions are however still rather uncertain (see Otoniel *et al.*, 2019).

Figure 5 shows the mass–radius relationships of non-rotating and rotating white dwarfs for a magnetic field of $B = 10^{12}$ G. The frequency of the rotating stars is the Keplers frequency given by Eq. (32). The EoS of these stars accounts for bcc lattice contributions. The dashed horizontal

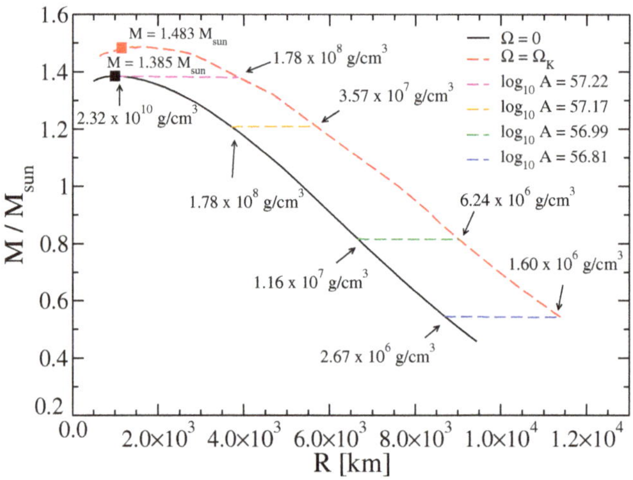

Fig. 5. Mass-radius relationships of non-rotating (sold black curve) and rotating (dashed red curve) bcc ^{12}C WDs with a magnetic field of $B = 1.0 \times 10^{12}$ G. Several stellar paths are shown (dashed blue, green, orange and magenta lines) which isolated rotating WDs (fixed baryon numbers, $\log_{10} A$) would follow when spinning down from the Kepler frequency, $\Omega = \Omega_K$, to zero, $\Omega = 0$. The numerical values next to the arrows denote the mass densities in the cores of these stars.

lines connect non-rotating and rotating WDs with the same baryon number ($\log_{10} A$) with each other. These lines represent potential evolutionary tracks of isolated WDs spinning down from high to low frequencies, as the baryon numbers of such stars would remain unchanged. As can be read off from the figures associated with each evolutionary path, the change in the central mass densities of these WDs ranges from a factor ~2, for the lighter stars ~0.5 $M_\odot$ of the sequence, to ~100 for the most massive members of the sequence.

In Fig. 6, we show the mass–radius relationships of non-rotating (black curve) and rotating (dashed red curve) ^{12}C (50%) + ^{16}O (50%) WDs carrying a magnetic field of $B = 10^{12}$ G. As already mentioned in connection with Fig. 5, the mass increase due to rotation at the Kepler frequency, Ω_K, is around 10%. We also note that the central densities of the rotating WDs are smaller than those of the non-rotating stars, since less pressure is to be provided by the EoS in the rotating case, and that the maximum-mass limit slightly decreases when lattice contributions are taken into account, which soften the EoS mildly (see Fig. 1).

Figure 7 shows the rotational deformation of a non-rotating ^{12}C + ^{16}O WD (with 1.384 $M_\odot$) caused by rotation at the Kepler frequency. The baryon number of both stars is the same, $\log_{10} A = 57.22$, but the their gravitational masses differ by a small amount because of rotation. The

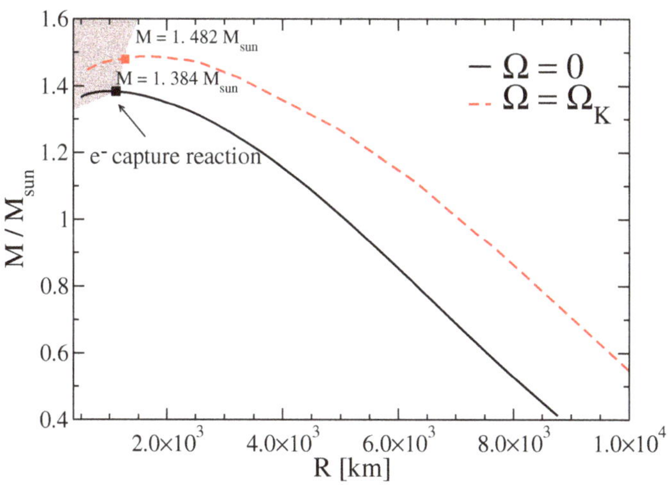

Fig. 6. Mass–radius relationship of rotating and non-rotating magnetic ^{12}C + ^{16}O WDs with bcc lattice contributions included. The colored area represents the regime where instabilities due to pycnonuclear reactions, electron-capture reactions, and general relativity are present.

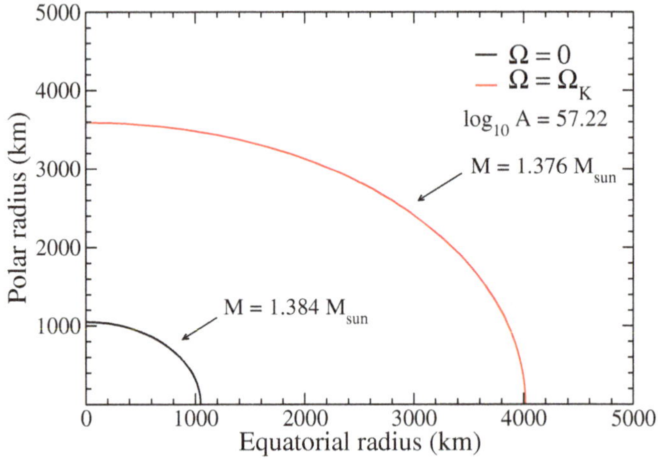

Fig. 7. Radial profiles of two magnetized $^{12}C + ^{16}O$ WDs with a same baryon number of $\log_{10} A = 57.22$. The end points of these curves show the polar and equatorial radii, R_p and R_{eq}, respectively. The red line is for a WD rotating at the Kepler frequency ($\Omega_K = 1.952$ Hz), the black curve is for the same star but at zero rotation. The respective gravitational masses are $M = 1.376M_\odot$ and $M = 1.384M_\odot$.

non-rotating WD has a mass of $M = 1.384\ M_\odot$ and a corresponding central density of $\rho = 2.02 \times 10^{10}$ g/cm^3, while the mass of the rotating WD is $M = 1.376\ M_\odot$ and $\rho = 2.14 \times 10^8$ g/cm^3. Hence this rotating white dwarf is around 94 times less dense than its non-rotating counterpart. The entire set of parameters of mass and radii taking into account different possibilities of instabilities to theses stars, constituted by ^{12}C and $^{12}C + ^{16}O$, is given in Table 5.

Figure 8 shows the Kepler period, $P_K = 2\pi/\Omega_K$ versus gravitational mass of selected magnetic WD models with and without bcc lattice contributions. As can be seen, the presence of lattice contributions $^{12}C + ^{16}O$ WDs reduces the Kepler periods (increases the Kepler frequencies) of WDs mildly, which has its origin in the softening of the EoS (see Fig. 2) caused by the lattice. The softening leads to WDs with somewhat smaller radii and therefore smaller Kepler periods. Overall, one sees that very massive magnetic $^{12}C + ^{16}O$ WDs have rotational Kepler periods on the order 0.5 s, which corresponds to a rotational frequency of 2.0 Hz. Pycnonuclear reactions are triggered in these WDs at masses that are markedly smaller than the maximum WD masses. The corresponding rotational periods are turn out to be in the 5 s (~0.2 Hz) range.

A possible mechanism for the formation of SChWDs was proposed by Das and Mukhopadhyay (2013) (and further extended in Das *et al.*, 2013),

Table 5. Maximum gravitational masses of non-rotating ($\Omega = 0$) and rotating (at Kepler frequency, Ω_K) magnetic white dwarfs experiencing different interior physical processes (i.e., pycnonuclear (pyc) reactions, electron capture (inverse-β) reactions).

Quantity	Units	^{12}C $B = 10^{12}$ G (with bcc lattice)	^{12}C $B = 10^{12}$ G (without lattice)	^{12}C (50%) + ^{16}O (50%) $B = 10^{12}$ G (with bcc lattice)	^{12}C (50%) + ^{16}O (50%) $B = 10^{12}$ G (without lattice)
$M^{\beta}(\Omega = 0)$	$M_{\odot}$	1.385	1.412	1.384	1.415
$R_{eq}(\Omega = 0)$	km	920.733	948.865	1131.792	1194.157
$M^{\beta}(\Omega = \Omega_K)$	$M_{\odot}$	1.484	1.501	1.479	1.486
$R_{eq}(\Omega = \Omega_K)$	km	1069.151	1220.946	1306.720	1401.261
$M(\Omega = 0)$	$M_{\odot}$	1.385	1.412	1.482	1.509
$R_{eq}(\Omega = 0)$	km	1004.845	1015.734	930.473	1034.863
$M(\Omega = \Omega_K)$	$M_{\odot}$	1.483	1.514	1.384	1.416
$R_{eq}(\Omega = \Omega_K)$	km	1163.572	1277.795	1080.723	1270.691
$M^{pyc}(\Omega = 0)$	$M_{\odot}$	1.361	1.389	1.478	1.495
$R_{eq}(\Omega = 0)$	km	1836.84	1907.881	3782.966	4323.396
$M^{pyc}(\Omega = \Omega_K)$	$M_{\odot}$	1.487	1.519	1.338	1.320
$R_{eq}(\Omega = \Omega_K)$	km	2108.868	2190.988	4413.811	5319.416

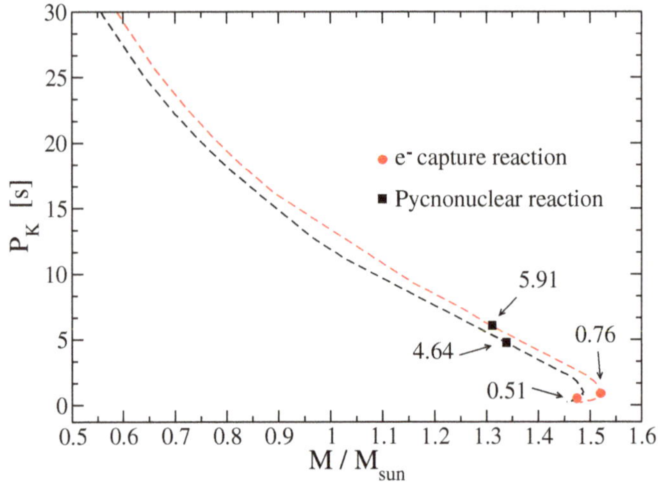

Fig. 8. Kepler period, given by Eq. (34), versus mass of magnetic $^{12}C+^{16}O$ white dwarfs ($B = 10^{12}$ G) with (black) and without (red) bcc lattice contributions. The numerical values next to the arrows display the rotational periods (in seconds) of a few selected WDs.

who suggested that such stars could be formed by mass accretion onto a HMWD. As pointed out in Coelho *et al.* (2014), electron-capture reactions and pycnonuclear fusion reactions should play a significant role for the evolution of such a star from its low-mass configuration to its high-mass termination point. On the other hand, it is interesting to note that the most stringent limit on the magnetic fields of WDs is possibly set by type-Ia supernovae (Das and Mukhopadhyay, 2013). It the traditional picture, such an explosion is triggered when a white dwarf reaches its maximum-possible mass, the Chandrasekhar mass, but that no such explosion can occur before this mass is reached. In the traditional thermonuclear run-away model, however, a type-Ia supernova is triggered when the critical density $\rho^C_{\rm ign}$ for carbon ignition inside of a white dwarf is reached. This value is not known precisely, but theoretical studies predict values for $\rho^C_{\rm ign}$ in the range of 2–3×10^9 g/cm^3 (see, e.g., Lesaffre *et al.*, 2006, and references therein). These $\rho^C_{\rm ign}$ values are way lower than the critical densities at which instability against electron capture reactions and pycnonuclear reaction (discussed above) set in. Hence the latter may severely impact the evolution of WDs as well.

An upper limit to the magnetic field can be obtained by equating the carbon ignition density $\rho^C_{\rm ign}$ to the central stellar density of the magnetic field, $\rho_B = B^2_{\text{center}}/8\pi c^2$. From this one obtains for the maximum magnetic field $B^{\max}_{\text{central}} = c\sqrt{8\pi\rho^{\text{C}}_{\text{ign}}} \approx 6.7 \times 10^{15}$ G, which indicates that in order to prevent a premature type-Ia supernova explosion, lower magnetic fields need to exist at the center of a WD in order to prevent a premature type-Ia supernova explosion.

Stellar rotation could be a possible explanation for sChWDs with central densities lower than $\rho^{\text{C}}_{\text{ign}}$ (Boshkayev *et al.*, 2013a). Besides, rotation could also be related to some superluminous type-Ia supernovae and supernova delay time distributions (see Ilkov and Soker, 2012, and references therein). However, this possibility does not seem to be plausible for the SChWDs model of Das and Mukhopadhyay (2013), since stellar configurations with large masses have densities up to three orders of magnitude larger than $\rho^{\text{C}}_{\text{ign}}$.

When this chapter was almost finalized, in 2021, for the first time, an isolated and very fast white dwarf (WD) designated ZTF J190132.9+145808.7 was discovered with a spin period of 6.94 minutes (see Caiazzo *et al.*, 2021). This nearby (~40 pc) WD is not in a binary system, as all the other rapid rotating WDs are, and seems to have originated from a merger of two WDs. The ZTF J190132.9+145808.7 was observed with a mass of 1.35 $M_\odot$ and a radius of 2140 km as small as the Moon. The star's mass needs to be close to the Chandrasekhar mass limit in order to have such a small radius. Furthermore, this WD is highly magnetized with a magnetic field ranging between 600 megagauss and 900 megagauss, which supports its WDs merger origin. In the same year, Kilic *et al.*, 2021 report the discovery of an isolated white dwarf with a spin period of 70 s, named as SDSS J2211+1136, which has a mass of 1.27 $M_\odot$, and surface magnetic field strength of 15 MG. These rotation rates are consistent with the theoretical predictions for double white dwarf (DWD) merger remnants, in agreement with the many works published in the last decade about the theory of highly magnetic, massive, and fast WDs from DWD mergers. It is important to point out that these sources may be the missing link to support a recent claim made by Malheiro *et al.* (2012) and Coelho and Malheiro (2012, 2014) that some of the anomalous X-ray pulsars (AXPs; see Olausen and Kaspi 2014; Kaspi and Beloborodov 2017) — understood as neutron stars that are isolated stars spinning with a period of ~10 s — could be in fact very massive, magnetic, and fast WDs that resulted from DWD mergers.

In summary, we have demonstrated here that the occurrence of pycnonuclear reactions and inverse beta reactions in the cores of highly magnetized WDs depends strongly on the rotation rate (and thus on the mass) of a WD. Rapid rotation increases the mass of a WD by up to 10%, accompanied by an up to 100-fold drop in central stellar density. Pycnonuclear reactions, for instance, are therefore triggered in rotation WDs at masses that differ markedly from maximum WD masses. This study ought to lay the groundwork for future research on the multifaceted properties of fast, massive and highly magnetized WDs.

Acknowledgments

E.0. is grateful for the support of PRPI — EDITAL 01/2020/PRPI/UFCA. J.G.C. is grateful for support of CNPq (421265/2018-3 and 305369/2018-0). M.M. acknowledges financial support from FAPESP under the thematic project 13/26258-4, Capes, CNPq and INCT-FNA (Proc. No. 464898/2014-5). F.W. is supported by the National Science Foundation (USA) under GrantPHY-1714068 and PHY-2012152.

References

Althaus, L. G., García-Berro, E., Isern, J., and Córsico, A. H. (2005). Mass-radius relations for massive white dwarf stars, *Astronomy & Astrophysics*, **441**, pp. 689–694.

Althaus, L. G., García-Berro, E., Isern, J., Córsico, A. H., and Rohrmann, R. D. (2007). The age and colors of massive white dwarf stars, *Astronomy & Astrophysics*, **465**, pp. 249–255.

Angel, J. R. P. (1978). Magnetic white dwarfs, *Annual Review of Astronomy and Astrophysics*, **16**, pp. 487–519.

Ashley, R. P. *et al.* (2020). V1460 her: A fast spinning white dwarf accreting from an evolved donor star, *Monthly Notices of the Royal Astronomical Society*, **499**, pp. 149–160.

Audi, G., Wang, Meng Wapstra, A. H., Kondev, F. G., MacCormick, M., Xu, X., and Pfeiffer, B. The Ame2012 atomic mass evaluation, *Chinese Physics C*, **36**, p. 1287.

Barstow, M. A., Jordan, S., O'Donoghue, D., Burleigh, M. R., Napiwotzki, R., and Harrop-Allin, M. K. (1995). RE J0317-853: The hottest known highly magnetic DA white dwarf, *Monthly Notices of the Royal Astronomical Society*, **277**, pp. 971–985.

Becerra, L., Rueda, J. A., Loren-Aguilar, P., and García-Berro, E. (2018). The spin evolution of fast-rotating, magnetized super-chandrasekhar white dwarfs in the aftermath of white dwarf mergers, *The Astrophysical Journal*, **857**, p. 134.

Bera, P. and Bhattacharya, D. (2014). Mass-radius relation of strongly magnetized white dwarfs: Nearly independent of landau quantization, *Monthly Notices of the Royal Astronomical Society*, **445**, pp. 3951–3958.

Bera, P. and Bhattacharya, D. (2016). Mass-radius relation of strongly magnetized white dwarfs: Dependence on field geometry, gr effects and electrostatic corrections to the EOS, *Monthly Notices of the Royal Astronomical Society*, **456**, pp. 3375–3385.

Boshkayev, K., Rueda, J. A., Ruffini, R., and Siutsou, I. (2013a). On general relativistic uniformly rotating white dwarfs, *The Astrophysical Journal*, **762**, p. 117.

Boshkayev, K., Rueda, J., and Ruffini, R. (2013b). On the maximum mass and minimum rotation period of relativistic uniformly rotating white dwarfs, *International Journal of Modern Physics: Conference Series*, **23**, pp. 193–197.

Cáceres, D. L., de Carvalho, S. M., Coelho, J. G., de Lima, R. C. R., and Rueda, Jorge A. (2016). Thermal X-ray emission from massive, fast rotating, highly magnetized white dwarfs, *Monthly Notices of the Royal Astronomical Society*, **465**, pp. 4434–4440.

Caiazzo, I., Burdge, K. B., Fuller, J., Heyl, J., Kulkarni, S. R., Prince, T. A., Richer, H. B., Schwab, J., Andreoni, I., Bellm, E. C., Drake, A., Duev, D. A., Graham, M. J., Helou, G., Mahabal, A. A., Masci, F. J., Smith, R., and Soumagnac, M. T. (2021). A highly magnetized and rapidly rotating white dwarf as small as the Moon, *Nature*, **595**, pp. 39–42.

Camisassa, M. E., Althaus, L. G., Córsico, Alejandro H., De Gerónimo, Francisco C., Miller Bertolami, Marcelo, M., Novarino, María L., Rohrmann, René D., Wachlin, Felipe C., and García-Berro, Enrique (2019). The evolution of ultra-massive white dwarfs, *Astronomy & Astrophysics*, **625**, p. A87.

Canuto, V. and Chiu, H.-Y. (1968). Quantum theory of an electron gas in intense magnetic fields, *Physical Review*, **173**, pp. 1210–1219.

Carvalho, G. Jr, Rubens, M., and Malheiro, M. (2015). The importance of GR for the radius of massive white dwarfs, *AIP Conference Proceedings*, **1693**, p. 030004.

Carvalho, G. A., Marinho, R. M., and Malheiro, M. (2018). General relativistic effects in the structure of massive white dwarfs, *General Relativity and Gravitation*, **50**, p. 38.

Castanheira, B. G., Kepler, S. O., Kleinman, S. J., Nitta, A., and Fraga, L. (2013). Discovery of five new massive pulsating white dwarf stars, *Monthly Notices of the Royal Astronomical Society*, **430**, pp. 50–59.

Chamel, N., Fantina, A. F. and Davis, P. J. (2013). Stability of super-chandrasekhar magnetic white dwarfs, *Physical Review D*, **88**, p. 081301.

Chamel, N., Molter, E., Fantina, A. F., and Arteaga, D. P. (2014). Maximum strength of the magnetic field in the core of the most massive white dwarfs, *physical Review D*, **90**, p. 043002.

Chamel, N. and Fantina, A. F. (2015). Electron capture instability in magnetic and nonmagnetic white dwarfs, Chamel, N. and Fantina, A. F., *Physical Review D*, **92**, p. 023008.

Chamel, N. and Haensel, P. (2008). Physics of neutron star crusts, *Living Reviews in Relativity*, **11**. https://link.springer.com/content/pdf/10.12942/lrr-2008-10.pdf

Chandrasekhar, S. and Fermi, E. (1953). Problems of gravitational stability in the presence of a magnetic field, *The Astrophysical Journal*, **118**, p. 116.

Coelho, J. G., Marinho, R. M., Malheiro, M., Negreiros, R., Cáceres, D. L., Rueda, J. A. and Ruffini, R. (2014). Dynamical instability of white dwarfs and breaking of spherical symmetry under the presence of extreme magnetic fields, *The Astrophysical Journal*, **794**, p. 86.

Coelho, J. G. and Malheiro, M. (2014). Magnetic dipole moment of SGRs and AXPs described as massive and magnetic white dwarfs, *Publications of the Astronomical Society of Japan*, **66**, p. 14.

Curd, B., Gianninas, A., Bell, Keaton J., Kilic, Mukremin, Romero, A. D., Allende Prieto, Carlos, Winget, D. E., and Winget, K. I. (2017). Four new massive pulsating white dwarfs including an ultramassive DAV, *Monthly Notices of the Royal Astronomical Society*, **468**, pp. 239–249.

Das, U., Mukhopadhyay, B., and Rao, A. R. (2013). A possible evolutionary scenario of highly magnetized super-chandrasekhar white dwarfs: Progenitors of peculiar type ia supernovae, *The Astrophysical Journal*, **767**, p. L14.

Das, U. and Mukhopadhyay, B. (2014). Maximum mass of stable magnetized highly super-chandrasekhar white dwarfs: Stable solutions with varying magnetic fields, *Journal of Cosmology and Astroparticle Physics*, **2014**, p. 050.

Das, U. and Mukhopadhyay, B. (2012). Strongly magnetized cold degenerate electron gas: Mass-radius relation of the magnetized white dwarf, *Physical Review D*, **86**, p. 042001.

Das, U. and Mukhopadhyay, B. (2013). New mass limit for white dwarfs: Super-Chandrasekhar type ia supernova as a new standard candle, *Physical Review Letters*, **110**, p. 071102.

Das, U. and Mukhopadhyay, B. (2015). GRMHD Formulation of highly super-chandrasekhar magnetized white dwarfs: Stable configurations of non-spherical white dwarfs, *Journal of Cosmology and Astroparticle Physics*, **2015**, p. 016.

Debes, J. H., Hoard, D. W., Wachter, S., Leisawitz, D. T., and Cohen, M. (2011). The WIRED Survey. II. Infrared excesses in the SDSS DR7 white dwarf catalog, *APJS*, **197**, p. 38.

Eisenstein, D. J., Liebert, J., Harris, H. C., Kleinman, S. J., Nitta, A., Silvestri, N., Anderson, S. A., Barentine, J. C., Brewington, H. J., Brinkmann, J., Harvanek, M., Krzesiski, J., Neilsen, Jr., Eric H., Long, D., Schneider, D. P., and Snedden, S. A. (2016). A catalog of spectroscopically confirmed white dwarfs from the sloan digital sky survey data release 4, *The Astrophysical Journal Supplement Series*, **167**, pp. 40–58.

Ferrario, L., de Martino, D., and Gaensicke, B. T. (2015). Magnetic white dwarfs, *Space Science Reviews*, **191**, pp. 111–169.

Franzon, B. and Schramm (2015). Effects of magnetic fields and rotation on white dwarf structure, *Physical Review D*, **92**, p. 083006.

Friedman, J. L., Ipser, J. R., and Parker, L. (1986). Rapidly rotating neutron star models, *The Astrophysical Journal*, **304**, pp. 115–139.

García-Berro, E., Lorén-Aguilar, P., Aznar-Siguán, G., Torres, S., Camacho, J., Althaus, L. G., Córsico, A. H., Kúlebi, B., and Isern, J. (2012). Double degenerate mergers as progenitors of high-field magnetic white dwarfs, *The Astrophysical Journal*, **749**, p. 25.

Gasques, L. R., Afanasjev, A. V., Aguilera, E. F., Beard, M., Chamon, L. C., Ring, P., Wiescher, M., and Yakovlev, D. G. (2005). Nuclear fusion in dense matter: Reaction rate and carbon burning, *Physical Review C*, **72**, p. 025806.

Gentile F. N. P., Tremblay, P.-E., Gänsicke, B. T., Manser, C. J., Cunningham, T., Cukanovaite, E., Hollands, M., Marsh, T., Raddi, R., Jordan, S., Toonen, S., Geier, S., Barstow, M., and Cummings, J. D. (2018). A gaia data release 2 catalogue of white dwarfs and a comparison with SDSS, *Monthly Notices of the Royal Astronomical Society*, **482**, pp. 4570–4591.

Glendenning, N. K. and Weber, F. (1994). Impact of frame dragging on the kepler frequency of relativistic stars, *Physical Review D*, **50**, pp. 3836–3841.

Hartle, J. B. and Thorne, K. S. (1967). Slowly rotating relativistic stars. I. Equations of structure, *The Astrophysical Journal*, **150**, p. 1005.

Hartle, J. B. and Thorne, K. S. (1968). Slowly rotating relativistic stars. II. Models for neutron stars and supermassive stars, *The Astrophysical Journal*, **153**, p. 807.

Hermes, J. J., Kepler, S. O., Castanheira, B. G., Gianninas, A., Winget, D. E., Montgomery, M. H., Brown, W. R., and Harrold, Samuel T. (2013). Discovery of an ultramassive pulsating white dwarf, *Astrophysical Journal Letters*, **771**, p. L2.

Ilkov, M. and Soker, N. (2012). Type Ia supernovae from very long delayed explosion of core-white dwarf merger: SNe Ia from core-white dwarf merger, *Monthly Notices of the Royal Astronomical Society*, **419**, pp. 1695–1700.

Jiménez-Esteban, F. M., Torres, S., Rebassa-Mansergas, A., Skorobogatov, G., Solano, E., Cantero, C., and Rodrigo, C. (2018). A white dwarf catalogue from Gaia-DR2 and the virtual observatory, *Monthly Notices of the Royal Astronomical Society*, **480**, pp. 4505–4518.

Kalita, S. and Mukhopadhyay, B. (2019). Continuous gravitational wave from magnetized white dwarfs and neutron stars: Possible missions for LISA, DECIGO, BBO, ET detectors, *Monthly Notices of the Royal Astronomical Society*, **490**, pp. 2692–2705.

Kepler, S. O., Kleinman, S. J., Pelisoli, I., Peçanha, V., Diaz, M., Koester, D., Castanheira, B. G., and Nitta, A. (2010). Magnetic white dwarfs in the SDSS and estimating the mean mass of normal DA and DB WDs, *American Institute of Physics Conference Series*, **1273**, pp. 19–24.

Kepler, S. O., Pelisoli, I., Jordan, S., Kleinman, S. J., Koester, D., Külebi, B., Peçanha, V., Castanheira, B. G., Nitta, A., Costa, J. E. S., Winget, D. E., Kanaan, A., and Fraga, L. (2013). Magnetic white dwarf stars in the sloan digital sky survey, *Monthly Notices of the Royal Astronomical Society*, **429**, pp. 2934–2944.

Kepler, S. O., Pelisoli, I., Koester, D., Ourique, G., Kleinman, S. J., Romero, A. D., Nitta, A., Eisenstein, D. J., Costa, J. E. S., Külebi, B., Jordan, S., Dufour, P., Giommi, P., and Rebassa-Mansergas, A. (2015). *New white dwarf stars in the sloan digital sky survey data release 10*, *Monthly Notices of the Royal Astronomical Society*, **446**, pp. 4078–4087.

Kilic, M., Kosakowski, A., Moss, A. G., Bergeron, P., and Conly, A. A. (2021). An isolated white dwarf with a 70 s spin period, *The Astrophysical Journal Letters*, **923** p. L6.

Külebi, B., Jordan, S., Euchner, F., Gänsicke, B. T., and Hirsch, H. (2009). Analysis of hydrogen-rich magnetic white dwarfs detected in the sloan digital sky survey, *Astronomy and Astrophysics*, **506**, pp. 1341–1350.

Külebi, B., Jordan, S., Euchner, F., Gaensicke, B. T., and Hirsch, H. (2010a). Magnetic fields in white dwarfs (Kuelebi+, 2009), *VizieR Online Data Catalog*, **350**, p. 61341.

Külebi, B., Jordan, S., Nelan, E., Bastian, U., and Altmann, M. (2010b). Constraints on the origin of the massive, hot, and rapidly rotating magnetic white dwarf RE J 0317-853 from an HST parallax measurement, *Astronomy and Astrophysics*, **524**, p. A36.

Lesaffre, P., Han, Z., Tout, C. A., Podsiadlowski, P., and Martin, R. G. (2006). The C flash and the ignition conditions of type Ia supernovae, *Monthly Notices of the Royal Astronomical Society*, **368**, pp. 187–195.

Liccardo, V., Malheiro, M., Hussein, M. S., Carlson, B. V., and Frederico, T. (2018). Nuclear processes in astrophysics: Recent progress, *European Physical Journal A*, p. 221.

Liebert, J., Schmidt, G. D., Green, R. F., Stockman, H. S., and McGraw, J. T. (1983). Two hot, low-field magnetic DA white dwarfs, *The Astrophysical Journal*, **264**, pp. 262–272.

Livio, M. (1999). The spins of white dwarfs and pulsars, *Annapolis Workshop on Magnetic Cataclysmic Variables*, **157**, p. 247.

Livio, M. and Pringle, J. E. (1998). The rotation rates of white dwarfs and pulsars, *The Astrophysical Journal*, **505**, pp. 339–343.

Lobato, R. V., Malheiro, M., and Coelho, Jaziel G. (2016). Magnetars and white dwarf pulsars, *International Journal of Modern Physics D*, **25**, p. 1641025.

Lopes, d. O. R., Bruch, A., Rodrigues, C. V., Oliveira, A. S., and Mukai, K. (2020). CTCV J2056-3014: An X-Ray-faint intermediate polar harboring an extremely fast-spinning white dwarf, *Astrophysical Jornal Letters*, **898**, p. L40.

Malheiro, M., Otoniel, E., and Coelho, J. G. (2021). Relevance of dynamical nuclear processes in quantum complex systems of massive white dwarfs, *Brazilian Journal of Physics*, **51**, pp. 223–230.

Malheiro, M., Rueda, J. A. and Ruffini, R. (2012). SGRs and AXPs as rotation-powered massive white dwarfs, *Publications of the Astronomical Society of Japan*, **64**, p. 56.

Marsh, T. R., Gänsicke, B. T., Hümmerich, S., Hambsch, F.-J., Bernhard, K., Lloyd, C., Breedt, E., Stanway, E. R., Steeghs, D. T., Parsons, S. G., Toloza, O., Schreiber, M. R., Jonker, P. G., van Roestel, J., Kupfer, T., Pala, A. F., Dhillon, V. S., Hardy, L. K., Littlefair, S. P., Aungwerojwit, A., Arjyotha, S., Koester, D., Bochinski, J. J., Haswell, C. A., Frank, P., and Wheatley, P. J. (2016). A radio-pulsing white dwarf binary star, *Nature*, **537**, pp. 374–377.

Mereghetti, S., Tiengo, A., Esposito, P., La Palombara, N., Israel, G. L., and Stella, L. (2009). An ultramassive, fast-spinning white dwarf in a peculiar binary system, *Science*, **325**, p. 1222.

Mukhopadhyay B., Sarkar A., and Tout C. A. (2020). Modified virial theorem for highly magnetized white dwarfs, *Monthly Notices of the Royal Astronomical Society*, **500**, pp. 763–771.

Oppenheimer, J. R. and Volkoff, G. M. (1939). On massive neutron cores, *Physical Review*, **55**, pp. 374–381.

Ostriker, J. P. and Hartwick, F. D. A. (1969). Rapidly rotating stars.IV. magnetic white dwarfs, *The Astrophysical Journal*, **153**, p. 797.

Otoniel, E., Coelho, J. G., Malheiro, M., and Weber, F. (2020). Mass limits of the extremely fast-spinning white dwarf CTCV J2056-3014, *arXiv e-prints.*

Otoniel, E., Franzon, B., Carvalho, G. A., Malheiro, M., Schramm, S., and Weber, F. (2019). Strongly magnetized white dwarfs and their instability due to nuclear processes, *The Astrophysical Journal*, **879**, p. 46.

Otoniel, E., Malheiro, M., and Coelho, J. G. (2015). Fermionic matter under the effects of high magnetic fields and its consequences in white dwarfs, *Journal of Physics: Conference Series*, **630**, p. 012039.

Paret, D. M., Horvath, J. E., and Martínez, A. P. (2015a). Anisotropic stellar structure equations for magnetized strange stars, *Research in Astronomy and Astrophysics*, **15**, pp. 975–985.

Paret, D. M., Horvath, J. E., and Martínez, A. P. (2015b). Maximum mass of magnetic white dwarfs, *Research in Astronomy and Astrophysics*, **15**, pp. 1735–1741.

Pearson, J. M., Goriely, S. and Chamel, N. (2011). Properties of the outer crust of neutron stars from Hartree-Fock-Bogoliubov mass models, *Physical Review C*, **83**.

Potekhin, A. Y. and Chabrier, G. (2000). Equation of state of fully ionized electron-ion plasmas. II. Extension to relativistic densities and to the solid phase, *Physical Review E*, **62**, pp. 8554–8563.

Potekhin, A. Y. and Chabrier, G. (2013). Equation of state for magnetized coulomb plasmas, *Astronomy & Astrophysics*, **550**, p. A43.

Potekhin, A. Y. and Yakovlev, D. G. (2012). Comment on "Equation of state of dense and magnetized fermion system", *Physical Review C*, **85**, p. 039801.

Rotondo, M., Rueda, J. A., Ruffini, R., and Xue, S.-S. (2011). Relativistic Feynman-Metropolis-Teller theory for white dwarfs in general relativity, *Physical Review D*, **84**, p. 084007.

Rueda, J. A., Boshkayev, K., Izzo, L., Ruffini, R., Lorén-Aguilar, P., Külebi, B., Aznar-Siguán, G., and García-Berro, E. (2013). A white dwarf merger as progenitor of the anomalous X-Ray pulsar 4U 0142+61?, *The Astrophysical Journall*, **772**, p. L24.

Salpeter, E. E. (1961). Energy and pressure of a zero-temperature plasma, *The Astrophysical Journal*, **134**, p. 669.

Sarah, V. B., Claudia, V. R., Jaziel, G. C., Manuel Malheiro and Manuel Castro (2020). A magnetic white dwarf accretion model for the anomalous X-ray pulsar 4U 0142+61, *The Astrophysical Journal*, **895**, p. 26.

Schmidt, G. D., Bergeron, P., Liebert, J., and Saffer, R. A. (1992). Two ultramassive white dwarfs found among candidates for magnetic fields, *Astrophysical Journal*, **394**, p. 603.
Schmidt, G. D., West, S. C., Liebert, J., Green, R. F., and Stockman, H. S. (1986). The new magnetic white dwarf PG 1031 + 234 — polarization and field structure at more than 500 milion gauss, *The Astrophysical Journal*, **309**, p. 218.
Sousa, M. F., Coelho, J. G., and de Araujo, J. C. N. (2020a). Gravitational waves from fast-spinning white dwarfs, *Monthly Notices of the Royal Astronomical Society*, **492**, p. 5949.
Sousa, M. F., Coelho, J. G., and de Araujo, J. C. N. (2020b). Gravitational waves from SGRs and AXPs as fast-spinning white dwarfs, *Monthly Notices of the Royal Astronomical Society*, **498**, p. 4426.
Strickland, M., Dexheimer, V., and Menezes, D. P. (2012). Bulk properties of a fermi gas in a magnetic field, *Physical Review D*, **86**, p. 125032.
Subramanian, S. and Mukhopadhyay, B. (2015). GRMHD formulation of highly super-chandrasekhar rotating magnetized white dwarfs: Stable configurations of non-spherical white dwarfs, *Monthly Notices of the Royal Astronomical Society*, **454**, p. 752.
Suh, I. and Mathews, G. J. (2000). Mass-radius relation for magnetic white dwarfs, *The Astrophysical Journal*, **530**, p. 949.
Terada, Y., Hayashi, T., Ishida, M., Mukai, K., Dotani, T., Okada, S., Nakamura, R., Naik, S., Bamba, A., and Makishima, K. (2008). Suzaku discovery of hard X-ray pulsations from a rotating magnetized white dwarf, AEAquarii, *Publications of the Astronomical Society of Japan*, **60**, p. 387.
Tolman, R. C. (1939). Static solutions of Einstein's field equations for spheres of fluid, *Physical Review*, **55**, p. 364
Vishal, M. V. and Mukhopadhyay, B. (2014). Revised density of magnetized nuclear matter at the neutron drip line, *Physical Review C*, **89**, p. 065804.
Wang, M., Audi, G., Wapstra, A. H., Kondev, F. G., MacCormick, M., Xu, X., and Pfeiffer, B. (2012). The Ame2012 atomic mass evaluation, *Chinese Physics C*, **36**, p. 1603.
Weber, F. (2005). Strange quark matter and compact stars, *Progress in Particle and Nuclear Physics*, **54**, p. 193.
Weber, F. (2005). *Pulsars as Astrophysical Laboratories for Nuclear and Particle Physics*, Bristol, Philadelphia, USA.
Weber, F. and Glendenning, N. K. (2012). Application of the improved hartle method for the construction of general relativistic rotating neutron star models, *The Astrophysical Journal*, **390**, p. 541.
Weber, F., Glendenning, N. K., and Weigel, M. K. (1991). Structure and stability of rotating relativistic neutron stars, *The Astrophysical Journal*, **373**, p. 579.
Yakovlev, D. G., Gasques, L. R., Afanasjev, A. V., Beard, M., and Wiescher, M. (2006). Fusion reactions in multicomponent dense matter, *Physical Review C*, **74**, p. 035803.

https://doi.org/10.1142/9789811220944_0005

Chapter 5

Hyperonization in Compact Stars

Armen Sedrakian[*,§], Jia-Jie Li[†,¶] and Fridolin Weber[‡,‖]

[*] *Frankfurt Institute for Advanced Studies*
Ruth-Moufang-Str. 1, 60438 Frankfurt am Main, Germany
Institute of Theoretical Physics
University of Wrocław pl. M. Borna 9, 50-204 Wrocław, Poland
[†] *School of Physical Science and Technology*
Southwest University, Chongqing 400700, China
[‡] *Department of Physics, San Diego State University,*
San Diego, CA 92182, USA
Center for Astrophysics and Space Sciences
University of California at San Diego, La Jolla, CA 92093, USA
[§] *sedrakian@fias.uni-frankfurt.de*
[¶] *jiajieli@swu.edu.cn*
[‖] *fweber@sdsu.edu*

Abstract: We review the covariant density functional approach to the equation-of-state (EoS) of the dense nuclear matter in compact stars. The main emphasis is on the hyperonization of the dense matter, and the role played by the Δ-resonances. The implications of hyperonization for the astrophysics of compact stars, including the EoS, composition, and stellar parameters are examined. The mass–radius relation and tidal deformabilities of static and rapidly rotating (Keplerian) configurations are discussed in some detail. We briefly touch upon some other recent developments involving hyperonization in hot hypernuclear matter at high- and low-densities.

Keywords: equation of state; neutron stars; gravitational waves.

1. Introduction

Compact (or neutron) stars represent the endpoints of the evolution of ordinary stars. They provide a natural astrophysical laboratory for particle and nuclear physics under conditions that are largely different from and often inaccessible in terrestrial laboratories. For example, densities reached in compact stars are by a factor 5–10 larger than found in ordinary nuclei. In the 60s and 70s of the past century it has been conjectured that they may contain non-nucleonic constituents of matter, for example, hyperons (Ambartsumyan and Saakyan, 1960, 1961; Leung and Wang, 1971; Moszkowski, 1974; Pandharipande, 1971) or deconfined quark matter (Collins and Perry, 1975; Itoh, 1970). The state of (hyper)nuclear matter found in compact stars may feature some extraordinary facets, such as superfluidity and superconductivity, a trapped neutrino component at the early stages of evolution, and super-strong magnetic fields, for reviews see Glendenning (2012), Oertel *et al.* (2017), Sedrakian (2007), Shapiro and Teukolsky (1983), and Weber (1999).

One of the key theoretical challenges of the description of compact stars is the diversity of the possible phases at high densities. A compact star has a (conservatively estimated) mass in the range $1.1 \lesssim M/M_\odot \lesssim 2.3$, where $M_\odot$ is the solar mass, the central densities may reach up to about 10 times the nuclear saturation density ($\rho_{\rm sat} = 0.16\,{\rm fm}^{-3}$), and a radius in the range $12 \lesssim R \lesssim 14$ km. A compact star consists roughly of five major regions: the atmosphere, the outer and inner crust, and the outer and inner core, see Fig. 1. The outer and inner crusts are characterized by the presence of the nuclear clusters immersed in electron gas and a neutron sea in the inner crust. The outer core consists of neutrons, protons, and leptons (mainly electrons with some admixture of muons). The composition of the inner core starting at about twice the nuclear saturation density is not well established: the multitude of possibilities include hyperonization, the deconfinement phase transition to quark matter, the onset of meson condensation, etc. For a textbook discussion of the phases of compact stars see, for example Glendenning (2012), Shapiro and Teukolsky (1983), and Weber (1999).

During the last decade, there have been several breakthrough observational advances that reshaped our understanding of compact stars. These include the measurements of heavy pulsar masses in binaries of neutron stars with white dwarfs, the observation of gravitational waves from double neutron stars mergers, and the simultaneous mass and radius measurement of nearby X-ray emitting neutron stars. These advances provided crucial insights for the development of theoretical models of dense matter in recent years. Most importantly, they allowed us to narrow down substantially the theoretical range of parameters of the models of dense matter.

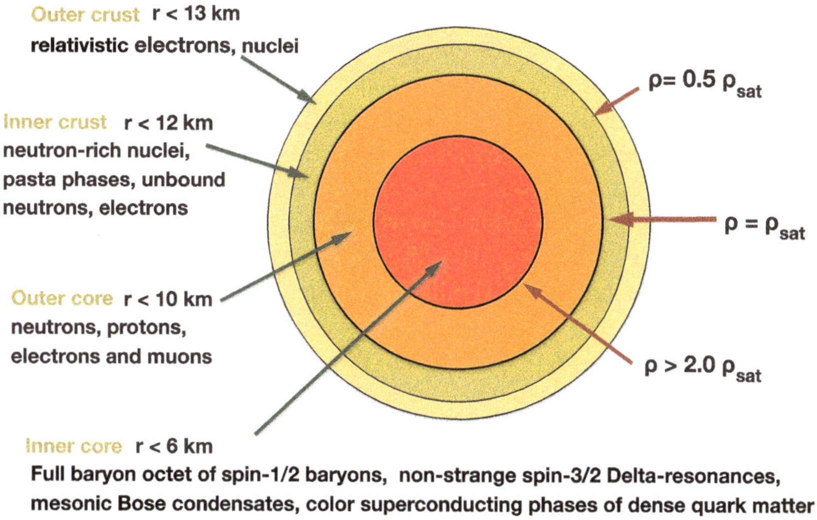

Fig. 1. Schematic illustration of the interior of a $M = 1.4M_{\odot}$ mass neutron star. The transition density between different regions (in units of nuclear saturation density $\rho_{\rm sat}$) and the corresponding radial coordinate are indicated. The particle content of each region are indicated as well. Note the possibility of mixed phases involving deconfined quark matter and hypernuclear matter.

This review aims to present recent advances in the studies of the EoS, composition, and physical manifestations of compact stars that contain hypernuclear cores. This will be done in the framework of the covariant density functional (hereafter CDF) methods (Serot and Walecka, 1997) for the description of dense hypernuclear matter — a method that provides a flexible enough approach to accommodate the information coming from astrophysics and laboratory on hypernuclei. The Lagrangian-based relativistic CDFs of nuclear systems (also known as nuclear relativistic mean-field models) operate with effective degrees of freedom - baryons and meson. They provide a well-motivated and straightforward way to obtain the energy density of matter. At the same time, there is no true fundamental field theory (in the sense of electroweak theory or quantum chromodynamics (QCD) associated with these models, i.e., the microscopic underpinning of these models is unknown. Therefore, they are best viewed as CDFs with parameters to be determined from available data in the sense of Kohn-Sham density functional theories which have been successfully applied in many complex many-body systems in various fields, including strongly correlated electronic systems, quantum chemistry, atomic and molecular systems, etc.

The relativistic mean-field model based CDFs have been applied to hypernuclear systems already in 80s and 90s (Glendenning, 1985; Glendenning and Moszkowski, 1991; Glendenning *et al.*, 1992; Huber *et al.*, 1994, 1998; Papazoglou *et al.*, 1998; Schaffner and Mishustin, 1996).

The motivation to study hypernuclear stars and the interest in hypernuclear CDFs resurged after the observations of two solar mass pulsars in a binary orbit with a white dwarf in 2010 (Demorest *et al.*, 2010) when, for the first time since the discovery of a pulsar in 1967, radioastronomy provided us with a significant constraint on the mass of a pulsar. The reason is that hyperons become energetically favourable once the Fermi energy of neutrons exceeds their rest mass. The onset of hyperons reduces the degeneracy pressure of a cold hypernuclear matter, therefore, its EoS (hereafter EoS) becomes softer than that of nucleonic matter. As a result, the maximum possible mass of a compact star with hyperons decreases to values below $2M_\odot$ (Baldo *et al.*, 2000; Schulze and Rijken, 2011), which is in direct contradiction with the observations. This contradiction is known as the "hyperon (or hyperonization) puzzle". As discussed below the modern density functionals for the hypernuclear matter not only resolve the hyperon puzzle but also account for the data coming from gravitational wave physics and X-ray observations of nearby pulsars (Banik *et al.*, 2014; Bednarek *et al.*, 2012; Bonanno and Sedrakian, 2012; Colucci and Sedrakian, 2013; Drago *et al.*, 2016; Fortin *et al.*, 2017, 2020; Gomes *et al.*, 2019, 2015; Gusakov *et al.*, 2014; Jiang *et al.*, 2012; Li *et al.*, 2018a,b; Lopes and Menezes, 2014; Maslov *et al.*, 2015, 2016; Massot *et al.*, 2012; Miyatsu *et al.*, 2015; Oertel *et al.*, 2016, 2015; Providencia and Rabhi, 2013; Tolos *et al.*, 2016; Torres *et al.*, 2017; Tsubakihara and Ohnishi, 2013; van Dalen *et al.*, 2014; Weissenborn *et al.*, 2012a). The possibility of nucleation of Δ-resonances has been considered along with the hyperonization based on essentially the same arguments which make heavy baryons preferable to highly energetic neutrons (Choudhury and Rakshit, 1993; Sawyer, 1972; Schürhoff *et al.*, 2010; Waldhauser *et al.*, 1987, 1988; Weber and Weigel, 1989b). Again, after the discovery of massive pulsars, the CDF methods have been invoked to treat Δ-resonance admixed (hyper)nuclear matter (Cai *et al.*, 2015; Dexheimer *et al.*, 2021b; Drago *et al.*, 2014; Kolomeitsev *et al.*, 2017; Li *et al.*, 2018b; Raduta *et al.*, 2020; Ribes *et al.*, 2019; Sahoo *et al.*, 2018; Spinella and Weber, 2020; Thapa *et al.*, 2020, 2021; Zhu *et al.*, 2016).

Because of space limitations, we will focus in this review on the CDF approach and do not cover the alternative many-body approaches as applied to hypernuclei and compact stars, which include lattice studies (Inoue and HAL QCD Collaboration, 2019; Nemura *et al.*, 2007; Sasaki *et al.*, 2020), many-body schemes based on G-matrix theory (Bombaci, 2017; Haidenbauer *et al.*, 2017; Yamamoto *et al.*, 2014), Monte-Carlo (Gandolfi and Lonardoni, 2017; Lonardoni *et al.*, 2013a,b, 2015) and variational (Shahrbaf *et al.*, 2020; Shahrbaf and Moshfegh, 2019; Togashi *et al.*, 2016) methods, see also the reviews (Blaschke and Chamel, 2018; Chatterjee and Vidaña, 2016; Providência *et al.*, 2019; Vidaña, 2018). An overview of the many-body methods with an emphasis on the compact stars can be found, for example, in Oertel *et al.* (2017), and Sedrakian (2007).

Compact star properties will be discussed below exclusively with Einstein's theory of general relativity. However, there has been substantial work in recent years on the interpretation of the astrophysics of compact stars within alternative theories of gravity, see, for example Altaha Motahar *et al.* (2017); Astashenok *et al.* (2014, 2015a,b); Blázquez-Salcedo *et al.* (2016) and references therein.

This chapter is organized as follows. Section 2 discusses the astrophysical constraints on compact star properties that became available in recent years. The key ideas of the CDF theory are presented in Section 3. Section 4 is devoted to the properties of hypernuclear stars, including the EoS, composition, and global properties. The recent results on rapidly rotating hypernuclear stars are discussed in Section 5. The interplay between clustering and heavy-baryon degrees of freedom in the warm and low-density nuclear matter is exposed in Section 6. The cooling of hypernuclear stars, the finite-temperature EoS, and universal relations are briefly touched upon in Sections 7 and 8, which is followed by concluding remarks in Section 9.

2. Astrophysical Constraints on Neutron Stars

Pulsar timing is among the methods that provide information on the masses of pulsars via measurements of the Keplerian parameters and the observed spin properties of pulsars within neutron star–neutron star and neutron star–white dwarf binaries. The Shapiro delay method of measurement of pulsar mass is based on the observation that electromagnetic radiation experiences a time delay as it passes in the vicinity companion compact object (neutron star or white dwarf) due to its gravitational field. The Shapiro delay method (Shapiro, 1964) has been successfully applied in binary systems involving millisecond pulsars, where the periodic pulsed signal from the pulsar covers tracks of different length within the space-time continuum depending on whether the pulsar passes in front or behind its binary companion relative to a distant observer. In the general theory of relativity, the Shapiro time delay depends on the companion mass and the degree of inclination of the binary system. The first high-mass pulsar measurement was carried out for PSR J1614-2230 which is a 3.2-ms pulsar in an 8.7-day orbit with a massive white-dwarf companion in a highly inclined orbit (Demorest *et al.*, 2010). The second massive pulsar with high-precision mass measurement is J0348+0432 which is 39 ms pulsar in a 2.46-hour orbit with a white dwarf. In this case, optical observation and modelling of companion white dwarf was used in addition to pulsar timing measurements of Keplerian parameters of binary to place the mass of this millisecond pulsar at $2.01 \pm 0.04 M_{\odot}$ (Antoniadis *et al.*, 2013). Improved measurements by the NANOGrav place the mass of this pulsar as $1.928(17)M_{\odot}$ (Fonseca *et al.*, 2016). The third, most massive (measured with high-precision) neutron star to date is PSR J0740+6620 (Cromartie *et al.*, 2020), which is a 2.89-ms pulsar in a 4.77-day orbit with a white dwarf. The timing analysis which measures Shapiro delay places the mass of this pulsar at $2.08 \pm 0.07 M_{\odot}$ at 68.3% credibility (Fonseca *et al.*, 2021). Thus, the timing observations of three millisecond pulsars J1614-2230, J0348+0432, and J0740+6620 indicate that compact stars with masses as large as $2M_{\odot}$ exist in Nature. On the other hand, general relativity predicts that stable compact star sequences end at a maximal mass independent of the EoS employed. (The stable configurations are determined by the Bardeen-Thorne-Meltzer criterion (Bardeen *et al.*, 1966), which implies that a star is stable only as long as its mass is increasing with the central density.) Therefore, we may conclude that this maximum mass predicted by any viable EoS must be at least as large as those observed in the timing observations. In other words, the millisecond pulsar observations place a *lower limit* on the maximum mass of a compact star.

With the advent of gravitational wave astronomy and the first measurement by the LIGO and Virgo collaboration (hereafter LVC) of the gravitational waves from a binary neutron star (hereafter BNS) merger in the GW170817 event (Abbott *et al.*, 2017) it became possible to constrain the properties of compact stars through the study of their tidal response. A weaker signal was detected later in the GW190425 event, which is likely to be a BNS coalescence (Abbott, 2020). These measurements, which were carried by the LVC second-generation ground-based gravitational observatories, are based on the idea that a neutron star is deformed in the tidal gravitational field of the companion. To the lowest order, such deformations are described through the star's induced quadrupole moment. The gravitational wave signal emitted before the merger of a binary carries direct information on the tidal properties of compact stars. The *tidal deformability* λ is defined as the coefficient which relates the induced quadrupole Q_{ij} to the perturbing tidal field $\mathcal{E}_{ij}$ that acts on a star perturbing its shape $Q_{ij} = -\lambda \mathcal{E}_{ij}$, where i and j label the space coordinates. It is related to the gravitational Love number k_2 via the relation

$$\lambda = \frac{2}{3} k_2 R^5, \tag{1}$$

which exhibits its sensitivity to the radius of the star R. Both k_2 and R depend on the EoS of matter (Baiotti, 2019). Frequently one uses dimensionless tidal deformability defined as

$$\Lambda = \frac{\lambda}{M^5} = \frac{2}{3} k_2 C, \tag{2}$$

where $C = M/R$ is the compactness of the star. Frequently one also defines an effective tidal deformability as (Flanagan and Hinderer, 2008)

$$\tilde{\Lambda} = \frac{16}{13} \frac{(M_1 + 12M_2)M_1^4 \Lambda_1 + (M_2 + 12M_1)M_2^4 \Lambda_2}{(M_1 + M_2)^5}, \tag{3}$$

involving the masses and tidal deformabilities of both stars. The analysis of the GW170817 and GW190425 events were carried out for high- and low-spin priors. We give below some characteristic numbers only for low-spin priors as suggested by galactic observations. The total binary mass values $2.73^{+0.04}_{-0.01} M_\odot$ and $3.4^{+0.3}_{-0.1} M_\odot$ were deduced from the GW170817 and GW190425 events. The masses of the components are in the range $1.16–1.6 M_\odot$ for GW170817 and $1.46 - 1.87 M_\odot$ for GW190425 event. The analysis of GW170817 event led to an upper limit $\Lambda_{1.4} < 700$ for a $1.4 M_\odot$ star (90% confidence). In the case of the GW190425 event, $\tilde{\Lambda} \leq 600$ has been inferred for the binary mass range indicated above.

Neutron star surfaces are emitting X-rays due to their thermal heating by currents associated with the particle flows in the magnetosphere. The location of the emitting hot spots reflects the structure and topology of the magnetosphere. Due to the rotation of the star, the hot spots are generating pulsed emission. The thermal X-ray emission pulse profile of millisecond pulsar J0030+0451 has been used by the NICER team to place limits on its mass *and* radius (Miller *et al.*, 2019; Riley *et al.*, 2019). The procedure involves modelling the soft X-ray pulses produced by the rotation of hot spots on the surface of the star and fitting them to the NICER waveform data. In doing so it was assumed that the star's emitting atmosphere is ionized hydrogen and that the magnetic field does not play any role. The two independent analyzes predict (68% credible interval) $M = 1.44^{+0.15}_{-0.14} M_\odot$, $R = 13.02^{+1.24}_{-1.06}$ km (Miller *et al.*, 2019) and $M = 1.34^{+0.15}_{-0.16} M_\odot$, $R = 12.71^{+1.14}_{-1.19}$ km (Riley *et al.*, 2019). The same collaboration has also measured the radius of PSR J0740+662 to be $R = 13.7^{+2.6}_{-1.5}$ km (Miller *et al.*, 2021) and $R = 12.39^{+1.3}_{-0.98}$ km (Riley *et al.*, 2021) (68% credible interval). In addition to the above constraints, the masses of neutron stars in binaries have been measured with high accuracy which lie in the range $1.2 \lesssim M/M_\odot \lesssim 1.6$ with a significant concentration around the value 1.4 $M_\odot$ (Lattimer, 2019; Özel and Freire, 2016). Furthermore, the moment of inertia of a neutron star is expected to be measured in the double pulsar system PSR J0737-3039 (Lattimer and Schutz, 2005), where both masses of stars are already accurately determined by observations.

3. Density Functionals for Hypernuclear Matter

Density functional theory offers a flexible framework to study the equilibrium thermodynamics of nuclear and neutron star matter. In nuclear physics, a particular class of density functionals can be obtained within the so-called *relativistic mean-field* models of nuclear matter, which by construction posses the main feature of density functional theory: the potential energy of the zero-temperature system is a functional of (energy) density alone. The self-energies in these theories are evaluated either in the Hartree or Hartree-Fock theories of many-body theory, the first approximation being the simplest realization of such scheme. The Hartree-Fock theories allow one to introduce the pion contribution explicitly in the density functional, which could be an advantage for a detailed treatment of the tensor force. Once the density functional is constructed the parameters of the relativistic Lagrangian are adjusted to reproduce the laboratory data within a range that is compatible with other constraints, such as those available from compact star observations. The microscopic *ab initio* many-body calculations are treated as data, i.e., they constrain the admissible range of parameters entering the functional. (In passing we note that density functionals have been derived directly from microscopic theories, but this will not be reviewed here.) An advantage of the approach taken here is the straightforward extension of the density functional from nuclear to hypernuclear and Δ-resonance matter. The fast numerical implementations allow us to scan a large parameter space associated with the density functional, which can become quite large as one includes the heavy baryons in the functional.

3.1. *Relativistic density functional with density-dependent couplings*

In this section, we briefly review the construction of the density functional of hypernuclear matter starting from a relativistic Lagrangian and adopting the Hartree approximation. The Hartree-Fock scheme is discussed in detail elsewhere (Li *et al.*, 2018a). We will use a particular class of such functionals that assign a density dependence to the coupling constants describing the meson-baryon interactions, which incorporate modifications of the interaction due to changes in the density of the medium in which baryons and mesons are embedded (Lalazissis *et al.*, 2005; Typel, 2018; Typel and Wolter, 1999). The scheme discussed below will include successively the $J = 1/2$ baryon octet and $J = 3/2$ resonances; see Fig. 2 for an illustration of the octet and decuplet arrangement of the heavy baryons and their quantum numbers. The Lagrangian of stellar matter with baryonic degrees of freedom can be written as

$$\mathscr{L} = \mathscr{L}_b + \mathscr{L}_m + \mathscr{L}_l + \mathscr{L}_{\mathrm{em}}, \tag{4}$$

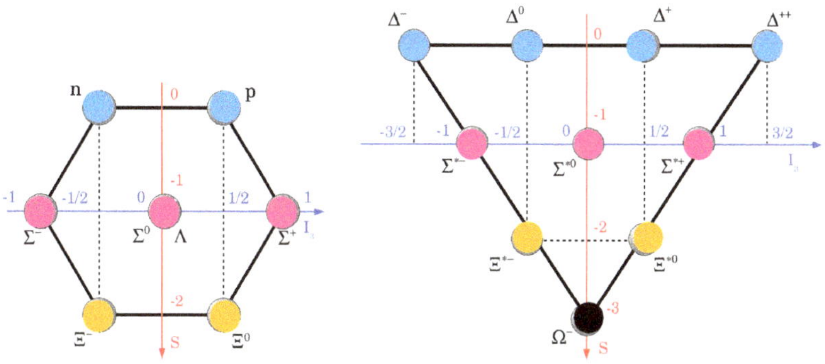

(a) (b)

Fig. 2. (a) An illustration of the spin-1/2 octet of baryons and (b) spin-3/2 decuplet of resonances (right), where the vertical axis is the strangeness and the horizontal axis is the third component of the isospin.

where the baryon Lagrangian is given by

$$\mathscr{L}_b = \sum_b \bar{\psi}_b \left[\gamma^\mu \left(i\partial_\mu - g_{\omega b}\omega_\mu - \frac{1}{2} g_{\rho B} \tau \cdot \rho_\mu \right) - (m_b - g_{\sigma b}\sigma) \right] \psi_b, \quad (5)$$

where the b-sum is over the $J^P = \frac{1}{2}^+$ baryon octet, ψ_b are the baryonic Dirac fields of baryons with masses m_b, and σ, ω_μ, and ρ_μ are the mesonic fields which mediate the interaction among baryon fields. The coupling constants g_{mb}, in general, are density-dependent. The mesonic part of the Lagrangian is given by

$$\mathscr{L}_m = \frac{1}{2}\partial^\mu \sigma \partial_\mu \sigma - \frac{m_\sigma^2}{2}\sigma^2 - \frac{1}{4}\omega^{\mu\nu}\omega_{\mu\nu} + \frac{m_\omega^2}{2}\omega^\mu \omega_\mu - \frac{1}{4}\rho^{\mu\nu} \cdot \rho_{\mu\nu} + \frac{m_\rho^2}{2}\rho^\mu \cdot \rho_\mu, \quad (6)$$

where m_σ, m_ω, and m_ρ are the meson masses and $\omega_{\mu\nu}$ and $\rho_{\mu\nu}$ represent the strength tensors of vector mesons

$$\omega_{\mu\nu} = \partial_\mu \omega_\nu - \partial_\mu \omega_\nu, \qquad \boldsymbol{\rho}_{\mu\nu} = \partial_\nu \boldsymbol{\rho}_\mu - \partial_\mu \boldsymbol{\rho}_\nu. \quad (7)$$

The leptonic Lagrangian is given by

$$\mathscr{L}_l = \sum_\lambda \bar{\psi}_\lambda (i\gamma^\mu \partial_\mu - m_\lambda)\psi_\lambda, \quad (8)$$

where ψ_λ are lepton fields (in stellar matter the λ-summation includes electron, muons, and at high temperatures the three flavours of neutrinos when neutrinos are trapped) and m_λ are their masses. Finally, if the stellar matter is permeated by sizable magnetic fields (which is the case, for example, in magnetars) then we need to include the electromagnetism via its gauge-field Lagrangian

$$\mathscr{L}_{\rm em} = -\frac{1}{4}F^{\mu\nu}F_{\mu\nu}, \quad (9)$$

where $F^{\mu\nu}$ is the electromagnetic field strength tensor and replace the partial derivatives by their gauge-invariant counterparts $D_\mu = \partial_\mu + ieQA_\mu$ where A_μ is the electromagnetic vector potential and eQ is the charge of the particle (Dexheimer *et al.*, 2021b; Thapa *et al.*, 2020).

The baryonic Lagrangian can be extended to include the non-strange $J = 3/2$ members of the baryons decuplet which is the quartet of Δ-resonances, see Fig. 2, by adding to the Lagrangian (4) the term

$$\mathscr{L}_d = \sum_d \bar{\psi}_d^\nu \left[\gamma^\mu \left(i\partial_\mu - g_{\omega d}\omega_\mu - \frac{1}{2} g_{\rho d} \tau \cdot \rho_\mu \right) - (m_d - g_{\sigma d}\sigma) \right] \psi_{d\nu}, \quad (10)$$

where the d-summation is over the resonances described by the Rarita-Schwinger fields $\psi_{d\nu}$. Furthermore, to describe the interactions between the strange particles the Lagrangian (4) can be extended to include the interactions in the hypernuclear sector via two additional (hidden strangeness) mesons σ^* and ϕ

$$\mathscr{L}_{m'} = \frac{1}{2}\partial^\mu \sigma^* \partial_\mu \sigma^* - \frac{m_\sigma^{*2}}{2}\sigma^{*2} - \frac{1}{4}\phi_{\mu\nu}\phi^{\mu\nu} + \frac{1}{2}m_\phi^2 \phi_\mu \phi^\mu, \tag{11}$$

with field strength tensor $\phi_{\mu\nu} = \partial_\nu \phi_\mu - \partial_\mu \phi_\nu$. These mesons do not couple to nucleons, i.e., $g_{\sigma^* N} = g_{\phi N} = 0$, see however Eq. (29) below.

Let us now turn to the coupling constants in the nucleonic sector. These are given by their values at the saturation density $\rho_{\rm sat}$ and by functional form describing their dependence on baryon density ρ_B

$$g_{iN}(\rho_B) = g_{iN}(\rho_{\rm sat}) h_i(x), \tag{12}$$

where $x = \rho_B/\rho_{\rm sat}$ and

$$h_i(x) = \frac{a_i + b_i(x+d_i)^2}{a_i + c_i(x+d_i)^2}, \quad i = \sigma, \omega, \quad h_\rho(x) = e^{-a_\rho(x-1)}. \tag{13}$$

The density dependence of the couplings implicitly takes into account many-body correlations that modify the interactions in the medium. In the following we will adopt the DDME2 parametrization (Lalazissis *et al.*, 2005); the values of the parameters are listed for completeness in Table 1.

The masses of ω and ρ mesons are taken to be their free values. The five constraints $h_i(1) = 1$, $h_i''(0) = 0$ and $h_\sigma''(1) = h_\omega''(1)$ allow one to reduce the number of free parameters in isocalar-scalar and iso-scalar-vector sector to three. Three additional parameters in this channel are $g_{\sigma N}(\rho_{\rm sat})$, $g_{\omega N}(\rho_{\rm sat})$ and m_σ — the mass of the phenomenological σ meson. With two additional parameter in the isovector-vector channel, the parameterization has in total eight parameters (seven entering the definition of the coupling constants and the mass of the σ-meson) which are adjusted to reproduce the

Table 1. The values of parameters of the DDME2 CDF.

Meson (i)	m_i (MeV)	a_i	b_i	c_i	d_i	g_{iN}
σ	550.1238	1.3881	1.0943	1.7057	0.4421	10.5396
ω	783	1.3892	0.9240	1.4620	0.4775	13.0189
ρ	763	0.5647				7.3672

properties of symmetric and asymmetric nuclear matter, binding energies, charge radii, and neutron skins of spherical nuclei.

The ground-state expectation values of mesons in the mean-field and infinite system approximations are given by

$$m_\sigma^2 \sigma = \sum_b g_{\sigma b} n_b^s + \sum_d g_{\sigma d} n_d^s, \quad m_{\sigma^*}^2 \sigma^* = \sum_b g_{\sigma^* b} n_b^s, \tag{14}$$

$$m_\omega^2 \omega_0 = \sum_b g_{\omega b} n_b + \sum_d g_{\omega d} n_d, \quad m_\phi^2 \phi_0 = \sum_b g_{\phi b} n_b, \tag{15}$$

$$m_\rho^2 \rho_{03} = \sum_b g_{\rho b} \tau_{b3} n_b + \sum_d g_{\rho d} \tau_{d3} n_d, \tag{16}$$

where the scalar and baryon (vector) number densities are defined for the baryon octet as $n_b^s = \langle \bar{\psi}_b \psi_b \rangle$ and $n_b = \langle \bar{\psi}_b \gamma^0 \psi_b \rangle$, respectively. For the Δ-resonances, these are defined as $n_d^s = \langle \bar{\psi}_{d\nu} \psi_d^\nu \rangle$ and $n_d = \langle \bar{\psi}_{d\nu} \gamma^0 \psi_d^\nu \rangle$, respectively. The effective (Dirac) baryon masses in the same approximation are given by

$$m_b^* = m_b - g_{\sigma b}\sigma - g_{\sigma^* b}\sigma^*, \quad m_d^* = m_d - g_{\sigma d}\sigma. \tag{17}$$

Given the Lagrangian density (5), the energy stress tensor can be constructed

$$T^{\mu\nu} = \frac{\partial \mathscr{L}}{\partial(\partial_\mu \varphi_i)} \partial^\nu \varphi_i - g^{\mu\nu} \mathscr{L}, \tag{18}$$

where φ_i stands generically for a boson or fermion field. Then, its diagonal elements define the energy density and pressure

$$\mathcal{E} = \langle T^{00} \rangle, \quad P = \frac{1}{3} \sum_i \langle T^{ii} \rangle, \tag{19}$$

where the brackets refer to statistical averaging. Explicitly one finds

$$\begin{aligned} P = & -\frac{m_\sigma^2}{2}\sigma^2 - \frac{m_\sigma^{*2}}{2}\sigma^{*2} + \frac{m_\omega^2}{2}\omega_0^2 + \frac{m_\rho^2}{2}\rho_{03}^2 + \frac{m_\phi^2}{2}\phi_0^2 \\ & + \frac{1}{3}\sum_{b,d} \frac{2J_{b,d}+1}{2\pi^2} \int_0^\infty \frac{dk\ k^4}{E_k^{b,d}} \left[f(E_k^{b,d} - \mu_{b,d}^*) + f(E_k^{b,d} + \mu_{b,d}^*) \right] \\ & + \frac{1}{3\pi^2} \sum_\lambda \int_0^\infty \frac{dk\ k^4}{E_k^\lambda} \left[f(E_k^\lambda - \mu_\lambda) + f(E_k^\lambda + \mu_\lambda) \right], \end{aligned} \tag{20}$$

and

$$\mathcal{E} = \frac{m_\sigma^2}{2}\sigma^2 + \frac{m_\sigma^{*2}}{2}\sigma^{*2} + \frac{m_\omega^2}{2}\omega_0^2 + \frac{m_\rho^2}{2}\rho_{03}^2 + \frac{m_\phi^2}{2}\phi_0^2$$
$$+ \sum_{b,d} \frac{2J_{b,d}+1}{2\pi^2} \int_0^\infty dk\ k^2 E_k^{b,d} \left[f(E_k^{b,d} - \mu_{b,d}^*) + f(E_k^{b,d} + \mu_{b,d}^*) \right]$$
$$+ \sum_\lambda \int_0^\infty \frac{dk}{\pi^2} k^2 E_k^\lambda \left[f(E_k^\lambda - \mu_\lambda) + f(E_k^\lambda + \mu_\lambda) \right], \tag{21}$$

where $2J_{b,d}+1$ is the baryon degeneracy factor with $J_b = 1/2$ for baryon octet and $J_d = 3/2$ for Δ-resonances,

$$\mu_b = \mu_b^* + g_{\omega b}\omega_0 + g_{\phi b}\phi_0 + g_{\rho b}\tau_{b3}\rho_{03} + \Sigma^r, \tag{22}$$

$$\mu_d = \mu_d^* + g_{\omega d}\omega_0 + g_{\rho d}\tau_{d3}\rho_{03} + \Sigma^r, \tag{23}$$

are the baryon chemical potentials, I_3 is the third component of baryon isospin, $E_k^{b,d} = \sqrt{k^2 + m_{b,d}^{*2}}$ and $E_k^\lambda = \sqrt{k^2 + m_\lambda^2}$ are the single particle energies of baryons and leptons respectively, and $f(E) = [1+\exp(E/T)]^{-1}$ is the Fermi distribution function at temperature T. The lepton mass m_λ can be taken equal to its free-space value. The self-energy Σ^r arises due to the density-dependence of the coupling constant and guarantees the thermodynamic consistency of the theory, i.e., the fact that the thermodynamic relation

$$P = \rho^2 \frac{\partial}{\partial\rho}\left(\frac{\mathcal{E}}{\rho}\right), \tag{24}$$

is fulfilled.

3.2. *Hyperonic and Δ-resonance couplings*

The lack of reliable information on the hyperon-nucleon and hyperon-hyperon interactions prevents a high-precision determination of the parameters entering the Lagrangian (5). The SU(3) flavour symmetric model allows one to determine the magnitude of the coupling constants appearing in the Lagrangian (5) based on symmetry argument and the principles of the "eight-fold way" of elementary particle classification in particle physics (de Swart, 1963). The SU(3) symmetry in flavour space is, of course, broken by the strange quark mass at the densities and temperatures relevant to compact stars. Nevertheless, it provides some guidance in instances where no or little information is available.

Within this model, the spin-1/2 baryons and mesons are arranged in octets, which is the lowest non-trivial irreducible representation of the symmetry group. The interaction part of the SU(3) invariant Lagrangian describing the coupling of baryons and mesons is constructed using matrix representations for the baryon $J^P = 1/2^+$ octet of baryons B and $J^P = 1^-$ meson octet (M_8) which is supplemented by meson singlet (M_1) which allows describing physical mesons via mixing mechanism. The Lagrangian contains linear combinations of the antisymmetric (F-type), symmetric (D-type), and singlet (S-type) scalar contributions (using the standard notations)

$$\begin{aligned}\mathscr{L}_{SU(3)} &= -g_8\sqrt{2}[\alpha \mathrm{Tr}([\bar{B}, M_8]B) + (1-\alpha)\mathrm{Tr}(\{\bar{B}, M_8\}B)] \\ &\quad - \frac{g_1}{\sqrt{3}}\mathrm{Tr}(\bar{B}B)\mathrm{Tr}(M_1), \end{aligned} \tag{25}$$

where g_8 and g_1 denote the meson octet and singlet coupling constant respectively and $\alpha = F/(F+D)$ with $0 \leq \alpha \leq 1$.

The physical mesons ω and ϕ then appear as a mixture of the ω_0 and ω_8 members of the vector meson nonet:

$$\begin{pmatrix}\omega_8 \\ \omega_0\end{pmatrix} = \begin{pmatrix}\cos\theta_V & \sin\theta_V \\ -\sin\theta_V & \cos\theta_V\end{pmatrix}\begin{pmatrix}\omega \\ \phi\end{pmatrix}, \tag{26}$$

where θ_V is the vector mixing angle. Within this mixing scheme the coupling of a baryon to the physical ω-meson is given by

$$g_{B\omega} = \cos\theta_V g_1 + \sin\theta_V \frac{g_8}{\sqrt{3}}\delta_B, \quad B \in \{N, \Xi, \Lambda, \Sigma\}, \tag{27}$$

where $\delta_N = 4\alpha_V - 1$, $\delta_\Xi = 1 + 2\alpha_V$ and $\delta_\Sigma = -\delta_\Lambda = 2(1-\alpha_V)$. It is convenient to express the coupling of hyperons to mesons by using the nucleonic couplings as normalization

$$R_{\omega B} = \frac{g_{\omega B}}{g_{\omega N}} = \frac{1 - \frac{g_8}{g_1\sqrt{3}}\delta_B \tan\theta_V}{1 - \frac{g_8}{g_1\sqrt{3}}(1-4\alpha_V)\tan\theta_V}. \tag{28}$$

The ϕ-meson couplings can be obtained from those for ω meson by the substitution $\cos\theta_V \to -\sin\theta_V$ and $\sin\theta_V \to \cos\theta_V$. In general, it is possible that ϕ meson couples to the nucleon with the coupling given by

$$R_{\phi N} = -\frac{\tan\theta_V - \frac{g_8}{g_1\sqrt{3}}\delta_N}{1 + \frac{g_8}{g_1\sqrt{3}}\delta_N \tan\theta_V}. \tag{29}$$

The coupling for the isovector ρ meson is given by

$$g_{\rho N} = g_8, \quad R_{\rho\Xi} = -(1 - 2\alpha_V), \quad R_{\rho\Sigma} = 2\alpha_V, \quad R_{\rho\Lambda} = 0. \tag{30}$$

Note that the ρ couplings vanish exponentially at high densities according to Eq. (13) and their effect on the properties of dense matter (beyond the threshold for the onset of hyperons) is small.

The approximate equality of the masses of ω and ρ-mesons implies that the mixing is ideal, in which case ϕ meson is a pure $\bar{s}s$ state and the mixing angle is given by the *ideal mixing* value $\tan\theta_V^* = 1/\sqrt{2}$. Since the nucleon does not couple to pure strange meson ϕ, Eq. (29) implies that this is the case when $g_1 = \sqrt{6}\, g_8$ and $\alpha_V = 1$, the latter being the universality assumption for the (electric) $F/(F+D)$ ratio, i.e., only F-type coupling is non-zero. In this case the coupling constants of heavy baryons are related to those of the nucleon as in the additive quark model.

The couplings in the case of the scalar mesons σ and σ^*, are obtained from those of ω and ϕ, respectively, with the replacements $\omega \to \sigma$, $\phi \to \sigma^*$, and changing the vector indices to scalar ones $V \to S$. In the case of σ-meson the coupling constants are given by

$$g_{B\sigma} = \cos\theta_S\; g_1 + \delta_B g_8 \sin\theta_S/\sqrt{3}, \tag{31}$$

where in the definitions of δ_B the scalar ratio α_S appears instead of its vector counterpart. It can be shown that the coupling defined in this manner obey the following relation (Colucci and Sedrakian, 2013)

$$2(g_{N\sigma} + g_{\Xi\sigma}) = 3g_{\Lambda\sigma} + g_{\Sigma\sigma}, \tag{32}$$

which is valid for arbitrary values of the four parameters α_S, g_1, g_8 and θ_S. It is easy to verify that Eq. (32) is satisfied for the coupling constants in the SU(6) spin-flavour symmetric quark model. Table 2 lists the couplings within this model.

The information on the couplings of hyperons to the scalar mesons can be obtained from the fits to their potentials in nuclear matter and nuclei within a particular model. For example, the single Λ-hypernuclei has been used to fix the value of $g_{\Lambda\sigma}$ coupling (van Dalen *et al.*, 2014). Similarly, the coupling to the σ^* is obtained from the fits to the binding of double

Table 2. The ratios of the couplings of hyperons in the SU(6) spin-flavour model.

$Y \backslash R$	$R_{\sigma Y}$	$R_{\sigma^* Y}$	$R_{\omega Y}$	$R_{\phi Y}$	$R_{\rho Y}$
Λ	2/3	$-\sqrt{2}/3$	2/3	$\sqrt{2}/3$	0
Σ	2/3	$-\sqrt{2}/3$	2/3	$-\sqrt{2}/3$	2
Ξ	1/3	$-2\sqrt{2}/3$	1/3	$-2\sqrt{2}/3$	1

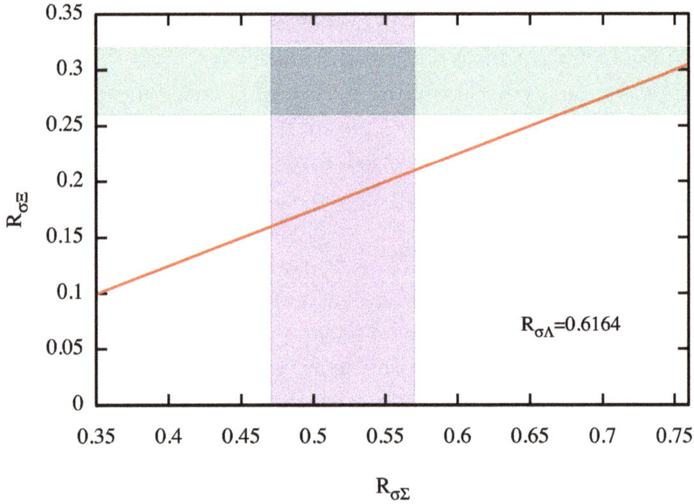

Fig. 3. The ranges of coupling constants $R_{\sigma\Sigma}$ and $R_{\sigma\Xi}$ (shaded areas) compatible with experimental/theoretical information and relation (32) (solid line) for DDME2 CDF and fixed $R_{\sigma\Lambda} = 0.6164$. The combined area of $R_{\sigma\Sigma}$ and $R_{\sigma\Xi}$ has no overlap with prediction based on the SU(3) flavour symmetry, indicating its breaking in the scalar-meson sector.

Λ hypernuclei (Fortin *et al.*, 2017). Furthermore, density functionals have been adapted to treat multi-strange nuclei hypernuclei (Güven *et al.*, 2018; Khan *et al.*, 2015; Margueron *et al.*, 2017).

Figure 3 shows the range of the (dimensionless) couplings $R_{\sigma\Sigma}$ and $R_{\sigma\Xi}$ which covers the potential depths ranges $U_\Sigma(\rho_{\rm sat}) = [-10 : +30]$ MeV and $U_\Xi(\rho_{\rm sat}) = [-24 : 0]$ MeV at nuclear saturation density. The value $U_\Xi(\rho_{\rm sat}) = -24$ MeV has been given in (Friedman and Gal, 2021) and is much deeper than the one expected from Lattice 2019 results (Inoue and HAL QCD Collaboration, 2019; Sasaki *et al.*, 2020). The fixed value of the $R_{\sigma\Lambda} = 0.6164$ (van Dalen *et al.*, 2014) has been used. It is seen that in contrast to the SU(6) model, the values of $R_{\sigma\Xi}$ predicted by relation (32) do not intercept the overlap area of $R_{\sigma\Sigma}$ and $R_{\sigma\Xi}$ couplings, which indicates the breaking of corresponding symmetry.

The results pertaining hyperonic compact stars below have been obtained with the parameters $R_{\sigma\Lambda} = 0.6106$, $R_{\sigma\Sigma} = 0.4426$, and $R_{\sigma\Xi} = 0.3024$ (Li *et al.*, 2018b) in the framework of the DDME2 CDF extension to hypernuclear sector.

Most studies of Δs in dense matter have been carried out in the relativistic mean-field approach (Cai *et al.*, 2015; Choudhury and Rakshit, 1993; Lavagno, 2010; Waldhauser *et al.*, 1987, 1988; Weber and Weigel, 1989b), the density-dependent CDF approach (Drago *et al.*, 2014; Kolomeitsev *et al.*, 2017; Li *et al.*, 2018b), the relativistic Hartree–Fock approach (Weber and Weigel, 1989a; Zhu *et al.*, 2016) and the quark–meson coupling model (Motta *et al.*, 2020). The interactions of Δ-resonance within the matter are not well known. Some information on the Δ-potential in the isospin symmetric nuclear matter is available from the analysis of scattering of electrons and pions off nuclei with Δ-excitation (Horikawa *et al.*, 1980; Koch, 1997; Nakamura *et al.*, 2010; O'Connell and Sealock, 1990; Wehrberger *et al.*, 1989) and photo-absorption (Alberico *et al.*, 1994; Riek *et al.*, 2009). The extracted value of the potential is (Drago *et al.*, 2014)

$$-30\ \text{MeV} + V_N(\rho_{\text{sat}}) \leq V_\Delta(\rho_{\text{sat}}) \leq V_N(\rho_{\text{sat}}), \tag{33}$$

where $V_N(\rho_{\text{sat}})$ is the nucleon isoscalar potential at the saturation density. The Δ-resonance production in heavy-ion collisions is another channel of information, where however collective dynamics of nuclear matter comes into play (Cozma, 2016; Cozma and Tsang, 2021; Ono *et al.*, 2019; Xu, 2019). Numerical simulations provide hints towards the values of the potential in the range (Kolomeitsev *et al.*, 2017)

$$V_N(\rho_{\text{sat}}) \leq V_\Delta(\rho_{\text{sat}}) \leq 2/3 V_N(\rho_{\text{sat}}). \tag{34}$$

The isovector meson-Δ-resonance couplings are not known. In the following we use the ratios $R_{m\Delta} = g_{m\Delta}/g_{mN}$ to describe the Δ-resonance couplings. In the recent work, these parameters have been varied in the range (Li and Sedrakian, 2019a; Li *et al.*, 2018b)

$$R_{\rho\Delta} = 1, \quad 0.8 \leq R_{\omega\Delta} \leq 1.6, \quad R_{\sigma\Delta} = R_{\omega\Delta} \pm 0.2, \tag{35}$$

to explore the consequences of the inclusion of Δ-resonance in the CDF of (hyper)nuclear matter.

Table 3. The coefficients of the expansion (36) for the DDME2 parametrization in MeV units, where we show the values of higher-order parameters not defined in (36), see Margueron *et al.* (2018a).

Density expansion					
ρ_{sat} [fm^{-3}]	m_N^*/m	E_{sat}	K_{sat}	Q_{sat}	Z_{sat}
0.152	0.57	−16.14	251.15	479	4448
Isospin-asymmetry expansion					
		E_{sym}	L_{sym}	K_{sym}	Q_{sym}
		32.31	51.27	−87	777

3.3. *Characteristics of nuclear matter close to saturation*

As is well known, the EoS of nuclear matter can be parametrized via an expansion in the vicinity of saturation density and isospin-symmetrical limit via a double-expansion in the Taylor series:

$$E(\chi,\delta) \simeq E_{\text{sat}} + \frac{1}{2!}K_{\text{sat}}\chi^2 + \frac{1}{3!}Q_{\text{sat}}\chi^3 + E_{\text{sym}}\delta^2 + L_{\text{sym}}\delta^2\chi + \mathcal{O}(\chi^4, \chi^2\delta^2), \tag{36}$$

where $\chi = (\rho - \rho_{\text{sat}})/3\rho_{\text{sat}}$, $\delta = (\rho_n - \rho_p)/\rho$ with ρ_n and ρ_p being the neutron and proton densities. The coefficients of the expansion are known as *incompressibility* K_{sat}, the *skewness* Q_{sat}, the *symmetry energy* E_{sym} and the *slope parameter* L_{sym}. The definitions of parameters are standard, see Margueron *et al.* (2018a). These four low-order coefficients can be constrained from experimental data on nuclear systems. The higher-order terms are less constrained (Margueron *et al.*, 2018a; Zhang and Li, 2019).

While the CDFs provide full access to the spectra of particles, the matter compositions, etc. the expansions of the type (36) provide only a more limited set of physical parameters, for example, the EoS. Since the uncertainties in the nuclear matter properties are easily characterized in terms of uncertainties in the characteristics entering the expansion (36), it is important to establish a one-to-one correspondence between the two descriptions.

The five macroscopic characteristics in Eq. (36) together with the preassigned values of saturation density ρ_{sat} and the nucleon Dirac mass m_N^*/m uniquely determine the seven adjustable parameters of the CDFs of the DDME2 type. This allows one to generate new parametrizations which reproduce desired values of the characteristics (Li and Sedrakian, 2019a), especially those that are associated with high-density and large-isospin behaviour. As finite nuclei do not probe these regimes the loss of accuracy of the new CDFs in reproducing the binding energies, charge radii and neutron skins of finite nuclei is marginal.

Having such a tool at the disposal one can now proceed to vary individually the characteristics within their acceptable ranges and extract the EoS of dense matter and, consequently, the properties of compact stars. This allows one to explore the correlation(s) between specific properties of nuclear matter and/or compact stars (Cai and Chen, 2017; Cai and Li, 2021; Li *et al.*, 2021; Li and Magno, 2020; Li and Sedrakian, 2019a,b). Detailed investigations of the EoS were carried out where the higher-order coefficients of the expansion, specifically, $Q_{\rm sat}$ and $L_{\rm sym}$ were varied since their values are weakly constrained by the conventional fitting protocol of CDF (Li and Sedrakian, 2019a; Margueron and Gulminelli, 2019; Margueron *et al.*, 2018a,b; Zhang *et al.*, 2018). It is interesting that the one-to-one mapping described above allows one to *predict* the higher-order terms in the expansion (36) (Li and Sedrakian, 2019a), which are highly model dependent (Dutra *et al.*, 2012, 2014).

4. Hypernuclear Stars: Static Properties

4.1. *EoS and composition*

Given a CDF one could compute the EoS of the stellar matter by implementing the additional conditions of weak equilibrium and change neutrality that prevail in compact star matter (except for the first instances after birth). The strangeness changing weak equilibrium conditions are given

$$\mu_\Lambda = \mu_{\Sigma^0} = \mu_{\Xi^0} = \mu_{\Delta^0} = \mu_n = \mu_B, \tag{37}$$

$$\mu_{\Sigma^-} = \mu_{\Xi^-} = \mu_{\Delta^-} = \mu_B - \mu_Q, \tag{38}$$

$$\mu_{\Sigma^+} = \mu_{\Delta^+} = \mu_B + \mu_Q, \tag{39}$$

$$\mu_{\Delta^{++}} = \mu_B + 2\mu_Q, \tag{40}$$

where μ_B and $\mu_Q = \mu_p - \mu_n$ are the baryon and charge chemical potentials, μ_i with $i \in \{\Lambda, \Sigma^{0,\pm}, \Xi^{0,\pm}, \Delta^{0,\pm,++}\}$ are the chemical potentials of the baryons. The baryonic charge is given by

$$n_p + n_{\Sigma^+} + 2n_{\Delta^{++}} + n_{\Delta^+} - (n_{\Sigma^-} + n_{\Xi^-} + n_{\Delta^-}) = n_Q,$$

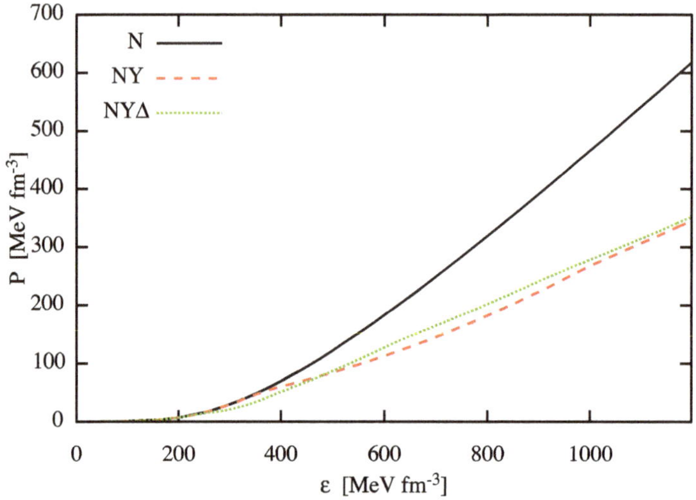

Fig. 4. The EoS of nucleonic (N), hyperonic (NY) and Δ-admixed hyperonic ($NY\Delta$) matter at zero temperature and in β-equilibrium. The Δ-potential is fixed by $V_\Delta(\rho_{\rm sat}) = V_N(\rho_{\rm sat})$.

where n_i are the baryon number densities. The electrical charge neutrality is then implemented by equating this quantity with the lepton charge

$$n_Q = n_e + n_\mu, \tag{41}$$

where $n_{e,\mu}$ are the number densities of electrons and muons. Figure 4 shows the EoS for nucleonic (N), hyperonic (NY) and Δ-admixed hyperonic ($NY\Delta$) matter at zero temperature and in weak β-equilibrium. The onset of hyperons and Δ particles are seen by the change in the slope of the pressure above the saturation density and significant softening of the EoS. In the case where Δ's are included in the composition, the EoS softens at low and stiffens at high densities compared to the purely hyperonic case. Figure 5 shows the particle abundances of hyperonic matter for the DDME2 plus hyperons parameterization. This EoS is smoothly matched at the density $\rho_{\rm sat}/2$ to that of the crust EoS (Baym *et al.*, 1971a,b). It is seen that the first hyperon to appear is the Λ, which is followed by the Ξ^- hyperon. The Σ^- hyperons do not appear because they are disfavoured by their repulsive potential at nuclear saturation density (Bart *et al.*, 1999; Dover and Gal, 1984; Gomes *et al.*, 2015; Lopes and Menezes, 2014; Maslov *et al.*, 2016; Miyatsu *et al.*, 2015). Similar ordering of hyperon thresholds was found with other hypernuclear CDFs (Fortin *et al.*, 2017; Li *et al.*, 2018a,b;

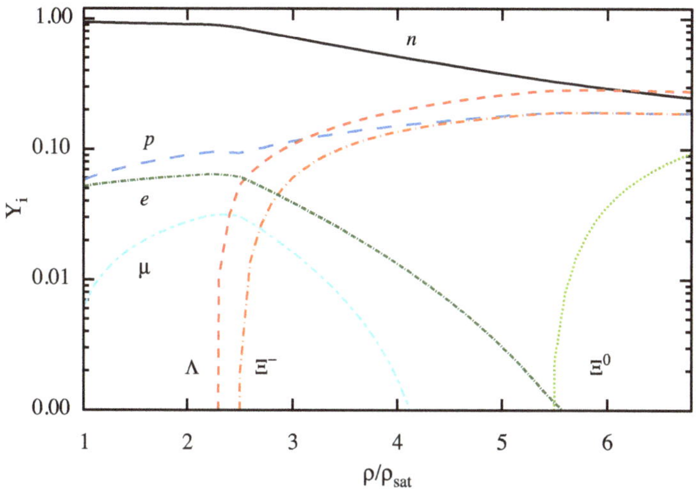

Fig. 5. Particle fractions $Y_i = n_i/n$, where n_i is the number density of species i and n is the total density, in zero-temperature hypernuclear matter according to the DDME2 model as a function of the ratio $\rho/\rho_{\rm sat}$.

Weissenborn *et al.*, 2012a,b); note also that this picture is in contrast to the case of free hyperonic gas, in which case Σ^- is, in fact, the first hyperon to nucleate (Ambartsumyan and Saakyan, 1960). The modification of the particle abundances in the cases when Δ's are included is as follows (Li *et al.*, 2018b). For strong enough Δ potential $V_\Delta^{(N)}$ the Δ-threshold density is considerably lower than that for the Λ hyperon. The larger $V_\Delta^{(N)}$ the lower is the onset densities of Δ's. Because of its negative charge, the first Δ resonance to appear is the Δ^-, which competes with negatively charged leptons, effectively eliminates the Σ^- and shifts the threshold for the Ξ^- to higher densities. The Λ hyperon abundance is weakly affected by the Δ's. For $V_\Delta \geq V_N$, the remaining $\Delta^{0,+,++}$ resonances also nucleate. By electric charge neutrality between the baryons and leptons, the appearance of negatively charged Δ^- and Ξ^- depletes the electron–muon population. Finally, for large ($V_\Delta \geq V_N$) potentials Δ's appear already in $1.4M_\odot$ compact stars.

4.2. *Global static properties*

The static properties of the compact stars in spherical symmetry (assuming no rotation and significant magnetic fields) are obtained from the integration of the Tolman–Oppenheimer–Volkoff equations (Oppenheimer and

Volkoff, 1939; Tolman, 1939), which represent the solution of Einstein's equation for a spherically symmetrical distribution of mass. It is often useful to compare the theoretical predictions with the observations on plots that contain only observable quantities, i.e., combinations of mass, radius, the moment of inertia, spin frequency, etc. As an example, we show in Fig. 6 the mass-radius (MR) relations for purely nucleonic, hyperonic, and hyperon-Δ admixed EoS presented in Fig. 4. The following observational constraints are included: (a) the 1σ constraints set by the NICER experiment on the mass *and* radius of PSR J0030+0451 (Miller *et al.*, 2019; Riley *et al.*, 2019); (b) the 2σ mass-limit on the largest millisecond pulsar MPS J0740+6620 mass measured (Cromartie *et al.*, 2020) in combination with the recent NICER teams' report of its 1σ radius measurement in the range $12 \leq R \leq 14$ km (Miller *et al.*, 2021; Riley *et al.*, 2021); and (c) the mass range inferred for the light companion of the binary observed in the GW190814 event to be discussed in detail in Section 5. The softening of the EoS when the heavy baryons are allowed is reflected in the significant reduction of the maximum mass of compact stars in both cases of only hyperons (NY) and hyperons and Δ-resonances ($NY\Delta$). The additional

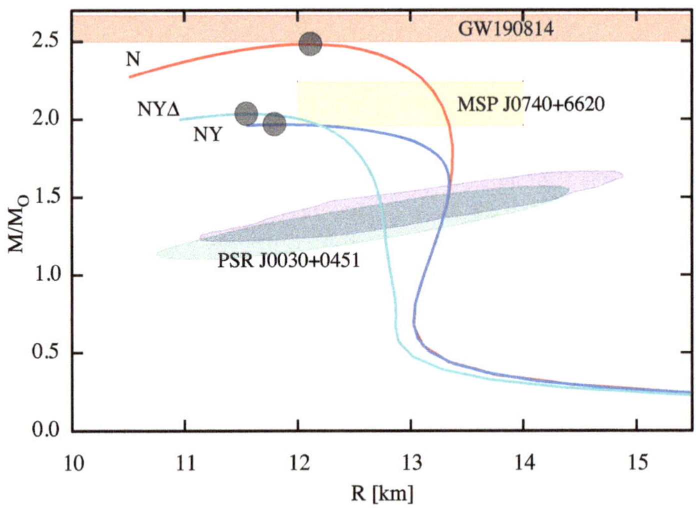

Fig. 6. MR relations for EoS of state with N, NY, and $NY\Delta$ compositions. The shaded areas labeled PSR J0030+0451 corresponds to the 1σ constraints set by the NICER experiment (Miller *et al.*, 2019; Riley *et al.*, 2019). We also show the 2σ range mass for MSPJ0740+6620 (Cromartie *et al.*, 2020) and the mass range extracted from the GW190814 event. The maximum mass for each stellar sequence is indicated by a solid dot.

softening of EoS at intermediate densities in the case when Δ-resonances are allowed results in a shift of the radius of the corresponding configuration to smaller values. Since in this case, the stiffness of the EoS is comparable to that of purely hyperonic matter at high densities (and even exceeds it asymptotically) the values of the maximum masses ($M_{\max} \gtrsim 2.0 M_\odot$) are comparable in both cases of pure and Δ-admixed hypernuclear matter, see Fig. 6.

To illustrate the effect of variations of the nuclear characteristics on the mass and radius of compact stars Fig. 7 shows the MR plot for the three cases N, NY and $NY\Delta$ where the isoscalar $Q_{\rm sat}$, and the isovector $L_{\rm sym}$ characteristics are varied, while the remaining parameters in Eq. (36) are fixed at their values predicted by the DDME2 functional (see Table 3). Note that the variation of $Q_{\rm sat}$ is achieved by varying the three density-dependent parameters, i.e., these variations do not impact the meson-hyperon and meson-Δ couplings at nuclear saturation density. It is seen that smaller values of $L_{\rm sym}$ and $Q_{\rm sat}$ imply smaller radius for a given mass. At the same time, a smaller value of $Q_{\rm sat}$ predicts smaller maximal mass, as the high-density asymptotics of Eq. (36) is reduced. The effect of varying the Δ-meson couplings are shown by using various values of Δ potential $R_{\Delta N} = V_\Delta(\rho_{\rm sat})/V_N(\rho_{\rm sat}) = 1; 4/3; 5/3$. It is seen that the larger is the value of

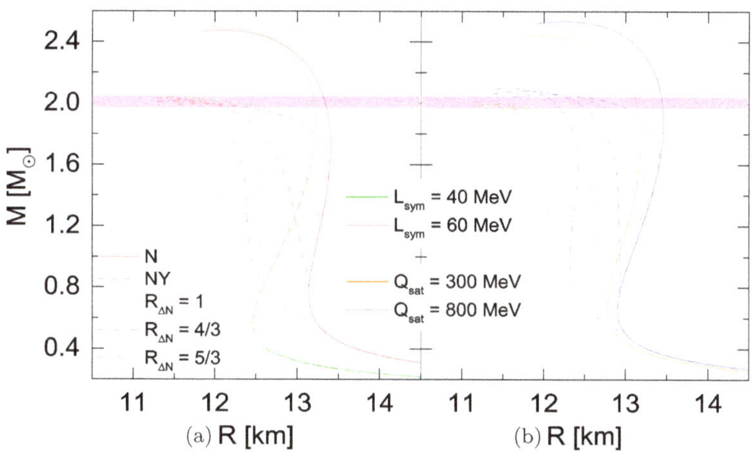

Fig. 7. Mass–radius relation for a set of EoS with varying $L_{\rm sym}$ (a) and $Q_{\rm sat}$ (b) and assuming purely nucleonic (N), hyperonic (NY), and hyperon-Δ admixed ($NY\Delta$) compositions of stellar matter (Li and Sedrakian, 2019b). Three values of the Δ-potential have been used: $R_{\Delta N} = V_\Delta/V_N = 1$, 4/3 and 5/3, where V_N is the nucleon potential in isospin-symmetrical matter at saturation density.

$V_\Delta(\rho_{\rm sat})$ the smaller is the radius of the predicted configuration, as can be anticipated from the discussion of Fig. 6. Furthermore, the overall trends are rather similar when varying individually the characteristics for NY and $NY\Delta$ matter, as these are related to the properties of nuclear matter itself and not the heavy baryon admixture. The differences between the two compositions (i.e., NY versus $NY\Delta$) seen in Fig. 7 (e.g., in the radius of a canonical neutron star) arise from the factors that have been discussed in the context of Fig. 6. We close by noting that the parameter space used in Fig. 7 implies maximum masses of configurations larger than 2 $M_\odot$ and radii within the range $12 \lesssim R \lesssim 14$ km independent of the composition.

4.3. *Tidal deformabilities*

Since the first observation of gravitational waves from a BNS merger — the GW170817 event — the tidal deformability of compact stars has become accessible observationally (Abbott *et al.*, 2017, 2019) and, thus, can be confronted with the theoretical predictions, as already indicated in Section 2. Figure 8 shows tidal deformabilities of two stars Λ_1 and Λ_2 as defined by Eq. (2) for N, NY, and $NY\Delta$ compositions, three values of the Δ-resonance potential in nuclear matter expressed through the ratio $R_{\Delta N} = V_\Delta/V_N$ and BNS member masses M_1 and M_2. A variation in the characteristics is also allowed by varying $Q_{\rm sat}$ and $L_{\rm sym}$. The diagonal line corresponds to the case of an equal-mass binary assumption for GW170817 event, in which case $M_1 = M_2 = 1.362 M_\odot$. The light and heavily shaded areas represent the 90% and 50% confidence limits extracted from the analysis of GW170817 event (Abbott *et al.*, 2019). It is seen that the data favours low values of $Q_{\rm sat} \simeq 300$ MeV and $L_{\rm sym} \simeq 40$ MeV for purely nucleonic compact stars, which are otherwise outside the allowed range. The inclusion of hyperons and Δ's reduces the tidal deformability, as seen clearly from the inset in Fig. 8. The most significant reduction arises from the inclusion of the Δ's with large attractive potential, i.e., large $R_{\Delta N}$ value. Thus, the appearance of Δ-resonances with reasonably attractive Δ-potential in the nuclear matter is needed to make the EoS models compatible with the data from GW170817. Purely hyperonic models are compatible with the 90% confidence limits for selected values of $Q_{\rm sat}$ and $L_{\rm sym}$. We note that the softening of EoS at intermediate densities caused by the onset of Δ's is similar in its effect to a first-order phase transition to a quark matter in which case again more compact configurations arise, see, for example, Alford *et al.* (2013), Alford and Sedrakian (2017), Alvarez-Castillo *et al.*

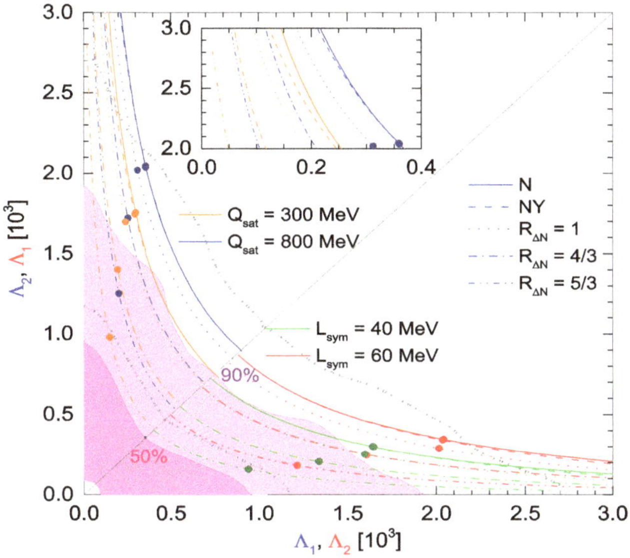

Fig. 8. Dimensionless tidal deformabilities extracted observationally from the GW170817 event (shaded areas) are compared with the predictions of EoS with various compositions and varying values of $L_{\rm sym}$, $Q_{\rm sat}$ and $R_{N\Delta}$ (Li and Sedrakian, 2019b). The light and heavy shadings correspond to the 50% and 90% credibility regions (Abbott *et al.*, 2019). The results of the earlier analysis by LVC (Abbott *et al.*, 2017) is shown by the gray dash-dotted curves. The dots correspond to the predictions for a BNS with mass ratio $q = 0.73$.

(2019), Alvarez-Castillo *et al.* (2019), Antić *et al.* (2021), Bonanno and Sedrakian (2012), Fukushima *et al.* (2020), Kojo (2020), Li *et al.* (2020a), Otto *et al.* (2020), and Zdunik and Haensel (2013). In closing, we note that the information gained from MR and $\Lambda_1 - \Lambda_2$ relations are complementary to each other: different EoS models predicting different MR relations could feature $\Lambda_1 - \Lambda_2$ relations that are rather close to each other.

5. Rapidly Rotating Hyperonic Stars

A generic feature of Einstein's gravity is the existence of the maximum mass for compact stars assuming that the pressure originates from "ordinary" baryonic matter. The exact value of this maximum mass is currently unknown. Rotating compact stars accommodate larger masses than their static (non-rotating) counterparts by about 20% because the centrifugal force provides additional support against the gravitational pull towards the centre of the star. There exist several public domain codes for computing stellar configurations of rapidly rotating compact stars (see the RNS code at Stergioulas (2003) and Gourgoulhon *et al.* (2021)). They are based on an iterative method of solution of Einstein's equations (Cook *et al.*, 1994; Nozawa *et al.*, 1998) in axial symmetry and use a tabulated EoS as an input. The method of solution is iterative: it starts with a "guess" configuration (density profile), integrates the stellar structure equations, and uses this result as an input for a new iteration. This procedure is repeated until convergence to the desired accuracy is reached at each point of the spatial grid. Rotating compact stars with hyperonic cores have been considered in recent years as well (Haensel *et al.*, 2016; Lenka *et al.*, 2019). The interest in the exploration of rapidly rotating analogous of the models discussed in the previous sections arose after the measurement by the LVC (Abbott *et al.*, 2020) of gravitational waves from a binary coalescence of a $24.3M_\odot$ black hole with a compact object in the mass the range of 2.50–$2.67M_\odot$ (the GW190814 event). The light member of this binary has a mass that lies in the so-called "mass gap" $2.5 \lesssim M/M_\odot \lesssim 5$ where neither a neutron star nor a black hole has been observed and their existence is not obvious from the point of view of stellar evolution scenarios. A natural question that arises in this context is the possible compact star nature of the light companion. Below we discuss the conjecture that it could be a compact object which is rotating at a frequency that is close to the mass-shedding (Keplerian) limit (Li *et al.*, 2020b; Sedrakian *et al.*, 2020) based on an EoS which contains hyperons and Δ-resonances; see also Dexheimer *et al.* (2021a). Figure 9 shows the mass–radius relations of compact stars with three compositions N, NY and $NY\Delta$ based on the EoS with underlying DDME2 parametrization. The nucleonic models cover the mass range $2.48 \leq M/M_\odot \leq 3$ in the spin frequency range $0 \leq \Omega \leq \Omega_K$, where Ω_K is the Kepler frequency. Therefore, the nucleonic models account for the mass of a compact star in GW190814 even without rotation. In the case of NY and $NY\Delta$ compositions, the maximum masses (but not the radii) are quite similar in the static case and this feature extends to the case of rapidly rotating stars. As already noticed, the softening of the EoS in the case of NY and $NY\Delta$ compositions imply a lower maximum mass compared to the nucleonic case. The maximum masses for these compositions are close to $2.0M_\odot$, therefore the corresponding EoS are inconsistent with the compact star interpretation of the light companion in the GW190814 event.

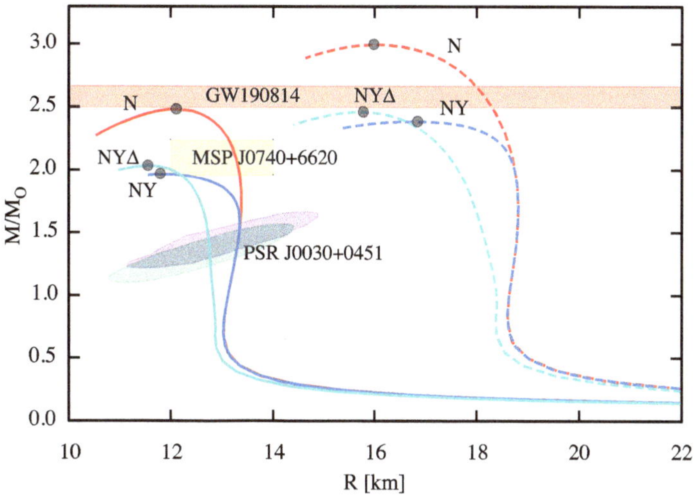

Fig. 9. The mass–radius relations for non-rotating (solid lines) and maximally rotating (dashed lines) nucleonic (N), hypernuclear (Y) and Δ-admixed-hypernuclear (Δ) stars. The colored areas show the constraints inferred from the most massive pulsar MSP J0740+6620 (Cromartie *et al.*, 2020), the mass–radius limits inferred from the NICER experiment (Miller *et al.*, 2019; Riley *et al.*, 2019) and the mass limits from GW190814 (Abbott *et al.*, 2020). The circles indicate the maximum masses of the sequences, to the left of which the stars are unstable.

Their maximally rotating Keplerian analogs on the other hand have maximum masses $\leq 2.4M_\odot$ suggesting that the maximal rotation is not sufficient to raise masses to the required value $2.5M_\odot$. Thus, there exists a significant tension between the hyperonization (with or without an admixture of Δ-resonances) and the interpretation of the light companion of GW190814 as a compact star. The results shown so far were obtained within a specific density functional and a valid question is whether the modifications of the CDF can alter this conclusion. A study of rapidly rotating stars in the case where the modifications of EoS of $NY\Delta$ matter were framed in the language of the characteristics appearing in Eq. (36), in particular the values of $Q_{\rm sat}$ and $L_{\rm sym}$ parameters, was carried out in Li *et al.* (2020b). The maximal value $R_{N\Delta} = 5/3$ was adopted although, as discussed above, the appearance of Δ-resonances increases the maximum mass of a configuration only slightly.

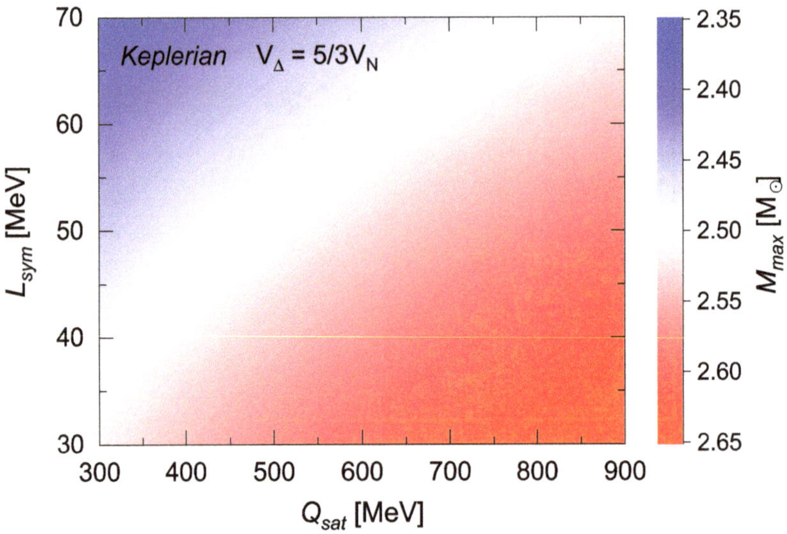

Fig. 10. The maximum masses of Keplerian sequences (color coded column on the right) as a function parameter space spanned by $Q_{\rm sat}$ and $L_{\rm sym}$. The Δ-resonance potential is fixed at assuming $R_{N\Delta} = 5/3$. The large-$Q_{\rm sat}$ and small-$L_{\rm sym}$ range corresponds to compact stars with masses exceeding $2.5\,M_{\odot}$.

Figure 10 shows the dependence of maximum masses of the Keplerian models on parameters $Q_{\rm sat}$ and $L_{\rm sym}$ for $R_{N\Delta} = 5/3$. It is seen that large masses, which are compatible with the compact star in GW190814, arise in the region of large $Q_{\rm sat}$ which imply larger energies at asymptotically large densities. Large masses are also favoured for smaller values of parameter $L_{\rm sym}$ (which implies smaller radii of stars and, therefore, more compact objects). One may compare the required values of $Q_{\rm sat}$ with other existing functionals (to avoid, in a first approximation, a full-scale computation with any given functional). A comparison shows that the range of $Q_{\rm sat}$ values compatible with a compact star in GW190814 has no overlap with the values predicted by large samples of non-relativistic and relativistic density functionals (Dutra *et al.*, 2012, 2014), exceptions being the DDME2 (Lalazissis *et al.*, 2005) and some of the recently proposed functionals (Fattoyev *et al.*, 2020; Taninah *et al.*, 2020).

One may therefore conclude that the secondary object in GW190814 event must have been a low-mass black hole. The compact star interpretation is in strong tension with the idea of hyperonization of dense matter (with and without Δ-resonances). Several extreme assumptions are required to state the contrary, such as maximally rapid rotation, as well as $Q_{\rm sat}$ and $L_{\rm sym}$ values that are outside the range covered by most of the functionals.

6. Hyperonization versus Clustering at Low Densities

So far we concentrated on the hyperonization in the cold and dense matter as it occurs in the mature compact stars, where the appearance of hyperons was made energetically favourable by the fact that in the compressed matter the neutron Fermi energy can exceed the in-medium rest masses of various hyperons. However, thermodynamic conditions may be favourable for the hyperonization in a different setting, such as in dilute and warm matter that may be (transiently) formed in supernova explosions and BNS mergers (Nakazato *et al.*, 2012; Peres *et al.*, 2013). Because under these conditions nuclear matter is composed of nucleons and clusters (characterized by a mass number and a charge) at certain isospin asymmetry one needs to resort to the ideas of nuclear statistical equilibrium (Avancini *et al.*, 2017; Botvina and Mishustin, 2010; Furusawa and Mishustin, 2017; Grams *et al.*, 2018; Gulminelli and Raduta, 2012, 2015; Hempel *et al.*, 2012, 2015; Pais *et al.*, 2020; Raduta and Gulminelli, 2010, 2019; Röpke, 2020; Typel *et al.*, 2010; Zhang and Chen, 2019). Additional facets are the pionization (at sufficiently high temperatures) (Colucci *et al.*, 2014; Fore and Reddy, 2020; Ishizuka *et al.*, 2008; Peres *et al.*, 2013) and formation of condensates of deuterons (Lombardo *et al.*, 2001; Sedrakian and Clark, 2006, 2019) and alpha-particles (Furusawa and Mishustin, 2020; Satarov *et al.*, 2019, 2021; Wu *et al.*, 2017; Zhang and Chen, 2017, 2019) at low temperatures.

An interesting question to be explored is the interplay between the clustering and heavy-baryon degrees of freedom in dilute, finite-temperature nuclear matter (Fortin *et al.*, 2018; Menezes and Providência, 2017; Sedrakian, 2020). The rich nucleon-Δ-pion dynamics, extensively explored in the heavy-ion context, suggests that one needs to include the quartet of Δ-resonances and the isotriplet of pions $\pi^{\pm,0}$ along with the nucleons and clusters (Sedrakian, 2020). In a first approximation, the full nuclear statistical ensemble can be approximated by the light clusters with the mass number $A \leq 4$ and a heavy nucleus, for example, ^{56}Fe. Since the hyperonic interactions are less important at low densities the lightest Λ-hyperon is expected to contribute dominantly to the strangeness content of matter. As an illustrative example, we show in Fig. 11 the composition of dilute nuclear matter in the temperature range $10 \leq T \leq 30$ MeV and

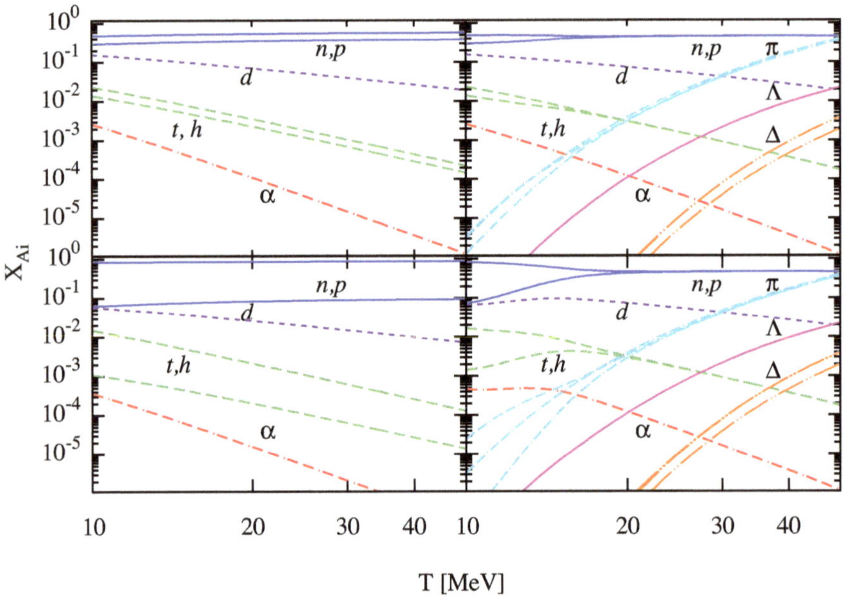

Fig. 11. Mass fractions of the constituents as a function of temperature at fixed density $10^{-2}\rho_{\rm sat}$ for charge fractions $Y_Q = 0.4$ (top panels) and 0.1 (bottom panels). In the two left panels include only nucleons and light clusters, whereas the two right panels the full composition. The composition includes neutrons n, protons p, deuterons d, triton t, helium h, α-particles, Δ resonances, Λ-hyperon and pions π.

density $10^{-2}\rho_{\rm sat}$ for charge fractions $Y_Q = 0.4$ (characteristic for supernova matter) and $Y_Q = 0.1$ (characteristic for BNS mergers) (Sedrakian, 2020). We show the *mass fraction* $X_j = A_j n_j/n$, where A_j is the mass number of a constituent, n_j is their number density, and n is the total density. The mass fraction of ^{56}Fe is not visible on the figure's scale. The main observation is that there is a transition from light baryons and clusters to heavy-baryon featuring matter at about temperature $T_{\rm tr} \sim 30$ MeV. Once heavy baryons and pions are included the isospin asymmetry in the neutron and proton components is reduced. This affects the abundances of the helion and triton which are then closer together. We believe that these are generic, model-independent features. The transition temperature itself is dependent on the treatment of the interactions and assumed composition; for alternatives see Fortin *et al.* (2018); Menezes and Providência (2017). Other generic aspects worth mentioning are: (a) in the presence of a heavy nucleus the abundances of light clusters are strongly suppressed at low temperatures; (b) the phase-space occupation suppresses completely the light clusters for densities $\geq 0.1\rho_{\rm sat}$ due to the Pauli blocking (Röpke, 2020); and (c) at low enough temperatures deuterons can cross-over from Bose-Einstein to Bardeen-Cooper-Schrieffer pair condensate (Lombardo *et al.*, 2001; Sedrakian and Clark, 2006, 2019).

7. Hypernuclear Stars: Pairing Patterns and Cooling

The long-term cooling of compact stars is a sensitive probe of the interior composition of such stars (Page, 2009; Page *et al.*, 2006; Potekhin, 2010; Schaab *et al.*, 1996; Sedrakian, 2007; Shapiro and Teukolsky, 1983). It is characterized by neutrino emission from the bulk of the stellar interior during the time-span $t \leq 10^5$ yr after the star's birth, which is followed by late-time photon cooling from its surface, assuming that there is no internal heating mechanism operating at any stage of evolution. With the advent of CDFs which were tuned to reproduce the available astrophysical and laboratory data, it became feasible to perform simulations of cooling of compact stars with hyperonization in a more constrained manner than was possible initially (Fortin *et al.*, 2021; Grigorian *et al.*, 2018; Negreiros *et al.*, 2018; Raduta *et al.*, 2017, 2019).

Hyperonic matter cools via the direct Urca processes (Prakash *et al.*, 1992). There exist density thresholds for these processes to operate, which are dictated by the kinematics involved in the reaction, but these densities are very low for hyperons, i.e., they start operating at a density that is slightly above the onset density for a hyperon participating in a reaction. Baryon pairing is known to suppress the rates of the Urca (and other baryonic) processes, therefore another unknown in the cooling simulations is the magnitude of the gaps in the spectra of various hyperons (Raduta *et al.*, 2017, 2019).

Nevertheless, with the input provided by any given CDF which can be supplemented by the solutions of the BCS equations for the hyperons, the simulations of the cooling of compact stars provided some generic insights (Raduta *et al.*, 2017, 2019). There appears to be a mass hierarchy with respect to the cooling behaviour. The lightest stars that contain hyperons with $M/M_\odot \gtrsim 1.5$ cool via the Urca process $\Lambda \to p + l + \bar{\nu}_l$, where l stands for lepton, $\bar{\nu}_l$ for the associated antineutrino. The appearance of Ξ^- in slightly more massive stars opens a competing channel via

$\Xi^- \to \Lambda + l + \bar{\nu}_l$, the degree of its efficacy depending on the pairing gaps of Ξ^-. For very massive stars with $M/M_\odot \sim 2$ the S-wave gaps of protons and Λ's will vanish at high densities (because the interactions will become repulsive at large energies) and therefore the Urca process involving Λ, p baryons will be the dominant one. Simulations show that the massive is the star the faster it cools. If Σ^- appears in the matter, the corresponding Urca process via Λ hyperon may be the dominant cooling agent, since Σ^- interaction is likely repulsive with no BCS gap in their spectrum. Given the mass hierarchy, the observations of the surface temperatures of neutron stars can be explained by the variation of their masses within their population (the light objects are hot, the heavier ones — cold). Note that the massive models may not develop normal cores of hyperons due to the possible pairing in the P-wave channel in the high-density matter in which case fast cooling will not take place (Raduta *et al.*, 2019). The pairing in the hyperonic sector remains the main unknown for cooling simulations of hypernuclear stars. Given this uncertainty, some studies neglect the hyperonic pairing altogether (Grigorian *et al.*, 2018; Negreiros *et al.*, 2018).

There are uncertainties in the studies of cooling of neutron stars, that are unrelated to the hyperonic component, which we list here for completeness. These include the composition of the atmosphere (Potekhin *et al.*, 2020) which substantially affects the surface temperature of the star and the pairing gaps of neutrons and protons in the domains where interactions are attractive (Sedrakian and Clark, 2019). Large magnetic fields are a factor, as they dissipate sufficient energy to heat up the star (Viganò *et al.*, 2021).

8. Universal Relations

In recent years, universal relations among the global properties of compact stars were established under various conditions, including stationary (non-rotating), rigidly rotating, magnetized, and finite-temperature stars, for a review, see Yagi and Yunes (2017). The universality of relations between global parameters refers to their independence of the EoS that has been input to compute them. These have proven to be rather useful for the interpretation of observational data because they reduce the uncertainties which are related to the EoS.

In the case of hypernuclear stars (with and without Δ-resonances) it has been established that the universality is maintained among the moment of inertia, quadruple moment and the Love number of the star (the so-called I-Love-Q relations) and the dependence of the moment of inertia, quadrupole moment, tidal deformability on the compactness in cold (Lenka *et al.*, 2017) and warm neutron stars (Raduta *et al.*, 2020). Their extension to finite temperature poses some challenges: it turns out that universalities are maintained if one considers stars at the same entropy per baryon S/A and lepton fraction (Raduta *et al.*, 2020), otherwise the universality is lost (Marques *et al.*, 2017; Maselli *et al.*, 2013). Clearly, constant entropy approximation will break down under realistic conditions to some degree. While we can be certain that there are no physical obstacles to maintaining the universality in the finite-temperature domain, a formulation that accounts for entropy gradients needs to be tested. The universalities so far established extend further to the rapidly rotating stars at finite temperature without loss of accuracy (Khadkikar *et al.*, 2021).

9. Concluding Remarks

The hyperonization of dense matter and the astrophysics of compact stars featuring hyperons and Δ-resonances is a vivid field of research with a strong relation to the observational multimessenger astronomy of compact stars. The recent advances on the observational front, which include the observation of gravitational waves from BNS mergers, notably GW170817, simultaneous measurement of masses and radii of neutron stars and the discovery of very massive pulsars motivate further improvements of the existing models and exploration of their new implications in novel astrophysical scenarios. The ongoing and future hypernuclear programs at CERN, BNL, JLab, GSI-FAIR, and J-PARC, and elsewhere will eventually provide us with insights on hypernuclei as they occur in terrestrial laboratories. These can be used to narrow down the parameters of the CDFs and constrain other theoretical models. The results achieved so far indicate that the combination of theoretical work with the observational advances has the potential of revealing the detailed features of the high-density matter found in compact stars in the observable future.

Acknowledgements

We are grateful to M. Alford, A. Harutyunyan, M. Oertel, A. Raduta, M. Sinha, V. Thapa for collaboration and insights on the topics related to this chapter. A. S. acknowledges the support by the Deutsche Forschungsgemeinschaft (Grant No. SE 1836/5-1) and the European COST Action CA16214 PHAROS "The multi-messenger physics and astrophysics of neutron stars". F. W. is supported through the U.S. National Science Foundation under Grants PHY-1714068 and PHY-2012152.

References

Abbott, B. P., Abbott, R., Abbott, T. D., Acernese, F., Ackley, K., *et al.* (2017). GW170817: Observation of gravitational waves from a binary neutron star inspiral, *Physical Review Letter*, **119**, p. 161101.

Abbott, B. P., Abbott, R., Abbott, T. D., Acernese, F., Ackley, K., *et al.* (2019). Properties of the binary neutron star merger GW170817, *Physical Review X*, **9**, p. 011001.

Abbott, B. P. *et al.* (2020). GW190425: Observation of a compact binary coalescence with total mass $\sim 3.4\ M_{\odot}$, *Astrophysical Journal Letters*, **892**, p. L3.

Abbott, R. *et al.* (2020). GW190814: Gravitational waves from the coalescence of a 23 solar mass black hole with a 2.6 solar mass compact object, *The Astrophysical Journal*, **896**, p. L44.

Alberico, W. M., Gervino, G., and Lavagno, A. (1994). Phenomenological approach to baryon resonance damping in nuclei, *Physics Letter B*, **321**, pp. 177–182.

Alford, M. G., Han, S., and Prakash, M. (2013). Generic conditions for stable hybrid stars, *Physical Review D*, **88**, p. 083013.

Alford, M. G. and Sedrakian, A. (2017). Compact stars with sequential qcd phase transitions, *Physical Review Letter*, **119**, p. 161104.

Altaha Motahar, Z., Blázquez-Salcedo, J. L., Kleihaus, B., and Kunz, J. (2017). Scalarization of neutron stars with realistic equations of state, *Physical Review D*, **96**, p. 064046.

Alvarez-Castillo, D. E., Antoniadis, J., Ayriyan, A., Blaschke, D., Danchev, V., Grigorian, H., Largani, N. K., and Weber, F. (2019). Accretion-induced collapse to third family compact stars as trigger for eccentric orbits of millisecond pulsars in binaries, *Astronomische Nachrichten*, **340**, pp. 878–884.

Alvarez-Castillo, D. E., Blaschke, D. B., Grunfeld, A. G., and Pagura, V. P. (2019). Third family of compact stars within a nonlocal chiral quark model equation of state, *Physical Review D*, **99**, p. 063010.

Ambartsumyan, V. A. and Saakyan, G. S. (1960). The degenerate superdense gas of elementary particles, *Soviet Astronomy*, **4**, p. 187.

Ambartsumyan, V. A. and Saakyan, G. S. (1961). On equilibrium configurations of superdense degenerate gas masses, *Astronomicheskii Zhurnal*, **38**, p. 785.

Antić, S., Shahrbaf, M., Blaschke, D., and Grunfeld, A. G. (2021). Parameter mapping between the microscopic nonlocal nambu jona-lasinio and constant sound speed models, http://arxiv.org/abs/2105.00029 (arXiv:2105.00029 [nucl-th]).

Antoniadis, J., Freire, P. C. C., Wex, N., Tauris, T. M., Lynch, R. S., van Kerkwijk, M. H., Kramer, M., Bassa, C., Dhillon, V. S., Driebe, T., Hessels, J. W. T., Kaspi, V. M., Kondratiev, V. I., Langer, N., Marsh, T. R., McLaughlin, M. A., Pennucci, T. T., Ransom, S. M., Stairs, I. H., van Leeuwen, J., Verbiest, J. P. W., and Whelan, D. G. (2013). A massive pulsar in a compact relativistic binary, *Science*, **340**, p. 448.

Astashenok, A. V., Capozziello, S., and Odintsov, S. D. (2014). Maximal neutron star mass and the resolution of the hyperon puzzle in modified gravity, *Physical Review D*, **89**, p. 103509.

Astashenok, A. V., Capozziello, S., and Odintsov, S. D. (2015a). Extreme neutron stars from extended theories of gravity, *JCAP*, **01**, p. 001.

Astashenok, A. V., Capozziello, S., and Odintsov, S. D. (2015b). Magnetic neutron stars in f(R) gravity, *Astrophysics and Space Science*, **355**, pp. 333–341.

Avancini, S. S., Ferreira, M., Pais, H., Providência, C., and Röpke, G. (2017). Light clusters and pasta phases in warm and dense nuclear matter, *Physical Review C*, **95**, p. 045804.

Baiotti, L. (2019). Gravitational waves from neutron star mergers and their relation to the nuclear equation of state, *Progress in Particle and Nuclear Physics*, **109**, p. 103714.

Baldo, M., Burgio, G. F., and Schulze, H.-J. (2000). Hyperon stars in the brueckner-bethe-goldstone theory, *Physical Review C*, **61**, p. 055801.

Banik, S., Hempel, M., and Bandyopadhyay, D. (2014). New hyperon equations of state for supernovae and neutron stars in density-dependent hadron field theory, *The Astrophysical Journal Supplement Series*, **214**, p. 22.

Bardeen, J. M., Thorne, K. S., and Meltzer, D. W. (1966). A catalogue of methods for studying the normal modes of radial pulsation of general-relativistic stellar models, *The Astrophysical Journal*, **145**, p. 505.

Bart, S., Chrien, R. E., Franklin, W. A., Fukuda, T., Hayano, R. S., Hicks, K., Hungerford, E. V., Michael, R., Miyachi, T., Nagae, T., Nakano, J., Naing, W., Omata, K., Sawafta, R., Shimizu, Y., Tang, L., and Wissink, S. W. (1999). Σ Hyperons in the nucleus, *Physical Review Letter*, **83**, pp. 5238–5241.

Baym, G., Bethe, H. A., and Pethick, C. J. (1971a). Neutron star matter, *Nuclear Physics A*, **175**, 2, pp. 225–271.

Baym, G., Pethick, C., and Sutherland, P. (1971b). The ground state of matter at high densities: equation of state and stellar models, *The Astrophysical Journal*, **170**, p. 299.

Bednarek, I., Haensel, P., Zdunik, J. L., Bejger, M., and Manka, R. (2012). Hyperons in neutron-star cores and two-solar-mass pulsar, *Astronomy & Astrophysics*, **543**, p. A157.

Blaschke, D. and Chamel, N. (2018). Phases of dense matter in compact stars, *Astrophysics and Space Science Library*, **457**, pp. 337–400.

Blázquez-Salcedo, J. L., González-Romero, L. M., Kunz, J., Mojica, S., and Navarro-Lérida, F. (2016). Axial quasinormal modes of einstein-gauss-bonnet-dilaton neutron stars, *Physical Review D*, **93**, p. 024052.

Bombaci, I. (2017). The hyperon puzzle in neutron stars, *JPS Conference Proceedings*, **17**, p. 101002.

Bonanno, L. and Sedrakian, A. (2012). Composition and stability of hybrid stars with hyperons and quark color-superconductivity, *Astronomy & Astrophysics*, **539**, p. A16.

Botvina, A. S. and Mishustin, I. N. (2010). Statistical approach for supernova matter, *Nuclear Physics A*, **843**, pp. 98–132.

Cai, B.-J. and Chen, L.-W. (2017). Constraints on the skewness coefficient of symmetric nuclear matter within the nonlinear relativistic mean field model, *Nuclear Science Technology*, **28**, p. 185.

Cai, B.-J., Fattoyev, F. J., Li, B.-A., and Newton, W. G. (2015). Critical density and impact of $\Delta(1232)$ resonance formation in neutron stars, *Physical Review C*, **92**, p. 015802.

Cai, B.-J. and Li, B.-A. (2021). Intrinsic correlations among characteristics of neutron-rich matter imposed by the unbound nature of pure neutron matter, *Physical Review C*, **103**, p. 034607.

Chatterjee, D. and Vidaña, I. (2016). Do hyperons exist in the interior of neutron stars? *European Physics Journal A*, **52**, p. 29.

Choudhury, S. K. and Rakshit, R. (1993). Δ-excited nuclear matter in the derivative scalar coupling model, *Physical Review C*, **48**, pp. 598–601.

Collins, J. C. and Perry, M. J. (1975). Superdense matter: neutrons or asymptotically free quarks? *Physical Review Letter*, **34**, pp. 1353–1356.

Colucci, G., Fraga, E. S., and Sedrakian, A. (2014). Chiral pions in a magnetic background, *Physics Letter B*, **728**, pp. 19–24.

Colucci, G. and Sedrakian, A. (2013). Equation of state of hypernuclear matter: impact of hyperon-scalar-meson couplings, *Physical Review C*, **87**, p. 055806.

Cook, G. B., Shapiro, S. L., and Teukolsky, S. A. (1994). Rapidly rotating neutron stars in general relativity: realistic equations of state, *The Astrophysical Journal*, **424**, p. 823.

Cozma, M. D. (2016). the impact of energy conservation in transport models on the π^-/π^+ multiplicity ratio in heavy-ion collisions and the symmetry energy, *Physics Letter B*, **753**, pp. 166–172.

Cozma, M. D. and Tsang, M. B. (2021). In-medium $\Delta(1232)$ potential, pion production in heavy-ion collisions and the symmetry energy, *European Physical Journal A*, **57**, p. 309.

Cromartie, H. T., Fonseca, E., Ransom, S. M., Demorest, P. B., Arzoumanian, Z., Blumer, H., Brook, P. R., DeCesar, M. E., Dolch, T., Ellis, J. A., Ferdman, R. D., Ferrara, E. C., Garver-Daniels, N., Gentile, P. A., Jones, M. L., Lam, M. T., Lorimer, D. R., Lynch, R. S., McLaughlin, M. A., Ng, C., Nice, D. J., Pennucci, T. T., Spiewak, R., Stairs, I. H., Stovall, K., Swiggum, J. K., and Zhu, W. W. (2020). Relativistic shapiro delay measurements of an extremely massive millisecond pulsar, *Nature Astronomy*, **4**, pp. 72–76.

de Swart, J. J. (1963). The octet model and its Clebsch-Gordan coefficients, *Reviews of Modern Physics*, **35**, pp. 916–939.

Demorest, P. B., Pennucci, T., Ransom, S. M., Roberts, M. S. E., and Hessels, J. W. T. (2010). A two-solar-mass neutron star measured using shapiro delay, *Nature*, **467**, pp. 1081–1083.

Dexheimer, V., Gomes, R. O., Klähn, T., Han, S., and Salinas, M. (2021a). GW190814 as a massive rapidly rotating neutron star with exotic degrees of freedom, *Physical Review C*, **103**, p. 025808.

Dexheimer, V., Marquez, K. D., and Menezes, D. P. (2021b). Delta baryons in neutron-star matter under strong magnetic fields, *European Physical Journal A*, **57**, p. 216.

Dover, C. and Gal, A. (1984). Hyperon-nucleus potentials, *Progress in Particle and Nuclear Physics*, **12**, pp. 171–239.

Drago, A., Lavagno, A., Pagliara, G., and Pigato, D. (2014). Early appearance of Δ isobars in neutron stars, *Physical Review C*, **90**, p. 065809.

Drago, A., Lavagno, A., Pagliara, G., and Pigato, D. (2016). The scenario of two families of compact stars: 1. Equations of state, mass-radius relations and binary systems, *European Physics Journal A*, **52**, p. 40.

Dutra, M., Lourenço, O., Sá Martins, J. S., Delfino, A., Stone, J. R., and Stevenson, P. D. (2012). Skyrme interaction and nuclear matter constraints, *Physical Review C*, **85**, p. 035201.

Dutra, M., Lourenço, O., Avancini, S. S., Carlson, B. V., Delfino, A., Menezes, D. P., Providência, C., Typel, S., and Stone, J. R. (2014). Relativistic mean-field hadronic models under nuclear matter constraints, *Physical Review C*, **90**, p. 055203.

Fattoyev, F. J., Horowitz, C. J., Piekarewicz, J., and Reed, B. (2020). GW190814: Impact of a 2.6 solar mass neutron star on the nucleonic equations of state, *Physical Review C*, **102**, p. 065805.

Flanagan, E. E. and Hinderer, T. (2008). Constraining neutron-star tidal love numbers with gravitational-wave detectors, *Physical Review D*, **77**, p. 021502.

Fonseca, E., Cromartie, H. T., Pennucci, T. T., Ray, P. S., Kirichenko, A. Y., Ransom, S. M., Demorest, P. B., Stairs, I. H., Arzoumanian, Z., Guillemot, L., Parthasarathy, A., Kerr, M., Cognard, I., Baker, P. T., Blumer, H., Brook, P. R., DeCesar, M., Dolch, T., Dong, F. A., Ferrara, E. C., Fiore, W., Garver-Daniels, N., Good, D. C., Jennings, R., Jones, M. L., Kaspi, V. M., Lam, M. T., Lorimer, D. R., Luo, J., McEwen, A., McKee, J. W., McLaughlin, M. A., McMann, N., Meyers, B. W., Naidu, A., Ng, C., Nice, D. J., Pol, N., Radovan, H. A., Shapiro-Albert, B., Tan, C. M., Tendulkar, S. P., Swiggum, J. K., Wahl, H. M., and Zhu, W. (2021). Refined mass and geometric measurements of the high-mass PSR J0740+6620, *Astrophysical Journal Letters*, **915**, p. L12.

Fonseca, E., Pennucci, T. T., Ellis, J. A., Stairs, I. H., Nice, D. J., Ransom, S. M., Demorest, P. B., Arzoumanian, Z., Crowter, K., Dolch, T., Ferdman, R. D., Gonzalez, M. E., Jones, G., Jones, M. L., Lam, M. T., Levin, L., McLaughlin, M. A., Stovall, K., Swiggum, J. K., and Zhu, W. (2016). The NANOGrav nine-year data set: mass and geometric measurements of binary millisecond pulsars, *The Astrophysical Journal*, **832**, p. 167.

Fore, B. and Reddy, S. (2020). Pions in hot dense matter and their astrophysical implications, *Physical Review C*, **101**, p. 035809.

Fortin, M., Avancini, S. S., Providência, C., and Vidaña, I. (2017). Hypernuclei and massive neutron stars, *Physical Review C*, **95**, p. 065803.

Fortin, M., Oertel, M., and Providência, C. (2018). Hyperons in hot dense matter: what do the constraints tell us for equation of state? *PASA*, **35**, p. 44.

Fortin, M., Raduta, A. R., Avancini, S., and Providência, C. (2020). Relativistic hypernuclear compact stars with calibrated equations of state, *Physical Review D*, **101**, p. 034017.

Fortin, M., Raduta, A. R., Avancini, S., and Providência, C. (2021). Thermal evolution of relativistic hyperonic compact stars with calibrated equations of state, *Physical Review D*, **103**, p. 083004.

Friedman, E. and Gal, A. (2021). Constraints on Ξ^- nuclear interactions from capture events in emulsion, *Physics Letters B*, **820**, p. 136555.

Fukushima, K., Kojo, T., and Weise, W. (2020). Hard-core deconfinement and soft-surface delocalization from nuclear to quark matter, *Physical Review D*, **102**, p. 096017.

Furusawa, S. and Mishustin, I. (2017). Self-consistent calculation of the nuclear composition in hot and dense stellar matter, *Physical Review C*, **95**, p. 035802.

Furusawa, S. and Mishustin, I. (2020). Degeneracy effects and bose condensation in warm nuclear matter with light and heavy clusters, *Nuclear Physics A*, **1002**, p. 121991.

Gandolfi, S. and Lonardoni, D. (2017). The EOS of neutron matter and the effect of Λ hyperons to neutron star structure, *JPS Conference Proceedings*, **17**, p. 101001.

Glendenning, N. (2012). Compact stars: nucl. phys., particle physics and general relativity, Astronomy and Astrophysics Library, Springer, New York, ISBN 9781468404913.

Glendenning, N. K. (1985). Neutron stars are giant hypernuclei? *The Astrophysical Journal*, **293**(1), pp. 470–493.

Glendenning, N. K. and Moszkowski, S. A. (1991). Reconciliation of neutron star masses and binding of the lambda in hypernuclei, *Physical Review Letter*, **67**, pp. 2414–2417.

Glendenning, N. K., Weber, F., and Moszkowski, S. A. (1992). Neutron and hybrid stars in the derivative coupling model, *Physical Review C*, **45**, pp. 844–855.

Gomes, R. O., Char, P., and Schramm, S. (2019). Constraining strangeness in dense matter with GW170817, *The Astrophysical Journal*, **877**, p. 139.

Gomes, R. O., Dexheimer, V., Schramm, S., and Vasconcellos, C. A. Z. (2015). Many-body forces in the equation of state of hyperonic matter, *The Astrophysical Journal*, **808**, p. 8.

Gourgoulhon, E., Grandclément, P, Marck, J.-A., Novak, J., and Taniguchi, K. (2021). *LORENE Code*, https://lorene.obspm.fr.

Grams, G., Giraud, S., Fantina, A. F., and Gulminelli, F. (2018). Distribution of nuclei in equilibrium stellar matter from the free-energy density in a wigner-seitz cell, *Physical Review C*, **97**, p. 035807.

Grigorian, H., Voskresensky, D. N., and Maslov, K. A. (2018). Cooling of Neutron Stars in "Nuclear Medium Cooling Scenario" with Stiff Equation of State Including Hyperons, *Nuclear Physics A*, **980**, pp. 105–130.

Gulminelli, F. and Raduta, A. R. (2012). Ensemble inequivalence in supernova matter within a simple model, *Physical Review C*, **85**, p. 025803.

Gulminelli, F. and Raduta, A. R. (2015). Unified treatment of subsaturation stellar matter at zero and finite temperature, *Physical Review C*, **92**, p. 055803.

Gusakov, M. E., Haensel, P., and Kantor, E. M. (2014). Physics input for modelling superfluid neutron stars with hyperon cores, *Monthly Notices of the Royal Astronomical Society*, **439**, pp. 318–333.

Güven, H., Bozkurt, K., Khan, E., and Margueron, J. (2018). $\Lambda\Lambda$ pairing in multi-strange hypernuclei, *Physical Review C*, **98**(1), p. 014318.

Haensel, P., Bejger, M., Fortin, M., and Zdunik, L. (2016). Rotating neutron stars with exotic cores: masses, radii, stability, *European Physics Journal A*, **52**, p. 59.

Haidenbauer, J., Meißner, U. G., Kaiser, N., and Weise, W. (2017). Lambda-nuclear interactions and hyperon puzzle in neutron stars, *European Physics Journal A*, **53**, p. 121.

Hempel, M., Fischer, T., Schaffner-Bielich, J., and Liebendörfer, M. (2012). New equations of state in simulations of core-collapse supernovae, *The Astrophysical Journal*, **748**, p. 70.

Hempel, M., Hagel, K., Natowitz, J., Röpke, G., and Typel, S. (2015). Constraining supernova equations of state with equilibrium constants from heavy-ion collisions, *Physical Review C*, **91**, p. 045805.

Horikawa, Y., Thies, M., and Lenz, F. (1980). The Δ-nucleus spin-orbit interaction in π-nucleus scattering, *Nuclear Physics A*, **345**, pp. 386–408.

Huber, H., Weber, F., and Weigel, M. K. (1994). Neutron star properties and the relativistic equation of state of asymmetric nuclear matter, *Physical Review C*, **50**, pp. R1287–R1291.

Huber, H., Weber, F., Weigel, M. K., and Schaab, C. (1998). Neutron star properties with relativistic equations of state, *International Journal of Modern Physics E*, **7**(03), pp. 301–339.

Inoue, T. and HAL QCD Collaboration (2019). Strange nuclear physics from qcd on lattice, in *the 13th international conference on hypernuclear and strange particle physics: HYP2018*, *American Institute of Physics Conference Series*, Vol. 2130, p. 020002.

Ishizuka, C., Ohnishi, A., Tsubakihara, K., Sumiyoshi, K., and Yamada, S. (2008). Tables of hyperonic matter equation of state for core-collapse supernovae, *Journal Nuclear Physics G*, **35**, p. 085201.

Itoh, N. (1970). hydrostatic equilibrium of hypothetical quark stars, *Progress of Theoretical Physics*, **44**, pp. 291–292.

Jiang, W.-Z., Li, B.-A., and Chen, L.-W. (2012). Large-mass neutron stars with hyperonization, *The Astrophysical Journal*, **756**, p. 56.

Khadkikar, S., Raduta, A. R., Oertel, M., and Sedrakian, A. (2021). Maximum mass of compact stars from gravitational wave events with finite-temperature equations of state, *Physical Review C*, **103**, p. 055811.

Khan, E., Margueron, J., Gulminelli, F., and Raduta, A. R. (2015). Microscopic evaluation of the hypernuclear chart with Λ hyperons, *Physical Review C*, **92**(4), p. 044313.

Koch, V. (1997). Aspects of chiral symmetry, *International Journal of Modern Physics E*, **6**, pp. 203–250.

Kojo, T. (2020). QCD equations of state and speed of sound in neutron stars, http://arxiv.org/abs/2011.10940 (arXiv:2011.10940 [nucl-th]).

Kolomeitsev, E. E., Maslov, K. A., and Voskresensky, D. N. (2017). Delta isobars in relativistic mean-field models with σ-scaled hadron masses and couplings, Nuclear Physics A, **961**, pp. 106–141.

Lalazissis, G. A., Nikšić, T., Vretenar, D., and Ring, P. (2005). New relativistic mean-field interaction with density-dependent meson-nucleon couplings, *Physical Review C*, **71**, p. 024312.

Lattimer, J. M. (2019). Neutron star mass and radius measurements, *Universe*, **5**, p. 159.

Lattimer, J. M. and Schutz, B. F. (2005). Constraining the equation of state with moment of inertia measurements, *The Astrophysical Journal*, **629**, pp. 979–984.

Lavagno, A. (2010). Hot and dense hadronic matter in an effective mean-field approach, *Physical Review C*, **81**, p. 044909.

Lenka, S. S., Char, P., and Banik, S. (2017). Critical mass, moment of inertia and universal relations of rapidly rotating neutron stars with exotic matter, *International Journal of Modern Physics D*, **26**, p. 1750127.

Lenka, S. S., Char, P., and Banik, S. (2019). Properties of massive rotating protoneutron stars with hyperons: structure and universality, *Journal of Physics G*, **46**, p. 105201.

Leung, Y. C. and Wang, C. G. (1971). Properties of hadron matter. II. dense baryon matter and neutron stars, *The Astrophysical Journal*, **170**, p. 499.

Li, B.-A., Cai, B.-J., Xie, W.-J., and Zhang, N.-B. (2021). Progress in constraining nuclear symmetry energy using neutron star observables since GW170817, *Universe*, **7**, p. 182.

Li, B.-A. and Magno, M. (2020). Curvature-slope correlation of nuclear symmetry energy and its imprints on the crust-core transition, radius and tidal deformability of canonical neutron stars, *Physical Review C*, **102**, p. 045807.

Li, J. J., Long, W. H., and Sedrakian, A. (2018a). Hypernuclear stars from relativistic hartree-fock density functional theory, *European Physics Journal A*, **54**, p. 133.

Li, J. J. and Sedrakian, A. (2019a). Constraining compact star properties with nuclear saturation parameters, *Physical Review C*, **100**, p. 015809.

Li, J. J. and Sedrakian, A. (2019b). Implications from GW170817 for Δ-isobar admixed hypernuclear compact stars, *Astrophysical Journal Letters*, **874**(2), p. L22.

Li, J. J., Sedrakian, A., and Alford, M. (2020a). Relativistic hybrid stars with sequential first-order phase transitions and heavy-baryon envelopes, *Physical Review D*, **101**, p. 063022.

Li, J. J., Sedrakian, A., and Weber, F. (2018b). Competition between delta isobars and hyperons and properties of compact stars, *Physics Letter B*, **783**, pp. 234–240.

Li, J. J., Sedrakian, A., and Weber, F. (2020b). Rapidly rotating δ-resonance-admixed hypernuclear compact stars, *Physics Letter B*, **810**, p. 135812.

Lombardo, U., Nozières, P., Schuck, P., Schulze, H. J., and Sedrakian, A. (2001). Transition from BCS pairing to Bose-Einstein condensation in low-density asymmetric nuclear matter, *Physical Review C*, **64**, p. 064314.

Lonardoni, D., Gandolfi, S., and Pederiva, F. (2013a). Effects of the two-body and three-body hyperon-nucleon interactions in λ-hypernuclei, *Physical Review C*, **87**, p. 041303.

Lonardoni, D., Pederiva, F., and Gandolfi, S. (2013b). Auxiliary field diffusion monte carlo study of the hyperon-nucleon interaction in λ-hypernuclei, *Nuclear Physics A*, **914**, pp. 243–247.

Lonardoni, D., Lovato, A., Gandolfi, S., and Pederiva, F. (2015). Hyperon puzzle: hints from quantum monte carlo calculations, *Physical Review Letter*, **114**, p. 092301.

Lopes, L. L. and Menezes, D. P. (2014). Hypernuclear matter in a complete su(3) symmetry group, *Physical Review C*, **89**, p. 025805.

Margueron, J. and Gulminelli, F. (2019). Effect of high-order empirical parameters on the nuclear equation of state, *Physical Review C*, **99**, p. 025806.

Margueron, J., Hoffmann Casali, R., and Gulminelli, F. (2018a). Equation of state for dense nucleonic matter from metamodeling. I. Foundational aspects, *Physical Review C*, **97**, p. 025805.

Margueron, J., Hoffmann Casali, R., and Gulminelli, F. (2018b). Equation of state for dense nucleonic matter from metamodeling. II. Predictions for neutron star properties, *Physical Review C*, **97**, p. 025806.

Margueron, J., Khan, E., and Gulminelli, F. (2017). Density functional approach for multi-strange hypernuclei: competition between λ and $\Xi^{0,-}$ hyperons, *Physical Review C*, **96**(5), p. 054317.

Marques, M., Oertel, M., Hempel, M., and Novak, J. (2017). New temperature dependent hyperonic equation of state: application to rotating neutron star models and I-Q relations, *Physical Review C*, **96**, p. 045806.

Maselli, A., Cardoso, V., Ferrari, V., Gualtieri, L., and Pani, P. (2013). Equation-of-state-independent relations in neutron stars, *Physical Review D*, **88**, p. 023007.

Maslov, K. A., Kolomeitsev, E. E., and Voskresensky, D. N. (2015). Solution of the hyperon puzzle within a relativistic mean-field model, *Physical Letter B*, **748**, pp. 369–375.

Maslov, K. A., Kolomeitsev, E. E., and Voskresensky, D. N. (2016). Relativistic mean-field models with scaled hadron masses and couplings: hyperons and maximum neutron star mass, *Nuclear Physics A*, **950**, pp. 64–109.

Massot, E., Margueron, J., and Chanfray, G. (2012). On the maximum mass of hyperonic neutron stars, *EPL*, **97**(3), p. 39002.

Menezes, D. P. and Providência, C. (2017). Hyperons in the nuclear pasta phase, *Physical Review C*, **96**, p. 045803.

Miller, M. C., Lamb, F. K., Dittmann, A. J., Bogdanov, S., Arzoumanian, Z., Gendreau, K. C., Guillot, S., Harding, A. K., Ho, W. C. G., Lattimer, J. M., Ludlam, R. M., Mahmoodifar, S., Morsink, S. M., Ray, P. S., Strohmayer, T. E., Wood, K. S., Enoto, T., Foster, R., Okajima, T., Prigozhin, G., and Soong, Y. (2019). PSR J0030+0451 mass and radius from NICER data and implications for the properties of neutron star matter, *Astrophysical Journal Letters*, **887**, p. L24.

Miller, M. C. *et al.* (2021). The radius of PSR J0740+6620 from NICER and XMM-newton data, *Astrophysical Journal Letters*, **918**, p. L28.

Miyatsu, T., Cheoun, M.-K., and Saito, K. (2015). Equation of state for neutron stars with hyperons and quarks in the relativistic Hartree-Fock approximation, *The Astrophysical Journal*, **813**, p. 135.

Moszkowski, S. A. (1974). Energy of neutron-star matter, *Physical Review D*, **9**, pp. 1613–1625.

Motta, T. F., Thomas, A. W., and Guichon, P. (2020). Do delta baryons play a role in neutron stars? *Physics Letter B*, **80**(2), p. 135266.

Nakamura, S. X., Sato, T., Lee, T. S. H., Szczerbinska, B., and Kubodera, K. (2010). Dynamical model of coherent pion production in neutrino-nucleus scattering, *Physical Review C*, **81**, p. 035502.

Nakazato, K., Furusawa, S., Sumiyoshi, K., Ohnishi, A., Yamada, S., and Suzuki, H. (2012). Hyperon matter and black hole formation in failed supernovae, *The Astrophysical Journal*, **745**, p. 197.

Negreiros, R., Tolos, L., Centelles, M., Ramos, A., and Dexheimer, V. (2018). Cooling of small and massive hyperonic stars, *The Astrophysical Journal*, **863**, p. 104.

Nemura, H., Ishii, N., Aoki, S., and Hatsuda, T. (2007). Hyperon-nucleon potentials from lattice QCD, *PoS* LATTICE2007, p. 156.

Nozawa, T., Stergioulas, N., Gourgoulhon, E., and Eriguchi, Y. (1998). Construction of highly accurate models of rotating neutron stars — comparison of three different numerical schemes, *Astronomy and Astrophysics Supplement Series*, **132**, pp. 431–454.

O'Connell, J. S. and Sealock, R. M. (1990). Phenomenological Δ-nucleus potential from inclusive electron-nucleus scattering data, *Physical Review C*, **42**, pp. 2290–2294.

Oertel, M., Gulminelli, F., Providência, C., and Raduta, A. R. (2016). Hyperons in neutron stars and supernova cores, *European Physics Journal A*, **52**, p. 50.

Oertel, M., Hempel, M., Klähn, T., and Typel, S. (2017). Equations of state for supernovae and compact stars, *Review Modern Physics*, **89**, p. 015007.

Oertel, M., Providência, C., Gulminelli, F., and Raduta, A. R. (2015). Hyperons in neutron star matter within relativistic mean-field models, *Journal of Physics G*, **42**(7), p. 075202.

Ono, A., Xu, J., Colonna, M., Danielewicz, P., Ko, C. M., Tsang, M. B., Wang, Y.-J., Wolter, H., Zhang, Y.-X., Chen, L.-W., Cozma, D., Elfner, H., Feng, Z.-Q., Ikeno, N., Li, B.-A., Mallik, S., Nara, Y., Ogawa, T., Ohnishi, A., Oliinychenko, D., Su, J., Song, T., Zhang, F.-S., and Zhang, Z.

(2019). Comparison of heavy-ion transport simulations: collision integral with pions and Δ resonances in a box, *Physical Review C*, **100**, p. 044617.
Oppenheimer, J. R. and Volkoff, G. M. (1939). On massive neutron cores, *Physical Review*, **55**, pp. 374–381.
Otto, K., Oertel, M., and Schaefer, B.-J. (2020). Nonperturbative quark matter equations of state with vector interactions, *European Physics Journal*, **229**, pp. 3629–3649.
Özel, F. and Freire, P. (2016). Masses, radii, and the equation of state of neutron stars, *Annual Review of Astronomy and Astrophysics*, **54**, pp. 401–440.
Page, D. (2009). Neutron Star Cooling: I, in W. Becker (ed.), *Astrophysics and Space Science Library*, Vol. **357**, (Springer), p. 247.
Page, D., Geppert, U., and Weber, F. (2006). The cooling of compact stars, *Nuclear Physics A*, **777**, pp. 497–530, special Issue on Nuclear Astrophysics.
Pais, H., Bougault, R., Gulminelli, F., Providência, C., Bonnet, E., Borderie, B., Chbihi, A., Frankland, J. D., Galichet, E., Gruyer, D., Henri, M., Le Neindre, N., Lopez, O., Manduci, L., Parlôg, M., and Verde, G. (2020). Low density in-medium effects on light clusters from heavy-ion data, *Physical Review Letter*, **125**, p. 012701.
Pandharipande, V. R. (1971). Hyperonic matter, *Nuclear Physics A*, **178**, pp. 123–144.
Papazoglou, P., Schramm, S., Schaffner-Bielich, J., Stoecker, H., and Greiner, W. (1998). Chiral lagrangian for strange hadronic matter, *Physical Review C*, **57**, pp. 2576–2588.
Peres, B., Oertel, M., and Novak, J. (2013). Influence of pions and hyperons on stellar black hole formation, *Physical Review D*, **87**, p. 043006.
Potekhin, A. Y. (2010). The physics of neutron stars, *Physics Uspekhi*, **53**, pp. 1235–1256.
Potekhin, A. Y., Zyuzin, D. A., Yakovlev, D. G., Beznogov, M. V., and Shibanov, Y. A. (2020). Thermal luminosities of cooling neutron stars, *Monthly Notices of the Royal Astronomical Society*, **496**, pp. 5052–5071.
Prakash, M., Prakash, M., Lattimer, J. M., and Pethick, C. J. (1992). Rapid cooling of neutron stars by hyperons and delta isobars, *Astrophysical Journal Letters*, **390**, p. L77.
Providência, C., Fortin, M., Pais, H., and Rabhi, A. (2019). Hyperonic stars and the symmetry energy, *Frontiers in Astronomy and Space Sciences*, **6**, p. 13.
Providencia, C. and Rabhi, A. (2013). Interplay between the symmetry energy and the strangeness content of neutron stars, *Physical Review C*, **87**, p. 055801.
Raduta, A. R. and Gulminelli, F. (2010). Statistical description of complex nuclear phases in supernovae and proto-neutron stars, *Physical Review C*, **82**, p. 065801.
Raduta, A. R. and Gulminelli, F. (2019). Nuclear statistical equilibrium equation of state for core collapse, *Nuclear Physics A*, **983**, pp. 252–275.
Raduta, A. R., Sedrakian, A., and Weber, F. (2017). Cooling of hypernuclear compact stars, *Monthly Notices of the Royal Astronomical Society*, **475**(4), pp. 4347–4356.

Raduta, A. R., Li, J. J., Sedrakian, A., and Weber, F. (2019). Cooling of hypernuclear compact stars: Hartree–Fock models and high-density pairing, *Monthly Notices of the Royal Astronomical Society*, **487**, pp. 2639–2652.

Raduta, A. R., Oertel, M., and Sedrakian, A. (2020). Proto-neutron stars with heavy baryons and universal relations, *Monthly Notices of the Royal Astronomical Society*, **499**, pp. 914–931.

Ribes, P., Ramos, A., Tolos, L., Gonzalez-Boquera, C., and Centelles, M. (2019). Interplay between delta particles and hyperons in neutron stars, *The Astrophysical Journal*, **883**, p. 168.

Riek, F., Lutz, M. F. M., and Korpa, C. L. (2009). Photoabsorption off nuclei with self-consistent vertex corrections, *Physical Review C*, **80**, p. 024902.

Riley, T. E., Watts, A. L., Bogdanov, S., Ray, P. S., Ludlam, R. M., Guillot, S., Arzoumanian, Z., Baker, C. L., Bilous, A. V., Chakrabarty, D., Gendreau, K. C., Harding, A. K., Ho, W. C. G., Lattimer, J. M., Morsink, S. M., and Strohmayer, T. E. (2019). A NICER view of PSR J0030+0451: Millisecond pulsar parameter estimation, *Astrophysical Journal Letters*, **887**, p. L21.

Riley, T. E. *et al.* (2021). A NICER view of the massive pulsar PSR J0740+6620 informed by radio timing and XMM-newton spectroscopy, *Astrophysical Journal Letters*, **918**, p. L27.

Röpke, G. (2020). Light P-shell nuclei with cluster structures ($4 \leq A \leq 16$) in nuclear matter, *Physical Review C*, **101**, p. 064310.

Sahoo, H. S., Mitra, G., Mishra, R., Panda, P. K., and Li, B.-A. (2018). Neutron star matter with δ isobars in a relativistic quark model, *Physical Review C*, **98**, p. 045801.

Sasaki, K. *et al.* (2020). $\Lambda\Lambda$ and $N\Xi$ interactions from lattice QCD near the physical point, *Nuclear Physics A*, **998**, p. 121737.

Satarov, L. M., Mishustin, I. N., Motornenko, A., Vovchenko, V., Gorenstein, M. I., and Stoecker, H. (2019). Phase transitions and Bose-Einstein condensation in α-nucleon matter, *Physical Review C*, **99**, p. 024909.

Satarov, L. M., Poberezhnyuk, R. V., Mishustin, I. N., and Stoecker, H. (2021). Phase diagram of α matter with a skyrme-like scalar interaction, *Physical Review C*, **103**, p. 024301.

Sawyer, R. F. (1972). Energy shifts of excited nucleons in neutron-star matter, *The Astrophysical Journal*, **176**, p. 205.

Schaab, C., Weber, F., Weigel, M. K., and Glendenning, N. K. (1996). Thermal evolution of compact stars, *Nuclear Physics A*, **605**, pp. 531–565.

Schaffner, J. and Mishustin, I. N. (1996). Hyperon rich matter in neutron stars, *Physical Review C*, **53**, pp. 1416–1429.

Schulze, H.-J. and Rijken, T. (2011). Maximum mass of hyperon stars with the nijmegen ESC08 model, *Physical Review C*, **84**, p. 035801.

Schürhoff, T., Schramm, S., and Dexheimer, V. (2010). Neutron stars with small radii — the role of Δ resonances, *The Astrophysical Journal*, **724**(1), p. L74.

Sedrakian, A. (2007). The physics of dense hadronic matter and compact stars, *Progress in Particle and Nuclear Physics*, **58**, pp. 168–246.

Sedrakian, A. (2020). Light clusters in dilute heavy-baryon admixed nuclear matter, *European Physics Journal A*, **56**, p. 258.

Sedrakian, A. and Clark, J. W. (2006). Pair condensation and bound states in fermionic systems, *Physical Review C*, **73**, p. 035803.

Sedrakian, A. and Clark, J. W. (2019). Superfluidity in nuclear systems and neutron stars, *European Physics Journal A*, **55**, p. 167.

Sedrakian, A., Weber, F., and Li, J. J. (2020). Confronting GW190814 with hyperonization in dense matter and hypernuclear compact stars, *Physical Review D*, **102**, p. 041301.

Serot, B. D. and Walecka, J. D. (1997). Recent progress in quantum hadrodynamics, *International Journal of Modern Physics E*, **6**(04), pp. 515–631.

Shahrbaf, M., Blaschke, D., and Khanmohamadi, S. (2020). Mixed phase transition from hypernuclear matter to deconfined quark matter fulfilling mass-radius constraints of neutron stars, *Journal of Physics G*, **47**, p. 115201.

Shahrbaf, M. and Moshfegh, H. R. (2019). Appearance of hyperons in neutron stars within LOCV method, *Annals Physics*, **402**, pp. 66–77.

Shapiro, I. I. (1964). Fourth test of general relativity, *Physical Review Letter*, **13**, pp. 789–791.

Shapiro, S. L. and Teukolsky, S. A. (1983). Black holes, white dwarfs, and neutron stars: The physics of compact objects, ISBN 978-0-471-87316-7.

Spinella, W. M. and Weber, F. (2020). Dense baryonic matter in the cores of neutron stars, in: Topics on strong gravity (World Scientific), ISBN 978-981-3277-33-5, pp. 85–152.

Stergioulas, N. (2003). Rapidly rotating neutron star code (RNS), *Living Reviews in Relativity*, **6**, Article number 3 (https://link.springer.com/article/10.12942/lrr-2003-3) (www.gravity.phys.uwm.edu/rns/).

Taninah, A., Agbemava, S. E., Afanasjev, A. V., and Ring, P. (2020). Parametric correlations in energy density functionals, *Physics Letter B*, **800**, p. 135065.

Thapa, V. B., Sinha, M., Li, J. J., and Sedrakian, A. (2020). Equation of state of strongly magnetized matter with hyperons and δ-resonances, *Particles*, **3**, pp. 660–675.

Thapa, V. B., Sinha, M., Li, J. J., and Sedrakian, A. (2021). Massive Δ-resonance admixed hypernuclear stars with antikaon condensations, *Physical Review D*, **103**, p. 063004.

Togashi, H., Hiyama, E., Yamamoto, Y., and Takano, M. (2016). Equation of state for neutron stars with hyperons by the variational method, *Physical Review C*, **93**, p. 035808.

Tolman, R. C. (1939). Static solutions of Einstein's field equations for spheres of fluid, *Physical Review*, **55**, pp. 364–373.

Tolos, L., Centelles, M., and Ramos, A. (2016). Equation of State for Nucleonic and Hyperonic Neutron Stars with Mass and Radius Constraints, *The Astrophysical Journal*, **834**(1), p. 3.

Torres, J. R., Gulminelli, F., and Menezes, D. P. (2017). Examination of strangeness instabilities and effects of strange meson couplings in dense strange hadronic matter and compact stars, *Physical Review C*, **95**, p. 025201.

Tsubakihara, K. and Ohnishi, A. (2013). Three-body couplings in RMF and its effects on hyperonic star equation of state, *Nuclear Physics A*, **914**, pp. 438–443.

Typel, S. (2018). Relativistic mean-field models with different parametrizations of density dependent couplings, *Particles*, **1**, pp. 3–22.

Typel, S., Röpke, G., Klähn, T., Blaschke, D., and Wolter, H. H. (2010). Composition and thermodynamics of nuclear matter with light clusters, *Physical Review C*, **81**, p. 015803.

Typel, S. and Wolter, H. H. (1999). Relativistic mean-field calculations with density-dependent meson-nucleon coupling, *Nuclear Physics A*, **656**(3-4), pp. 331–364.

van Dalen, E., Colucci, G., and Sedrakian, A. (2014). Constraining hypernuclear density functional with Λ-hypernuclei and compact stars, *Physics Letter B*, **734**, pp. 383–387.

Vidaña, I. (2018). Hyperons: The strange ingredients of the nuclear equation of state, *Procedings of Royal Society of London A*, **474**, p. 0145.

Viganò, D., Garcia-Garcia, A., Pons, J. A., Dehman, C., and Graber, V. (2021). Magneto-thermal evolution of neutron stars with coupled ohmic, hall and ambipolar effects via accurate finite-volume simulations, *Computer Physics Communications*, **265**, 108001.

Waldhauser, B. M., Maruhn, J. A., Stöcker, H., and Greiner, W. (1987). Sensitivity of pion production to the nuclear eos and to the delta's coupling constant, *Z. für Phys. A Atomic Nuclei*, **328**(1), pp. 19–22.

Waldhauser, B. M., Maruhn, J. A., Stöcker, H., and Greiner, W. (1988). Nuclear equation of state from the nonlinear relativistic mean-field theory, *Physical Review C*, **38**, pp. 1003–1009.

Weber, F. (1999). Pulsars as astrophysical laboratories for nuclear and particle physics, Series in High Energy Physics, Cosmology and Gravitation (Taylor & Francis), ISBN 9780750303323.

Weber, F. and Weigel, M. (1989a). Baryon composition and macroscopic properties of neutron stars, *Nuclear Physics A*, **505**, 3, pp. 779–822.

Weber, F. and Weigel, M. K. (1989b). Equation of state of dense baryonic matter, *Journal of Physics G*, **15**(6), pp. 765–792.

Wehrberger, K., Bedau, C., and Beck, F. (1989). Electromagnetic excitation of the delta-baryon in quantum hadrodynamics, *Nuclear Physics A*, **504**, pp. 797–817.

Weissenborn, S., Chatterjee, D., and Schaffner-Bielich, J. (2012a). Hyperons and massive neutron stars: the role of hyperon potentials, *Nuclear Physics A*, **881**, pp. 62–77.

Weissenborn, S., Chatterjee, D., and Schaffner-Bielich, J. (2012b). Hyperons and massive neutron stars: vector repulsion and SU(3) symmetry, *Physical Rcvicw C*, **85**, p. 065802.

Wu, X.-H., Wang, S.-B., Sedrakian, A., and Röpke, G. (2017). Composition of nuclear matter with light clusters and Bose-Einstein condensation of α particles, *Journal of Low Temperature Physics*, **189**, pp. 133–146.

Xu, J. (2019). Transport approaches for the description of intermediate-energy heavy-ion collisions, *Progress in Particle and Nuclear Physics*, **106**, pp. 312–359.

Yagi, K. and Yunes, N. (2017). Approximate universal relations for neutron stars and quark stars, *Physics Reports*, **681**, pp. 1–72.

Yamamoto, Y., Furumoto, T., Yasutake, N., and Rijken, T. A. (2014). Hyperon mixing and universal many-body repulsion in neutron stars, *Physical Review C*, **90**, p. 045805.

Zdunik, J. L. and Haensel, P. (2013). Maximum mass of neutron stars and strange neutron-star cores, *Astronomy & Astrophysics*, **551**, p. A61.

Zhang, N.-B. and Li, B.-A. (2019). Extracting nuclear symmetry energies at high densities from observations of neutron stars and gravitational waves, *European Physics Journal A*, **55**(3), p. 39.

Zhang, N.-B., Li, B.-A., and Xu, J. (2018). Combined constraints on the equation of state of dense neutron-rich matter from terrestrial nuclear experiments and observations of neutron stars, *The Astrophysical Journal*, **859**(2), p. 90.

Zhang, Z.-W. and Chen, L.-W. (2017). Low density nuclear matter with light clusters in a generalized nonlinear relativistic mean-field model, *Physical Review C*, **95**, p. 064330.

Zhang, Z.-W. and Chen, L.-W. (2019). Cold dilute nuclear matter with α-particle condensation in a generalized nonlinear relativistic mean-field model, *Physical Review C*, **100**, p. 054304.

Zhu, Z.-Y., Li, A., Hu, J.-N., and Sagawa, H. (2016). $\Delta(1232)$ effects in density-dependent relativistic HArtree-FOck theory and neutron stars, *Physical Review C*, **94**, p. 045803.

https://doi.org/10.1142/9789811220944_0006

Chapter 6

Learning from the Frequency Content of Continuous Gravitational Wave Signals

David Ian Jones
Mathematical Sciences and STAG Research Centre
University of Southampton, Southampton SO17 1BJ, UK
D.I.Jones@soton.ac.uk

Abstract: Spinning neutron stars can emit long-lived gravitational waves. There are several mechanisms that can produce such *continuous wave* emission. These mechanisms relate to the strains in the elastic crust, the star's magnetic field, superfluidity of the neutron fluid, and bulk oscillations of the entire star. In this chapter, we describe how the frequency content of the gravitational wave signal, and its relation to any electromagnetically observed spin frequency, can be used to constrain the mechanism producing the gravitational waves. These ideas will be of use in the event of the first detections of such signals, and help convert a detection into useful physical insight.

Keywords: neutron stars; gravitational waves.

1. Overview

Spinning neutron stars can emit gravitational waves, as long as their mass distribution is deformed away from axisymmetry in some way (Andersson, 2019; Misner *et al.*, 1973). Such gravitational wave signals could potentially last for very long periods of time. The spin-down timescales of known pulsars give us an indication of these timescales, ranging from thousands of year for young pulsars, through to billions of years for millisecond pulsars (MSPs) (see e.g., Lyne and Graham–Smith (2012)). Such durations are far longer than the duration of any gravitational wave observation, typically of length weeks to years, hence the designation of *continuous waves* (CWs).

The gravitational wave signal itself may however change on timescales much shorter than this, if, for instance, there is a sudden release of strain within the star. This interplay of long and (possibly) short timescales makes searching for CWs computationally and theoretically challenging.

There has nevertheless been a sustained and vigorous effort to detect such CWs. So far, no detections have been made, but upper limits have been put on such emission from both knows sources (pulsars and low-mass X-ray binaries), and from currently unknown sources. See the following for recent examples: Abbott *et al.* (2021, 2020, 2019a, 2019d).

There are several mechanisms that could produce such deformations. The two main scenarios are non-axisymmetries in rigidly rotating stars, colloquially referred to as *mountains*, and *oscillations* of the star; see (Glampedakis and Gualtieri, 2018) for a review. However, within these two cases are multiple possibilities, relating to free precession (Jones and Andersson, 2002; Zimmermann and Szedenits, 1979), multi-component stars (Abbott *et al.*, 2008), and superfluidity (Jones, 2010), that can determine the strength and frequency spectrum of the emission.

In this article, we look forward to the first detections, and ask the question as to what one can learn following such detections. The physics relating to mountains and fluid oscillations combines many interesting areas, including nuclear physics, fluid dynamics (including magnetohydrodynamics), elasticity, superfluidity, superconductivity, and, of course, general relativity; see Glampedakis and Gualtieri (2018). There is clearly a lot at stake, in terms of turning a detection into physical insight. This is a complex problem, and various facets of it can be found scattered over the extensive literature on CW emission. Here we narrow this question down a little, and reflect upon the fact that (in some cases) different emission mechanisms produce gravitational waves at different frequencies. We therefore pose and attempt to answer the following question:

> *In the event of the detection of a continuous gravitational wave source, to what extend can we use the frequency content of the signal to constrain the emission mechanism?*

Note that there are other candidate mechanisms for producing long-lived CW signals that have nothing to do with neutron stars, but these are of more speculative nature. These include emission from boson clouds outside of black holes (D'Antonio *et al.*, 2018), emission from low mass primordial

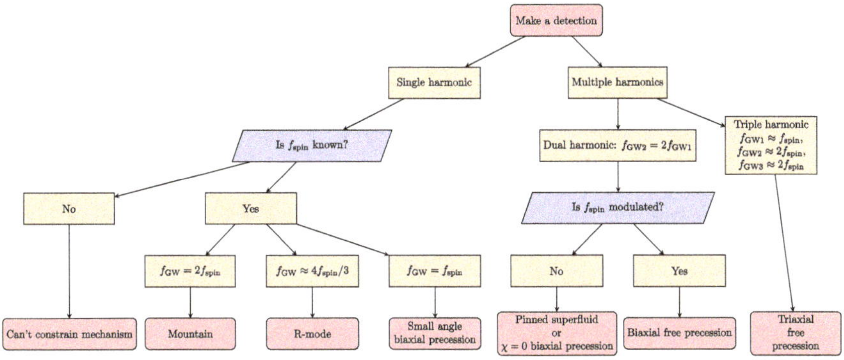

Fig. 1. Flow chart illustrating possible emission mechanisms, and how they relate the the frequency content of a CW detection.

black holes (Miller *et al.*, 2021), and the direct interaction of dark photon dark matter with the gravitational wave detectors (The LIGO Scientific Collaboration *et al.*, 2021). We will not consider such scenarios here.

The structure of this chapter is as follows. In Section 2, I give a brief review of the main gravitational wave mechanisms that I will consider, describing their frequency content, and how it relates to an electromagnetically observed spin frequency. In Section 3, I consider the scenario where a gravitational wave detection has been made, consisting of a single frequency, and describe what can be inferred, depending upon whether a radio pulsar spin frequency is know or not, and how that frequency compares with the gravitational wave frequency. In Section 4, I pose a similar question, but for the case where the are multiple frequency components in the gravitational wave signal. I briefly summarize in Section 5.

I have produced a flow chart to accompany this chapter; see Fig. 1. The reader is invited to follow this flow chart when reading this article, to help see the relations between the various scenarios.

2. Continuous Wave Emission Mechanisms

The simplest possible CW emission mechanism is that of a steadily and rigidly rotating neutron star with a non-axisymmetric mass distribution, i.e., a mountain (Shapiro and Teukolsky, 1983). For a star with angular velocity $\Omega_{\rm spin} = 2\pi f_{\rm spin}$, at distance r from Earth, with moment of inertia I_3 about the spin axis, and moments of inertia $I_1 \neq I_2$ about the two orthogonal axes, the gravitational wave amplitude can be parameterized by

$$h_0 = \frac{16\pi^2 G}{c^4} \frac{I_3 \epsilon_{21} f_{\text{spin}}^2}{r}, \tag{1}$$

where the asymmetry in the moment of inertia is parameterized by

$$\epsilon_{21} = \frac{I_2 - I_1}{I_3}, \tag{2}$$

a dimensionless quantity often known as the *ellipticity*.

A rotating perfect fluid star in equilibrium would have $\epsilon_{21} = 0$. Non-zero ellipticities require some sort of deforming force, either from elastic strain in the solid crust (and/or solid core, if such a component exists), or from magnetic strains (Glampedakis and Gualtieri, 2018). In either case, the gravitational wave emission is at twice the spin frequency, i.e., $f_{\text{GW}} = 2f_{\text{spin}}$. In the case of elastic strains, the ellipticity is proportional to the strain u in the crust, and is given approximately by (Ushomirsky *et al.*, 2000):

$$\epsilon_{21} \sim 10^{-7} \frac{u}{10^{-2}}. \tag{3}$$

In the case of magnetic deformations, the situation depends upon whether or not the interior proton fluid is in the *normal* or the *superconducting state*. If normal, the ellipticity is of order the magnetostatic energy divided by the gravitational binding energy:

$$\epsilon_{21} \sim 10^{-12} \left(\frac{B_{\text{int}}}{10^{12}\,\text{G}} \right)^2, \tag{4}$$

where B_{int} is the *internal* magnetic field strength; see e.g., Jones (2002); Lander and Jones (2009). However, a few hundred years after birth, a neutron star will have cooled sufficiently for its protons to go into the superconducting state, in which case the ellipticity instead scales linearly in the internal magnetic field strength:

$$\epsilon_{21} \sim 10^{-8} \frac{B_{\text{int}}}{10^{12}\,\text{G}}; \tag{5}$$

see (Lander, 2014; Lander *et al.*, 2012).

The next most commonly considered emission mechanism is from fluid oscillations, specifically *r-mode oscillations* (Provost *et al.*, 1981). The r-modes are a type of oscillation that only exist in rotating stars, with the restoring force being the Coriolis force. They are favoured over other modes as they are considered to be the most likely mode to undergo the Chandrasekhar–Friedman-Schutz instability, where gravitational radiation reaction acts to amplify the mode (Andersson, 1998; Lindblom *et al.*, 1998). The mode can be parameterized by a dimensionless amplitude α, so that the gravitational waves can be parameterized by (Owen, 2010; Owen *et al.*, 1998):

$$h_0 = \sqrt{\frac{8\pi}{5}} \frac{G}{c^5} \frac{1}{r} \alpha \omega^3 M R^3 \tilde{J}\,, \tag{6}$$

where ω is the gravitational wave (angular) frequency, M and R the mass and radius of the star, and $\tilde{J}$ a dimensionless constant hat depends upon stellar structure (see Owen *et al.* (1998) for details).

In the simplest approximation, where the star is modelled as a slowly rotating perfect fluid in Newtonian gravity, the oscillation frequency of the relevant r-mode is simply 4/3 times the spin frequency (Provost *et al.*, 1981). This is also the gravitational wave frequency, so we have $f_{\rm GW} = 4 f_{\rm spin}/3$. In more realistic treatments, other effects, principally rotation, elasticity and general relativity, combine to give a gravitational wave frequency slightly less than this (Caride *et al.*, 2019; Idrisy *et al.*, 2015; Lockitch *et al.*, 2003).

These are the two most commonly considered emission mechanisms. But there are others. If a steadily spinning star is perturbed, it can be set into *free precession*. In this case, even a biaxial ($I_1 = I_2 \neq I_3$) star will radiate. As first computed by Zimmermann and Szedenits (1979), there is then emission at two harmonically-related frequencies, which we can write as $\dot{\phi}$ and $2\dot{\phi}$:

$$h(\dot{\phi}) \sim \frac{2\dot{\phi}^2}{r} I_3 \epsilon_{31} \sin\theta \cos\theta, \tag{7}$$

$$h(2\dot{\phi}) \sim \frac{2\dot{\phi}^2}{r} I_3 \epsilon_{31} \sin^2\theta, \tag{8}$$

where it is now a different sort of ellipticity that sources the emission:

$$\epsilon_{31} \equiv \frac{I_3 - I_1}{I_3}. \tag{9}$$

Here, $\dot{\phi}$ is the angular velocity at which the star's 3-axis moves in a cone around the invariant angular momentum axis $\mathbf{J}$, and θ is the opening angle of this cone, sometimes known as the *wobble angle*; see Jones and Andersson (2001) for a discussion. We have stopped short of writing down proper equalities as we have not included various inclination angle factors that determine the polarisation content, as they are not important for our purposes (see Zimmermann and Szedenits (1979) for the full equalities).

In this case, if the star is also visible as a radio pulsar, the relation between the frequency of the radio pulsations and the gravitational wave frequencies $(\dot{\phi}, 2\dot{\phi})$ is slightly complicated, as described in Jones and Andersson (2002, 2001). The relation depends upon χ, the angle between the star's 3-axis and the "magnetic" axis $\mathbf{m}$ along which the pulsar beam is emitted. If $\chi = 0$, the pulsar beam simply follows the 3-axis, and the radio pulsations are emitted at the constant $\dot{\phi}$ frequency. However, if $\chi \neq 0$, the situation is more complicated. This is because the free precession of a biaxial body consists not just of the motion of the 3-axis about $\mathbf{J}$; there is a slow superimposed rotation of the body about the (instantaneous position of the) 3-axis at a rate $\dot{\psi}$, known as the "body frame precession frequency" or simply the precession frequency. This precession frequency will be small in the realistic case of a nearly spherical ($|\epsilon_{31}| \ll 1$) star:

$$\dot{\psi} = -\epsilon_{31}\dot{\psi}\cos\theta. \tag{10}$$

This produces a rather complex motion of the magnetic axis $\mathbf{m}$, as its position is determined by the sum of these two rotations: the fast rotation of the 3-axis about $\mathbf{J}$ at angular velocity $\dot{\phi}$, and the slow superimposed rotation of $\mathbf{m}$ about the 3-axis at the rate $\dot{\psi}$. The observed frequency of the radio pulsations is then modulated once per free precession period $P_{\rm fp} = 2\pi/\dot{\psi}$, as is the observer's latitudinal cut through the pulsar beam. Importantly for our purposes, the time-averaged radio pulsation frequency $\langle\nu_{\rm radio}\rangle$ will in general differ from the (constant) gravitational wave frequency $\dot{\phi}$, according to (Jones and Andersson, 2001):

$$2\pi\langle\nu_{\rm radio}\rangle = \dot{\phi} + \dot{\psi} \quad \text{for} \quad \theta < \chi, \tag{11}$$

$$2\pi\langle\nu_{\rm radio}\rangle = \dot{\phi} \quad \text{for} \quad \theta > \chi. \tag{12}$$

This will play a role in the discussion below.

All of this applies to the free precession of a biaxial star. In the more general case of a triaxial star $I_1 \neq I_2 \neq I_3$ the situation is more complex still, with both the radio pulsar and the gravitational wave frequencies being functions of time (Zimmermann, 1980). In the case of the small angle precession of a nearly spherical star, there is gravitational wave emission at three frequencies, which can be loosely identified with the $\dot{\phi}$, $2\dot{\phi}$ and $2\pi\langle\nu_{\rm radio}\rangle$ frequencies above (Van Den Broeck, 2005; Zimmermann, 1980).

As described above, the free precession of a star leads to the emission of more than one gravitational wave harmonic. Also, in general, it produces modulation of the radio pulsar frequency. The lack of observations of modulation in radio pulsar emission suggests that free precession is, at best, a rare phenomenon, which is why relative few gravitational wave searches have looked for such dual harmonic signals.

There are two special cases where there is no modulation in the radio emission, but there is nevertheless dual harmonic gravitational wave emission. The first was mentioned above: if the pulsar beam of a biaxial precessing star points exactly along the symmetry axis of the body, and if additionally the beam itself is symmetric about this axis, there will be no observed modulation. This is clearly quite a special case.

There is however one other scenario, where a steadily rotating star, with constant radio pulsation frequency, can produce dual harmonic gravitational waves. It is thought that all but the very youngest of neutron stars will be cool enough for some component of their neutron to be in a superfluid state (Chamel and Haensel, 2008). Such a superfluid rotates by forming an array of vortices, which can "pin" to the solid crust, preventing the associate angular momentum vector from moving relative to the crust. The effect is as if a spinning gyroscope has been sewn into the crust (Shaham, 1977). If this vortex array is misaligned with the principle axes of the crust's moment of inertia tensor, the non-precessing state of the star is such that, even though the rotation is steady, gravitational waves are produced at two harmonics, $f_{\rm spin}$ and $2f_{\rm spin}$ (Jones, 2010).

In recognition of these two emission modes, a number of gravitational wave searches have been performed looking for precisely this dual harmonic emission, even in known pulsars with no sign of precession in their radio emission; see Abbott *et al.* (2019c) for a recent example.

Another simpler but more speculative scenario can apply if the star has a solid core. Such a core might consist of a solid quark phase, and have a very high shear modulus (Glampedakis *et al.*, 2012; Owen, 2010). If it is this core that is deformed, rather than the elastic crust, it will be the rotation rate of the core rather than the crust that fixed the gravitational wave frequency. In principle, the two may rotate at different rates, thus opening up a new degree of freedom in the frequency spectrum (Abbott *et al.*, 2008).

In recognition of the fact that both free precession and independently spinning deformed cores might induce a small off-set between the observed pulsar spin frequency and (a component of) the gravitational weave frequency, a number of so-called *narrow band* gravitational wave searches have been performed; see Abbott *et al.* (2019b) for a recent example.

3. Detection of a Single GW Harmonic

A CW detection might reveal the presence of just one gravitational wave component ("single harmonic"), or it may consist of multiple frequency components ("multi harmonic"). Let us consider the single harmonic case first.

3.1. *Single GW harmonic: $f_{\rm spin}$ unknown*

Suppose a CW detection consists of just a single frequency component, and there is no associated electromagnetic counterpart, so we have no knowledge of $f_{\rm spin}$. This is the worst-case scenario, in terms of learning about the emission mechanism. We may have a mountain, or an r-mode. In the mountain case, there is no way of inferring if it is of elastic or magnetic nature.

There is one possible avenue for progress here, if the evolution in gravitational wave frequency of the spinning down star is fast enough for both the first and second time derivatives of $f_{\rm GW}$ to be measured. In this case, assuming that the gravitational wave frequency is proportional to the spin frequency, one can take the definition of the *braking index*, written in terms of the spin frequency:

$$n \equiv \frac{f_{\rm spin}\ddot{f}_{\rm spin}}{\dot{f}_{\rm spin}^2}, \tag{13}$$

and evaluate it making use of the observed values of $f_{\rm GW}, \dot{f}_{\rm GW}, \ddot{f}_{\rm GW}$ in place of $f_{\rm spin}, \dot{f}_{\rm spin}, \ddot{f}_{\rm spin}$.

The key consideration is this. For a star spinning down entirely due to gravitational wave emission from a mountain of constant ellipticity ϵ_{21}, we have $n = 5$. Similarly, for a star spinning down entirely due to gravitational wave emission from an r-mode of constant amplitude α, we have $n = 7$ (Glampedakis and Gualtieri, 2018). It follows that a measured value of n close to one or other of these two values is strongly suggestive of (but does not prove) emission from the corresponding mechanism.

Furthermore, in such a case, Sieniawska and Jones (2021) have shown in detail that one can go some way to breaking the three-way degeneracy between the quantities (I_3, r, ϵ_{21}) of Eq. (1), or between the quantities (I_3, r, α) of Eq. (6), to infer combinations of only two of these quantities. Given that I_3 is known to within a factor of two or so from the modelling on neutron star structure for realistic equations-of-state (EoS) (see discussion in Woan *et al.* (2018), this allows estimation of ϵ_{21} or α to within a factor of $\sqrt{2}$ or so. The distance d can be similarly constrained, hence the "standard siren" analysis of Sieniawska and Jones (2021). In this way, an estimate of ϵ_{21} or α can be obtained. However, in the mountain case, there is still no way of distinguishing between elastically and magnetically supported deformation.

There is one further possibility. As described above, in general biaxial free precession leads to emission at a pair of harmonically-related frequencies, which we denoted as $\dot{\phi}$ and $2\dot{\phi}$; see Eqs. (7) and (8). However, in the (possibly most plausible) case of small wobble angle free precession, i.e., $\theta \ll 1$, the amplitudes of the two components scale as θ and θ^2, respectively; see Eqs. (7) and (8), i.e., the emission at the lower harmonic dominates. In such a case only a single component $(h_0(\dot{\phi}))$ might be detected. So, we need to add small angle free precession of a biaxial star to our list of possibilities.

3.2. *Single GW harmonic: $f_{\rm spin}$ known*

In the event of a single harmonic CW detection where $f_{\rm spin}$ is also known, one can clearly use the relative values of $f_{\rm GW}$ and $f_{\rm spin}$ to make progress. We will consider the most obvious possibilities.

3.2.1. *Single GW harmonic: $f_{\rm spin}$ known: $f_{\rm GW} \approx 2 f_{\rm spin}$*

In the case $f_{\rm GW} \approx 2 f_{\rm spin}$ we clearly have emission from a mountain. If $f_{\rm GW} = 2 f_{\rm spin}$ exactly, or at least to within the accuracy of the gravitational wave measurement, so would infer that the gravitational wave-producing deformation is tightly locked to the radio pulsation mechanism. This is in fact what seems most likely from a theoretical standpoint. The radio pulsations are almost certainly produced by particle acceleration in the (external) magnetic field; see e.g., Lyne and Graham–Smith (2012). The magnetic field itself is likely to remain tightly locked to the neutron star, as neutron star matter is thought to be of sufficiently high electrical conductivity that diffusive processes would act on timescales long compared with gravitational wave observations (see e.g., Goldreich and Reisenegger (1992)).

Given this expectation, a small offset between $f_{\rm GW}$ and $f_{\rm spin}$ would be of great interest. We can parameterize this offset in terms of a dimensionless quantity δ:

$$f_{\rm GW} = 2f_{\rm spin}(1+\delta), \tag{14}$$

where $f_{\rm GW}/2$ is to be identified with the spin of the solid core, and $f_{\rm spin}$ with the spin of the crust.

A small positive value of δ could be interpreted as gravitational wave emission from a spinning down sold core, that is imperfectly coupled to the rest of the star (Abbott *et al.*, 2008). The picture would be that the external electromagnetic spin-down torque acts directly on the crust, and whatever is coupled tightly to the crust, but not directly on the solid core. We will denote the moments of inertia of these two components as $I_{\rm crust}$ and $I_{\rm core}$, so that the total is $I = I_{\rm crust} + I_{\rm core}$. The coupling torque between the crust and core may be of frictional nature, acting so as to restore co-rotation between the two on timescales short compared with the spin-down timescale, with a magnitude proportional to the spin frequency difference. We could then write the coupling torque acting on the crust as

$$T_{\rm coupling} = -A(f_{\rm crust} - f_{\rm core}), \tag{15}$$

where A is some positive constant that parameterizes the strength of the coupling torque. There will be an equal but opposite coupling torque $-T_{\rm coupling}$ acting on the core. In such a situation, if the spin-down torque due to the gravitational wave emission were to be neglected, one can easily show that the fractional frequency off-set between the crust and core is given by the ratio of the crust-core coupling timescale to the spin-down timescale:

$$\frac{f_{\rm GW}/2 - f_{\rm spin}}{f_{\rm spin}} = \delta \approx \frac{\tau_{\rm coupling}}{\tau_{\rm spin-down}} \frac{I}{I_{\rm crust}}. \tag{16}$$

In this way, one could infer not only the existence of the sold core, but also the strength of its coupling with the rest of the star.

A variation on this scenario would be that the angular momentum exchange between the spinning down crust and solid core is not perfectly smooth, but rather has a random component to it, that might be related to the seemingly ubiquitous phenomenon of pulsar *timing noise* (Lyne and Graham–Smith, 2012). In this case, small upward fluctuations in $f_{\rm spin}$ would, by angular momentum conservation, be accompanied by small downward fluctuations in $f_{\rm GW}$ (and *vice versa*), with a constant of proportionality related to the relative moments of inertia of the two components (Jones, 2004). Identification of such a correlation by comparison of the electromagnetic and gravitational wave data streams would provide a novel insight into the interior physics of neutron stars.

Alternatively, a small but non-zero value of δ could be interpreted as due to free precession (Abbott *et al.*, 2008; Jones and Andersson, 2002). However, in this case, emission at other gravitational harmonics is to be expected, and the radio pulsar frequency should be modulated in time. The absence of such features would favour the solid core interpretation.

Regardless of these issues, it is of interest to ask if there is any way of discriminating between elastically and magnetically supported mountains. This is a tricky degeneracy to break. One possibility applies if the distance to the source is known, say from electromagnetic means (e.g., radio pulsar dispersion measure (Lyne and Graham–Smith, 2012)). In that case, the value of the ellipticity ϵ_{21} can be estimated (again, only up the uncertainty in I_3, about a factor of 2), and the total contribution of gravitational wave emission to the total spin-down rate inferred. If this is small, so that most of the spin-down is from electromagnetic emissions, one can use the standard magnetic dipole formula to estimate the strength of the *external* magnetic field $B_{\rm ext}$ (Shapiro and Teukolsky, 1983). This could be compared with the strength that would be inferred from the *internal* magnetic field $B_{\rm int}$, required to support the mountain, if of magnetic origin, as per Eqs. (4) or (5) above. One could then compute the ratio

$$r_{\rm mag} \equiv \frac{B_{\rm ext}}{B_{\rm int}} \,. \tag{17}$$

The actual value of this ratio in neutron stars is not known, but modelling suggests that the internal field is likely to be somewhat stronger than the external one, see e.g., Lander and Jones (2009). It follows that a value of $r_{\rm mag}$ greater than unity, or a value much less than unity, would make the magnetic mountain interpretation less plausible, favouring the elastic strain interpretation.

3.2.2. *Single GW harmonic:* $f_{\rm spin}$ *known:* $f_{\rm GW} \approx 4f_{\rm spin}/3$

In the case of a single gravitational wave component with $f_{\rm GW} \approx 4f_{\rm spin}/3$ we clearly have r-mode emission. In this case the departure of $f_{\rm GW}$ from $4f_{\rm spin}/3$ encodes useful information, as is discussed in some detail in Caride *et al.* (2019), Idrisy *et al.* (2015), and Lockitch *et al.* (2003). The largest contribution to this frequency shift is likely to be due to the effects of general relativity, and to a good approximation allows one to infer the star's compactness M/R. However, care must be taken when making this argument. There are likely to be narrow bands in $f_{\rm spin}$ where there is a near resonance between the r-mode frequency and the frequencies of other normal modes, and this can have an even larger effect of the r-mode (and hence gravitational wave) frequency than the relativistic effects. This has been shown explicitly for the resonance of the r-mode with torsional oscillations of the elastic crust (Levin and Ushomirsky, 2001). Resonances with superfluid modes can also play a role, and may be important in determining the spin frequencies of some accreting neutron stars (Kantor *et al.*, 2016).

3.2.3. *Single GW harmonic: $f_{\rm spin}$ known: $f_{\rm GW} \approx f_{\rm spin}$*

In the case $f_{\rm GW} \approx f_{\rm spin}$ neither mountains nor r-modes can be at work. As described in Section 3.1, the natural interpretation would be small-angle free precession of a biaxial body, i.e., the $\theta \ll 1$ case of Eqs. (7) and (8). Given that $\theta \ll 1$, we would presumably be in the regime where $\theta < \chi$, where χ is the angle between the biaxial body's symmetry axis and the magnetic dipole axis $\mathbf{m}$. In this case, Eq. (11) applies, and the fractional difference between $f_{\rm GW}$ ($= \dot{\phi}/(2\pi)$ in the notation of Section 2), and $f_{\rm spin}$ can be used to infer the relevant ellipticity ϵ_{31}, as follows from combining Eqs. (10) and (11) to give:

$$\epsilon_{31} \approx \frac{f_{\rm GW} - f_{\rm spin}}{f_{\rm GW}}, \tag{18}$$

where we have used the small θ, small ϵ_{31} approximations.

4. Detection of Multiple GW Harmonics

In the case of emission at multiple gravitational wave harmonics, it is easier to make progress, even in the absence of an observed pulsar spin frequency. We will label the frequencies of the gravitational wave components as $f_{\rm GW1}, f_{\rm GW2}, \ldots$.

4.1. *Dual GW harmonic: $f_{\rm GW1}$, $f_{\rm GW2} = 2f_{\rm GW1}$*

In the case of dual harmonic emission, with the upper harmonic at twice the frequency of the lower ($f_{\rm GW2} = 2f_{\rm GW1}$), biaxial free precession is indicated, as described by Eqs. (7) and (8) (Jones and Andersson, 2002; Zimmermann and Szedenits, 1979). The relative amplitude of the harmonics contains information on the wobble angle θ. In the absence of a measurement of $f_{\rm spin}$, one cannot go further. However, if $f_{\rm spin}$ is measured, then one can draw different conclusions depending upon whether it is modulated in time or not.

4.1.1. *Dual GW harmonic: $f_{\rm GW1} = f_{\rm spin}$, $f_{\rm GW2} = 2f_{\rm spin}$, with $f_{\rm spin}$ not modulated*

If the radio pulsation is measured to be steady, and equal to the lower of the two (harmonically-related) gravitational wave frequencies, then there are two possibilities. The first is that biaxial free precession is taking place, with $\chi = 0$, i.e., the radio emission is taking place along the symmetry axis of the biaxial body, and so the magnetic axis is simply carried around at the frequency $\dot{\phi}$ that appears in Eqs. (7) and (8) (Jones and Andersson, 2001). The second possibility is that superfluid pinning is at work, misaligned with the principle axes of the star's moment of inertia tensor, giving a perfectly steady $f_{\rm spin}$, but nevertheless emitting gravitational waves at both $f_{\rm spin}$ and $2f_{\rm spin}$ (Jones, 2010).

4.1.2. *Dual GW harmonic: $f_{\rm GW1} = f_{\rm spin}$, $f_{\rm GW2} = 2f_{\rm spin}$, with $f_{\rm spin}$ modulated*

In the case of emission at two harmonically related frequencies, if the radio pulsation frequency is modulated, free precession of a biaxial body with $\chi \neq 0$ is indicated. The time-averaged radio frequency is then given by Eq. (11) or (12) above, depending upon the relative values of θ and χ. Either way, the observed periodicity of the modulations in $f_{\rm spin}$ gives the free precession frequency $P_{\rm fp}$, which in turn can be used to estimate the degree of biaxiality ϵ_{31} via Eq. (10).

4.2. *Triple GW harmonic: $f_{\rm GW1} \approx f_{\rm spin}$, $f_{\rm GW2} \approx 2f_{\rm spin}$, $f_{\rm GW3} \approx 2f_{\rm spin}$*

In the case of gravitational wave emission from a triplet of frequencies, of the approximate form $(f_{\rm spin}, 2f_{\rm spin}, 2f_{\rm spin})$, small angle ($\theta \ll 1$) free precession of a triaxial star is indicated, for a star with a small degree of biaxiality ($|\epsilon_{31}| \ll 1$) and and even smaller degree of triaxiality ($|\epsilon_{21}| \ll |\epsilon_{21}|$); see Zimmermann (1980) and Van Den Broeck (2005). In this case, the emission is close to a sum of what one would expect from biaxial precession, plus a mountain, with f_{GW1} and (say) f_{GW2} sourced mainly by ϵ_{31}, and f_{GW3} sourced by ϵ_{21}. In this case, the relative amplitudes of the three harmonics, and their separation in frequency, would give information on the degree of triaxiality; see Zimmermann (1980) and Van Den Broeck (2005) for discussion.

5. Summary and Reflections

Clearly, there are many different scenarios when it comes to the frequency content of a CW signal. I have sketched out here what I believe to be some of the more plausible possibilities, and attempted to indicate how we can use this information to draw conclusions about some parameters of the star. I hope that in the event of an eventual, and, let us hope not far off, first detection, one or other of these many possibilities can be identified, allowing us to extract as much information about the physics of the neutron star as possible.

Acknowledgements

I would like to acknowledge many useful conversations with colleagues in the Southampton Gravity Group, the LIGO Scientific Collaboration, and the Virgo Collaboration, on many of the ideas discussed here, including comments on this chapter from Bryn Haskell. I also acknowledge financial support from the Science and Technology Facilities Council (STFC, UK) via grant no. ST/R00045X/1.

References

Abbott, R. and 1371 colleagues. (2021). All-sky search in early O3 LIGO data for continuous gravitational-wave signals from unknown neutron stars in binary systems, *Physical Review D*, **103**, p. 064017.

Abbott, R. and 1338 colleagues. (2020). Gravitational-wave constraints on the equatorial ellipticity of millisecond pulsars, *The Astrophysical Journal Letters*, **902**(1), p. L21.

Abbott, B. P. and 1136 colleagues. (2019). Search for gravitational waves from scorpius X-1 in the second advanced LIGO observing run with an improved hidden markov model, *Physical Review D*, **100**, p. 122002.

Abbott, B. P. and 1153 colleagues. (2019). Searches for gravitational waves from known pulsars at two harmonics in 2015-2017 LIGO data, *The Astrophysical Journal*, **879**(1), p. 10.

Abbott, B. P. and 1137 colleagues. (2019). Narrow-band search for gravitational waves from known pulsars using the second LIGO observing run, *Physical Review D*, **99**, p. 122002.

Abbott, B. P. and 1132 colleagues. (2019). Searches for continuous gravitational waves from 15 supernova remnants and fomalhaut b with advanced LIGO, *The Astrophysical Journal*, **875**(2), p. 122.

Abbott, B. and 443 colleagues. (2008). Beating the spin-down limit on gravitational wave emission from the crab pulsar, *The Astrophysical Journal*, **683**, p. L45.

Andersson, N. (2019). *Gravitational-Wave Astronomy: Exploring the Dark Side of the Universe*, Oxford Scholarship Series, Oxford University Press, Oxford, UK. ISBN: 9780198568032, 688pp.

Andersson, N. (1998). A new class of unstable modes of rotating relativistic stars, *The Astrophysical Journal*, **502**, p. 708.

Caride, S., Inta, R., Owen, B. J., and Rajbhandari, B. (2019). How to search for gravitational waves from r-modes of known pulsars, *Physical Review D*, **100**, p. 064013.

Chamel, N. and Haensel, P. (2008). Physics of neutron star crusts, *Living Reviews in Relativity*, **11**(10).

D'Antonio, S. and 11 colleagues. (2018). Semicoherent analysis method to search for continuous gravitational waves emitted by ultralight boson clouds around spinning black holes, *Physical Review D*, **98**, p. 103017.

Glampedakis, K. and Gualtieri, L. (2018). *Gravitational Waves from Single Neutron Stars: An Advanced Detector Era Survey*, in *The Physics and Astrophysics of Neutron Stars* (Astrophysics and Space Science Library 673). Luciano Rezzolla, Pierre Pizzochero, David Ian Jones, Nanda Rea, Isaac Vidaña (editors), Springer Nature, Cham, Switzerland.

Glampedakis, K., Jones, D. I., and Samuelsson, L. (2012). Gravitational waves from color-magnetic "mountains" in neutron stars, *Physical Review Letters*, **109**, p. 081103.

Goldreich P., and Reisenegger, A. (1992). Magnetic field decay in isolated neutron stars, *The Astrophysical Journal*, **395**, p. 250.

Idrisy, A., Owen, B. J., and Jones, D. I. (2015). R-mode frequencies of slowly rotating relativistic neutron stars with realistic equations of state, *Physical Review D*, **91**, p. 024001.

Jones, D. I. (2010). Gravitational wave emission from rotating superfluid neutron stars, *Monthly Notices of the Royal Astronomical Society*, **402**(4), p. 2503.

Jones, D. I. (2004). Is timing noise important in the gravitational wave detection of neutron stars?, *Physical Review D*, **70**, p. 042002.

Jones, D. I. (2002). Gravitational waves from rotating strained neutron stars, *Classical and Quantum Gravity*, **19**(7), p. 1255.

Jones, D. I. and Andersson, N. (2002). Gravitational waves from freely precessing neutron stars, *Monthly Notices of the Royal Astronomical Society*, **331**(1), p. 203.

Jones, D. I. and Andersson, N. (2001). Freely precessing neutron stars: model and observations, *Monthly Notices of the Royal Astronomical Society*, **324**(4), p. 811.

Kantor, E. M., Gusakov, M. E., and Chugunov, A. I. (2016). Observational signatures of neutron stars in low-mass X-ray binaries climbing a stability peak, *Monthly Notices of the Royal Astronomical Society*, **455**, p. 739.
Lander, S. K. (2014). The contrasting magnetic fields of superconducting pulsars and magnetars, *Monthly Notices of the Royal Astronomical Society*, **437**, p. 424.
Lander, S. K., Andersson N., and Glampedakis, K. (2012). Magnetic neutron star equilibria with stratification and type II superconductivity, *Monthly Notices of the Royal Astronomical Society*, **419**, p. 732.
Lander, S. K. and Jones, D. I. (2009). Magnetic fields in axisymmetric neutron stars, *Monthly Notices of the Royal Astronomical Society*, **395**, p. 2162.
Levin, Y. and Ushomirsky, G. (2001). Crust-core coupling and r-mode damping in neutron stars: A toy model, *Monthly Notices of the Royal Astronomical Society*, **324**, p. 917.
The LIGO Scientific Collaboration and 1629 colleagues. (2021). Constraints on Dark Photon Dark Matter Using Data from LIGO's and Virgo's Third Observing Run. (arXiv e-prints).
Lindblom, L., Owen, B. J., and Morsink, S. M. (1998). Gravitational radiation instability in hot young neutron stars, *Physical Review Letters*, **80**, p. 4843.
Lockitch, K. H., Friedman, J. L., and Andersson, N. (2003). Rotational modes of relativistic stars: numerical results, *Physical Review D*, **68**, p. 124010.
Lyne, A. and Graham–Smith, F. (2012). *Pulsar Astronomy*, Cambridge University Press, Cambridge, UK.
Miller, A. L., Clesse, S., De Lillo, F., Bruno, G., Depasse, A., and Tanasijczuk, A. (2021). Probing planetary-mass primordial black holes with continuous gravitational waves, *Physics of the Dark Universe*, **32**, p. 100836.
Misner, C. W., Thorne, K. S., and Wheeler, J. A. (1973). *Gravitation*, W.H. Freeman and Co., San Francisco, USA.
Owen, B. J. (2010). How to adapt broad-band gravitational-wave searches for r-modes, *Physical Review D*, **82**, p. 104002.
Owen, B. J., Lindblom, L., Cutler, C., Schutz, B. F., Vecchio, A., and Andersson, N. (1998). Gravitational waves from hot young rapidly rotating neutron stars, *Physical Review D*, **58**, p. 084020.
Provost, J., Berthomieu, G., Rocca, A. (1981). Low frequency oscillations of a slowly rotating star — quasi toroidal modes, *Astronomy and Astrophysics*, **94**, p. 126.
Shaham, J. (1977). Free precession of neutron stars: role of possible vortex pinning, *The Astrophysical Journal*, **214**, p. 251.
Shapiro, S. L. and Teukolsky, S. A. (1983). *Black Holes, White Dwarfs, and Neutron Stars : The physics of compact objects*, Wiley, New York, USA.
Sieniawska, M. and Jones, D. I. (2021). Gravitational waves from spinning neutron stars as not-quite-standard Sirens, (To be submitted).
Ushomirsky, G., Cutler, C., and Bildsten, L. (2000). Deformations of accreting neutron star crusts and gravitational wave emission, *Monthly Notices of the Royal Astronomical Society*, **319**, p. 902.

Van Den Broeck, C. (2005). The gravitational wave spectrum of non-axisymmetric, freely precessing neutron stars, *Classical and Quantum Gravity*, **22**, p. 1825.

Woan, G., Pitkin, M. D., Haskell, B., Jones, D. I., and Lasky, P. D. (2018). Evidence for a minimum ellipticity in millisecond pulsars, *The Astrophysical Journal*, **863**, p. L40 .

Zimmermann, M. and Szedenits, E. (1979). Gravitational waves from rotating and precessing rigid bodies: simple models and applications to pulsars, *Physical Review D*, **20**, p. 351.

Zimmermann, M. (1980). Gravitational waves from rotating and precessing rigid bodies. II. general solutions and computationally useful formulas, *Physical Review D*, **21**, p. 891.

https://doi.org/10.1142/9789811220944_0007

Chapter 7

Insights Into the Physics of Neutron Star Interiors from Pulsar Glitches

Marco Antonelli[*,‡], Alessandro Montoli[†,§] and Pierre M. Pizzochero[†,¶]

[*]*Nicolaus Copernicus Astronomical Center of the Polish Academy of Sciences,*
ul. Bartycka 18, 00-716 Warszawa, Poland
CNRS/in2p3 and LPC-Caen, 1400 Caen, France
[†]*Dipartimento di Fisica*
Università degli Studi di Milano
Via Celoria 16, 20133 Milano, Italy
Istituto Nazionale di Fisica Nucleare
sezione di Milano, Via Celoria 16, 20133 Milano, Italy
[‡]*mantonelli@camk.edu.pl*
[§]*alessandro.montoli@unimi.it,*
[¶]*pierre.pizzochero@mi.infn.it*

Abstract: The presence of superfluid phases in the interior of a neutron star affects its dynamics, as it allows for neutrons to flow relative to the non-superfluid (normal) components of the star with little or no viscosity. A probe of superfluidity comes from pulsar glitches, sudden jumps in the observed rotational period of radio pulsars. Most models of glitches build on the idea that a superfluid component of the star is decoupled from the spin-down of the normal component, and its sudden re-coupling leads to a glitch. This transition in the strength of the hydrodynamic coupling is explained in terms of quantum vortices (long-lived vortices that are naturally present in the neutron superfluid at the microscopic scale). After introducing some basic ideas, we derive (as a pedagogical exercise) the formal scheme shared by many glitch studies. Then, we apply these notions to present some progress in understanding and modelling pulsar glitches, notably the possibility to constrain the nuclear physics input and the extraction of the coupling strength from observations of the elusive spin-up phase of a glitch.

Keywords: stars: neutron; stars: rotation; pulsars: general; dense matter; superfluidity.

1. Introduction

This chapter provides an introduction to the current understanding and theoretical modelling of the glitch phenomenon, a sudden change in the stable rotation period of pulsars. The field is far from being settled, as most of the results are qualitative and based on (Newtonian) phenomenological models. Nonetheless, the glitch phenomenon is probably the most clear observational evidence for the existence of an extended (kilometre-sized) superfluid region in the inner layers of a neutron star (NS). Nucleon superfluidity in the core of NSs has found additional support from the real time monitoring of the rapid cooling of the young (300-year-old) NS in Cassiopeia A (Page *et al.*, 2011; Shternin *et al.*, 2011), but this observational result and its interpretation still have to be firmly assessed (Posselt and Pavlov, 2018; Wijngaarden *et al.*, 2019).

Pulsar glitches are sudden increases in the rotation frequency of a spinning down NS (Espinoza *et al.*, 2011; Fuentes *et al.*, 2017; Lyne *et al.*, 2000; Manchester, 2018). Soon after the first glitch observations in the Vela pulsar (Radhakrishnan and Manchester, 1969; Reichley and Downs, 1969), Baym *et al.* (1969) associated the long post-glitch relaxation timescales of order days to months to the presence of a superfluid component in the internal layers.

The existence of superfluids in NSs was conjectured even before the discovery of pulsars. Neutron superfluidity is theoretically expected in NS interiors as most of the star will be cold enough for neutrons to undergo Cooper pairing (Sauls, 1989; Sedrakian and Clark, 2019). Depending on the age of the NS, the interior temperatures are in the range 10^7–10^9K, which are low on the scale defined by the typical Fermi energies of about 10–100MeV (equivalent to 10^{11}–10^{12}K), so that the ratio of temperature to Fermi energy in a NS is comparable to that of laboratory superfluid ^{3}He in the milli-kelvin temperature range.

For what concerns the glitch phenomenon, the presence of a neutron superfluid significantly modifies the internal dynamics of a NS, allowing for the superfluid to flow relative to the "normal" component. This relative flow determines the angular momentum reservoir needed to explain the observed jumps of the rotation period.

1.1. *An analogy with type-II superconductors: pining and creep*

To date, the most promising explanation of the glitch mechanism is based on the seminal idea of Anderson and Itoh (1975) and is built on the analogy between the superfluid current of dripped neutrons in the inner crust of a NS and the electron current in a type-II superconductor (Anderson and Kim, 1964). In fact, a type-II superconductor immersed in a magnetic field can sustain electron currents in a non-dissipative state as long as the vortices carrying a quantum of magnetic flux are anchored to the crystalline structure of the specimen — i.e., the quantized flux-tubes are "pinned". On the contrary, the superconductor enters into a resistive state — so that the current is dissipated unless some external voltage is applied — when the flux-tubes unpin and start to move (Blatter *et al.*, 1994).

Similarly, in a NS a current of neutrons can flow without resistance as long as the quantized vortex lines, that are naturally present in a rotating superfluid, find energetically favourable to pin to some impurity in the medium. In the non-homogeneous environment of the inner crust the pinning impurities may be the ions that constitute the crustal lattice (Alpar, 1977; Bulgac *et al.*, 2013; Donati and Pizzochero, 2004; Seveso *et al.*, 2016). If the proton fluid in the outer core forms a type-II superconductor (which is the still debated picture established in the seminal work of Baym *et al.* (1969)), pinning of neutron vortices to the quantized flux-tubes may be a viable possibility as well (see, e.g., Alpar, 2017; Sourie and Chamel, 2021, and references therein). However, the pinning in the outer core is, possibly, even less studied than the already difficult issue of pinning in the inner crust. In fact, the nature of the proton superconductor may be more complex than that of a usual type-II (Haber and Schmitt, 2017; Wood *et al.*, 2020), and also the features of the vortices threading the spin-triplet neutron superfluid expected in most of the outer core can differ from the simpler spin-singlet Cooper pairing case realized in the crust (Brand and Pleiner, 1981; Leinson, 2020).

Only when the quantized vortices unpin, the neutron superfluid and the rest of the star can interact at the hydrodynamic level via a dissipative force (the so-called *vortex-mediated mutual friction*), which tends to drive the system towards a zero current equilibrium state (Andersson *et al.*, 2006; Gavassino *et al.*, 2021; Mendell, 1991).

To complete the above qualitative picture, it is necessary to provide a reasonable argument for how a relative current of neutrons should develop in the first place. The seminal Anderson's idea is based on the fact that a rotating superfluid can spin down only by expelling part of its quantized vorticity. However, if the vortex lines are somehow pinned to the nuclear clusters in the crust, their natural outward motion is hindered and the superfluid cannot exactly follow the spin-down of the crust (Pines and Alpar, 1985). In this way, the superfluid component naturally tends to rotate faster than the spinning-down normal component, storing an excess of angular momentum. This reservoir of momentum can then be suddenly released during a glitch when several vortex lines ($\sim 10^{12}$ out of $\sim 10^{18}$ in a typical pulsar) are expelled from the superfluid domain.[1] Clearly, the same set of ideas applies also in reverse to explain anti-glitches in spinning-up pulsars, namely sudden spin-up events during which the creep motion of vortices is directed inward.

1.2. *Collective vortex dynamics*

The glitch phenomenon is the result of a competition between pinning and the hydrodynamic lift — the so-called *Magnus force* — felt by pinned vortices because of the background flow of neutrons. Interestingly, vortex avalanches occur routinely in two dimensional simulations of many vortices (Howitt *et al.*, 2020), when chains of unpinning events are triggered collectively by vortex–vortex repulsion (Warszawski *et al.*, 2012), in accordance with smaller scale studies based on the Gross–Pitaevskii equation (Warszawski and Melatos, 2011).

This process is sketched in Fig. 1: in principle, we can imagine to associate a certain energy to each vortex configuration in the inner crust, by accounting for the vortex–vortex and the vortex–lattice interactions. This would generate an intricate energy landscape in the space of all possible configurations of vortices (reduced to only two dimensions in the figure). If no velocity lag between the components is present (as in Fig. 1(a)), then each configuration relaxes to the local minimum by rearranging the mutual positions and shape of the vortices. However, the external torque slowly builds a velocity lag, which tends to tilt the energy landscape. Whether this produces a continuous rearrangement of the vortex configuration, or a sudden slip to a new minimum (as in Fig. 1(b)) is still not clear and it really depends on the properties of the landscape. The idea that glitches may the

[1]More precisely, the unpinned vortices will tend to move outward — away from the NS's rotation axis — but their displacement is not necessarily uniform across the superfluid region (see, e.g., Alpar *et al.* (1984b); Antonelli and Pizzochero (2017) and Howitt *et al.* (2020)).

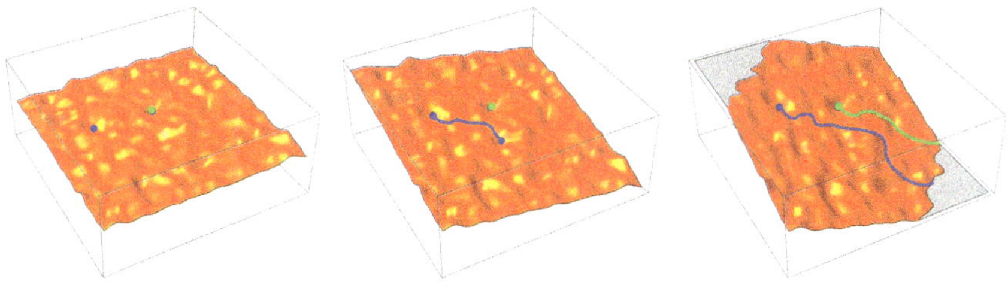

(a) (b) (c)

Fig. 1. Cartoon of the relaxation dynamics between different metastable vortex configurations. A particular configuration of vortex lines, here represented by the blue (or green) point, is immersed in the space of all possible configurations. The total energy associated to a configuration defines a disordered landscape over the configuration space. The external torque drives the system and continuously tilts such landscape. When a critical tilt is reached, a metastable configuration can relax to a new minimum (during the relaxation process the tilt angle decreases as the lag between the components is reduced). Since the energy landscape is extremely irregular, different configurations (like the green and the blue points) will start to relax at different tilt angles, following different paths. The final configuration depends on how quickly the tilt angle decreases during the process and on the initial configuration itself. In principle, the same idea applies in reverse for anti-glitches in spinning-up pulsars.

byproduct of a self-organized criticality of the vortex configuration (Melatos and Peralta, 2007; Morley and Garcia–Pelayo, 1993) is based on the latter possibility. Whether or not this process would be really a manifestation of self-organization in the strict sense (Watkins *et al.*, 2016), Anderson and Itoh (1975) themselves described this driven relaxation process of the vortex configuration as a *noisy vorticity creep process*, that may be thought of as a quake in the vortex configuration. Despite being just a cartoon, this abstract point of view on the glitch phenomenon allows one to recognize that the sequence of dynamical phases (explored in Section 4) experienced by a pulsar during its slowly driven evolution depends on the properties of an intricate energy landscape, so that the observed probability distributions of the avalanche sizes and waiting times are the direct manifestation of a stress-release model that naturally has multiple thresholds (Carlin and Melatos, 2021). A fast driven system (for which the tilt of the potential tends to be steep) is less likely to develop such intermittent fast-slow dynamics; on the contrary a smoother and continuous creep of vortices has to be expected (Alpar *et al.*, 1984b; Baym *et al.*, 1992).

The nature of the trigger for vortex unpinning is still unclear and proposals range from vortex avalanches (Anderson and Itoh, 1975; Melatos and Peralta, 2007; Pines, 1999) to hydrodynamical instabilities (Andersson *et al.*, 2003; Glampedakis and Andersson, 2009; Khomenko *et al.*, 2019) or crust quakes (Alpar *et al.*, 1994; Giliberti *et al.*, 2020; Ruderman, 1969). Recently, during the observation of the 2016 Vela glitch, changes in the pulsar magnetosphere have been observed[2] (Palfreyman *et al.*, 2018). This may be a consequence of one of the trigger processes listed above, in particular a quake (Bransgrove *et al.*, 2020), or it may even be a clue of the involvement of the magnetic field itself during the glitch phenomenon (Ruderman *et al.*, 1998). This interpretation seems to be also supported by the recent analysis of Ashton *et al.* (2020), that reveals an increased flickering activity of the magnetosphere around the glitch epoch. This result, if confirmed, is interesting, as the flickering activity may be linked to precursors and aftershocks, much like in terrestrial earthquakes. If this is the case, then in Fig. 1 the relaxation to a new metastable vortex configuration can be triggered before the local critical tilt of the energy landscape is reached, because the quake provides an initial kick to initiate the process.

Whatever the trigger mechanism, one of the key ingredients in the above picture is the pinning force that the crustal lattice can exert on a vortex. Similarly to what happens in type-II superconductors where pinning enhances the critical current value, this quantity determines the maximum amount of angular momentum that could be exchanged during a glitch (Antonelli *et al.*, 2018). The understanding of how much angular momentum can be stored in different regions of the star would, in fact, allow a comparison with observations of glitching pulsars, potentially constraining the internal structure of a NS (Ho *et al.*, 2015; Pizzochero *et al.*, 2017) when used in tandem with other methods based on the study of the average glitch activity (Andersson *et al.*, 2012; Carreau *et al.*, 2019; Chamel, 2013; Montoli *et al.*, 2021), dynamical fits of the spin-up phase (Ashton *et al.*, 2019; Graber *et al.*, 2018; Montoli *et al.*, 2020a) or the glitching history of pulsars (Gügercinoğlu and Alpar, 2020; Montoli *et al.*, 2020b).

In the following, we review some notions on glitch modelling and then move to discuss some recent advances in the field. In the last years, a number of studies have been devoted to demonstrating that the analysis of glitches across the pulsar population has the potential to put constraints on the dense-matter properties. These analyzes are only first steps, as more high-cadence pulsar timing observations are necessary before it will be possible to falsify some of the theoretical predictions and test the microscopic inputs.

---

[2] An even more dramatic change was observed in the high-magnetic field pulsar J1119-6127, where the pulse profile switched from single to double following large glitch (Weltevrede *et al.*, 2011).

2. Basic Formalism for Glitch Modelling

Glitch modelling is based on several fruitful analogies between spinning NSs and superfluid, or superconducting, systems that are studied in laboratories (see, e.g., Graber *et al.* 2017). In fact, the tools and ideas that are currently used to describe the rotational dynamics of a NS are the byproducts of an attempt to adapt and extend the Tisza–Landau two-fluid model for superfluid Helium to account for the several fluid species, and their mutual interactions, present in the interior of a NS, as it has recently been reviewed by Haskell and Sedrakian (2017), Chamel (2017b) and Andersson (2021).

2.1. *Phenomenological model with two rigid components*

The pulsar rotation frequency $\nu(t) = 1/P(t)$ can be measured at a given reference time t by means of pulsar timing techniques, namely the continuous recording of the time of arrival of each pulse at the telescope. The precise timing of rotation-powered pulsars reveals a steady and extremely slow increase of the pulse period, indicating that these objects lose angular momentum and kinetic energy due to some emission mechanism. However, the rotational frequency of a pulsar occasionally undergoes sudden jumps of amplitude $\Delta\nu > 0$, which are followed by a period of slow recovery that may last for days or months, the so-called glitches. The signature of a glitch in the timing residuals is rather clear for relative jumps bigger than $\Delta\nu/\nu \sim 10^{-7}$: when a glitch occurs, the model $P(t)$ for the slow period evolution before the glitch can no longer predict the time of arrivals, which have to be described by a different timing solution by including a discontinuity in ν and $\dot{\nu}$ around the inferred epoch of the glitch.

Since, the star is an isolated object, the observed sudden spin-up can be explained only in terms of a sudden change of the NS's moment of inertia, or by invoking the action of an internal torque (or both). In particular, the subsequent slow post-glitch relaxation of $\nu(t)$ observed in several pulsars is an indication that an internal torque is at work. In other words, there is a loose component inside the star whose rotation is not directly observable, but that is exchanging angular momentum with the observable component.

2.1.1. *A minimal linear model*

Baym *et al.* (1969) proposed what is probably the simplest phenomenological description of a glitch: the observable component (assumed to be rigidly coupled to the magnetosphere, where the pulsar signal originates) and the loose component in the interior are both modelled as rigid and non-precessing bodies rotating around a common axis and coupled by a linear viscous-like internal torque (see, e.g., Gügercinoğlu and Alpar (2017) and Celora *et al.* (2020) for recent nonlinear extensions),

$$I_o \dot{\Omega}_o + I_s \dot{\Omega}_s = -I|\dot{\Omega}_\infty|, \tag{1}$$

$$I_s \dot{\Omega}_s = -\frac{I_s\, I_o}{I\, \tau_{exp}}(\Omega_s - \Omega_o) = -\frac{I_s}{\tau}(\Omega_s - \Omega_o)\,, \tag{2}$$

where $\Omega_s(t)$ and $\Omega_o(t) = 2\pi\nu(t)$ are the angular velocities of the loose component (identified with the superfluid in an unknown internal region) and of the observable component, respectively. The other parameters are the moments of inertia of the two components, I_s and I_o, while $|\dot{\Omega}_\infty| > 0$ sets the intensity of the external spin-down torque, usually linked to electromagnetic radiation (its exact nature is unimportant here, but it is assumed to be constant on the timescale of interest). The total moment of inertia is $I = I_s + I_o$. Since a real NS is not uniform and the rotation of the loose component may be differential (its rotation rate may depend on the position inside the star), the quantities that appear in the above equations have to be understood as suitable spatial averages over some region in the interior. Finally, there are two common ways to parametrize the internal torque: in terms of the observed post-glitch exponential timescale τ_{exp} (the *relaxation* timescale), or in terms of the *coupling* timescale τ, that is more directly linked to the microscopic processes responsible for the interaction between the two components.

It is interesting to notice that the first Eq. (1) — representing the total angular momentum balance of the star — is a fundamental statement, while the modelling enters into the second Eq. (2). The latter, in fact, includes the interaction between the loose and the observable components which is responsible for the angular momentum exchange between the two components. For a given angular velocity lag Ω^0_{so} at $t = 0$ so that $\Omega_o(0) = \Omega^0_o$ and $\Omega_s(0) = \Omega^0_o + \Omega^0_{so}$, the solution to the system is

$$\Omega_o(t) = \Omega^\infty_o(t) + \Delta\Omega_o(t), \tag{3}$$

$$\Omega_s(t) = \Omega^\infty_o(t) + \Omega^0_{so} - (I_o/I_s)\,\Delta\Omega_o(t)\,, \tag{4}$$

where $\Omega^\infty_o(t) = \Omega^0_o - t|\dot{\Omega}_\infty|$ describes the steady spin-down, while the angular velocity residue $\Delta\Omega_o(t)$ reads

$$\Delta\Omega_o(t) = \frac{I_s}{I}(\Omega^0_{so} - \Omega^\infty_{so})\big(1 - e^{-t/\tau_{exp}}\big)\,. \tag{5}$$

Here $\Omega_{so}^{\infty} = \tau|\dot{\Omega}_{\infty}|$ is the constant value of the lag in the steady state, that is the asymptotic situation in which both components spin-down at the same rate $|\dot{\Omega}_{\infty}|$.

The general solution (3) may be used to fit the observed post-glitch evolution: assuming that the steady-state spin-down rate $|\dot{\Omega}_{\infty}|$ and the angular velocity Ω_o^0 are known (so that $\Omega_o^{\infty}(t)$ is given), we just have to fit the residue, an operation that would require the fitting of 2 parameters (the timescale τ_{exp} and the overall amplitude) that are related to the physical properties of the NS.[3]

2.1.2. *Spin-up due to a starquake*

In their original work, Baym and collaborators assumed that the unresolved spin-up is due to a change in the moments of inertia after a starquake, following the abrupt release of part of the stress accumulated into the crustal solid during the whole NS's life. The solid crust formed when the rotation rate was higher, so that its subsequent relaxation towards a less oblate shape (because of the spin-down) may be realized via a sequence of failures. Within this picture, the loose component, whose location and extension within the star is unknown, only serves the purpose to explain why the slow post-glitch relaxation proceeds over the timescale τ_{exp}.

We can use the linear system in (1) and (2) to study the relaxation after a spin-up induced by a starquake (see also Pines *et al.*, 1974; Shapiro and Teukolsky, 1983). We already have the general solution, so we just have to link the initial condition Ω_{so}^0 with the angular velocity jump due to a change in the moments of inertia. The starquake can be modelled as an instantaneous (i.e., unresolved) global change at $t = 0$, such that

$$\begin{aligned} I_{s,o} &\rightarrow (1 - \epsilon_{s,o}) I_{s,o}, \\ \Omega_{s,o}^0 &\rightarrow \Omega_{s,o}^0 / (1 - \epsilon_{s,o}), \\ |\dot{\Omega}_{\infty}| &\rightarrow (1 + \epsilon_{\infty}) |\dot{\Omega}_{\infty}|, \\ \tau &\rightarrow (1 + \epsilon_{\tau}) \tau, \end{aligned} \tag{6}$$

where $\epsilon_{s,o} > 0$ since the star should evolve towards a less oblate shape, while the sign of $\epsilon_{\tau,\infty}$ is undetermined. Assuming the star to be in the

[3] If the model is used to fit the post-glitch relaxation, then the initial conditions Ω_{so}^0 and Ω_o^0 refer to some time $t = 0$ after the unresolved spin-up. From the practical point of view, observations may not be frequent enough to pin the exact instant of the spin-up event.

steady state before the starquake, the post-glitch relaxation can be written in terms of the pre-glitch parameters and the relative variations ϵ,

$$\Omega_o(t) = (1+\epsilon_o)\Omega_o^0 - (1+\epsilon_\infty)|\dot{\Omega}_\infty|t$$
$$+\frac{I_s}{I}\left[(\epsilon_s-\epsilon_o)\Omega_o^0 - (\epsilon_\infty+\epsilon_\tau-\epsilon_s)\Omega_{so}^\infty\right]\left(1-e^{-t/\tau_{exp}}\right) + O(\epsilon^2), \quad (7)$$

where, again, $\Omega_{so}^\infty = |\dot{\Omega}_\infty|\tau$ is the pre-starquake steady-state lag (all the quantities but ϵ in (7) refer to some instant before the starquake, including Ω_o^0).

The observed glitch jump $\Delta\nu$ allows us to fix the parameter ϵ_o as $\epsilon_o = \Delta\nu/\nu$, but it is difficult to estimate the other ϵ parameters. However, Baym and collaborators noticed that it is possible to relate ϵ_s to the so-called healing parameter Q, which may be obtained from observations: to the first order in ϵ, Eq. (7) may be expressed as

$$\Omega_o(t) = \Omega_o^0 + \Delta\Omega_o - (1+\epsilon_\infty)|\dot{\Omega}_\infty|t - Q\Delta\Omega_o\left(1-e^{-t/\tau_{exp}}\right), \quad (8)$$

where $\Delta\Omega_o \approx \epsilon_o\Omega_o^0$ is the observed glitch amplitude. Hence,

$$Q = \frac{I_s}{I}\left(1-\frac{\epsilon_s}{\epsilon_o}+\frac{\epsilon_\infty+\epsilon_\tau-\epsilon_s}{\epsilon_o}\frac{\Omega_{so}^\infty}{\Omega_o^0}\right) \approx \frac{I_s}{I}\left(1-\frac{\epsilon_s}{\epsilon_o}\right) > \frac{I_s}{I}, \quad (9)$$

where the approximated expression, that is the one found by Baym *et al.* (1969), is obtained in the physical limit $\Omega_{so}^\infty \ll \Omega_o^0$. Crawford and Demiański (2003) collected the all the measured values of the healing parameter Q for Crab and Vela pulsars available in the literature at that time. They found that a weighted mean of the $\sim$ 20 measured values of Q for Crab glitches the values yields $Q \approx 0.7 \pm 0.05$. On the other hand, a mean value for the glitches in the Vela pulsar yields the smaller value $Q \approx 0.12 \pm 0.07$. To test the starquake model for the Crab and the Vela pulsars, they also estimated the healing parameter as $Q \approx I_s/I$ — which is valid in the limit of $I_o \ll I_s$ and/or $\epsilon_s I_s \ll \epsilon_o I_o$ — for different equations-of-state (EoS) dense nuclear matter covering a range of NS masses, and interpreting I_s as the moment of inertia of the neutrons in the core. They conclude that the results are consistent with the starquake model predictions for the Crab pulsar, but the much smaller values of $Q < 0.2$ of the Vela pulsar are inconsistent with the constraint in (9), since the implied Vela mass would be unreasonably small, about $\sim 0.5\,M_\odot$.

2.2. *Some insights from seismology*

As we have seen in the previous subsection, the starquake model is a plausible model for the glitches in the Crab pulsar, although some difficulty arises in the case of the Vela. Apart from the large amplitudes of the Vela glitches, that would require a consistent change in the moments of inertia of the order of $\delta\nu/\nu \approx \epsilon_o \sim 10^{-6}$, the problem is also due to the frequency of its large glitches (roughly one in 2–3 years): the spin-down is too slow for the crust to break because of the change in the centrifugal force between two glitches. Despite the situation is less severe for the Crab, the important point is that starquakes may still be a viable trigger of large Vela-like glitches (Akbal *et al.*, 2017), but only if (see, e.g., the discussion in Giliberti *et al.* (2020)):

- the crust never relieves all the accumulated stresses (which is also the case of earthquakes, see Fig. 2),
- and/or some internal loading force (possibly due to vortex pinning (Ruderman, 1976) or to the magnetic field (Ruderman *et al.*, 1998)) that is orders of magnitude stronger than the variation in the centrifugal force during the inter-glitch time is operating in the crust.

Despite the fact that crustal motion after failure is also invoked as a plausible mechanism to explain magnetar emission, it is possible that NS matter cannot exhibit brittle fracture at any temperature or magnetic field strength (Jones, 2003). Hence, the crust of pulsars can deform in a gradual plastic manner in spite of its extreme strength, leading to rarer starquakes than calculated for a rigid crust (Smoluchowski and Welch, 1970). On the other hand, the presence of defects and of crystal domains (Ruderman, 1991) and the fact that stresses are applied continuously for a long time (Chugunov and Horowitz, 2010) are all elements that point in the opposite direction, namely toward the possibility of more frequent crust failures. The situation is far from being settled and the very question of whether or not crustquakes actually can happen (and with what frequency) in radio pulsars is uncertain.

However, the link between quakes and glitches goes beyond the questions about the nature of the trigger and the magnitude of the change in the moment of inertia in (6). In fact, an earthquake is triggered when a mechanical instability occurs and a fracture (the sudden slip of a fault) occurs in the Earth's crust. Likewise, a glitch is is triggered by the same mechanism — in the case the trigger is a crustquake — or by a mechanical instability in the vortex configuration.

Both glitches and earthquakes are the byproduct of a slowly driven system that sporadically releases some sort of internal stress in a sequence of short bursts. This is the kind of dynamics expected to emerge in many complex systems, that have the tendency to exhibit rare and sudden transitions occurring over intervals that are short compared with the timescales of their prior evolution (Jensen, 1998). Such extreme events are expressions of the underlying forces (like the competing Magnus and pinning forces), or stresses, usually hidden by an almost perfect balance.

Despite there is an opinion that earthquakes — and so glitches, if the analogy between the two phenomena holds — could be inherently unpredictable, observations have shown that the process of seismogenesis is not completely random. The fact that earthquakes seem to be clustered in time more than would be expected for a random process (namely, the Omori's law, which describes the increased rate of small earthquakes after large earthquakes) is considered to be expression of a degree of determinism and predictability in the properties of the earthquake population (Sornette, 2002). In particular, Shimazaki and Nakata (1980) identified the three idealized scenarios shown in Fig. 2:

(1) *Time-predictable* — Assume that the stress $\sigma(t)$ along a fault builds up at a nearly constant rate $\dot{\sigma}$ and that a certain amount of stress — resulting in a slip of the fault — is released when a certain (constant in time) critical threshold σ^c is topped. If we observe a drop in the stress $\Delta\sigma$, then the time to the next quake will be $\Delta t \sim \Delta\sigma/\dot{\sigma}$. Even if $\dot{\sigma}$ is not directly observable, it can be estimated by observation of the previous waiting times between quakes and the amplitudes.
(2) *Slip-predictable* — The critical threshold σ^c may not be constant in time: the fault itself is spatially non-homogeneous and it may also rearrange in time (because of small rearrangements during the continuous stress build up or modifications induced by past quakes). However, if the stress in a quake is released till it drops below a certain level σ_{low} that is fairly constant in time, then it is possible to predict the size of the quake: $\Delta\sigma \sim \dot{\sigma}\,\Delta t$, where now Δt is the observed time elapsed since the preceding quake.
(3) *Periodic* — When both thresholds σ^c and σ^{low} remain almost constant in time, then the stress release works a clock, so that the distribution

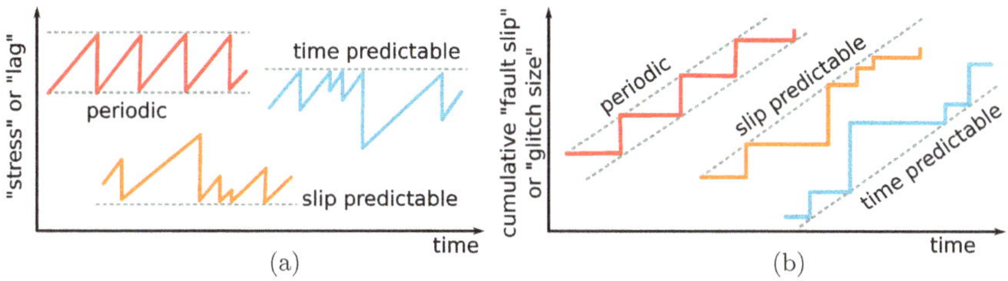

Fig. 2. (a) The three idealized scenarios of the temporal evolution of tectonic stresses (Shimazaki and Nakata, 1980): *periodic* (where the initial σ^c and final σ^{low} stresses of faulting are constant), *time-predictable* (only σ^c is constant, so the larger a quake the longer the following quiet period), *slip-predictable* (the final stress σ^{low} is constant, so the longer the quiet period the larger the quake that follows). The same scheme can be applied to the stress accumulated in the crust of a NS and/or in the vortex configuration because of the competition between the Magnus and pinning forces. (b) The cumulative observed slip (i.e., the cumulative quake size, or the cumulative glitch amplitude) as a function of time for the three scenarios.

of both waiting times between consecutive quakes and sizes should be peaked around a well defined value.

What is important is that observations of a sequence of events may reveal a tendency of the system for one of those scenarios (or for the most general one, in which neither σ^c or σ^{low} is constant), as shown in Fig. 2.

Glitch forecast may not be as socially important as earthquake forecast, but, given the fact that observing glitches requires a constant monitoring effort, it is interesting to try to give at least rough estimates of the time till the next glitch (Akbal *et al.*, 2017; Itoh, 1983; Middleditch *et al.*, 2006). To date, a statistical procedure for predicting a glitch in a given pulsar has not yet been developed, as we still have a lot to learn about the statistics of glitch occurrence in different objects.

While statistical analysis of the whole glitch population has been carried out by aggregating data from different pulsars (Fuentes *et al.*, 2017; Lyne *et al.*, 2000; Shemar and Lyne, 1996), studies of the glitch statistics for individual pulsars are considerably more difficult because of small number statistics. Nonetheless, as the number of detected glitches increases, it is becoming possible to have a rough idea of the probability density functions for the glitch sizes and the waiting times in single pulsars (Ashton *et al.*, 2017; Fuentes *et al.*, 2019; Howitt *et al.*, 2018). Most pulsars are roughly compatible with exponentially distributed waiting times and power-law distributed event sizes, which is the expected behaviour arising

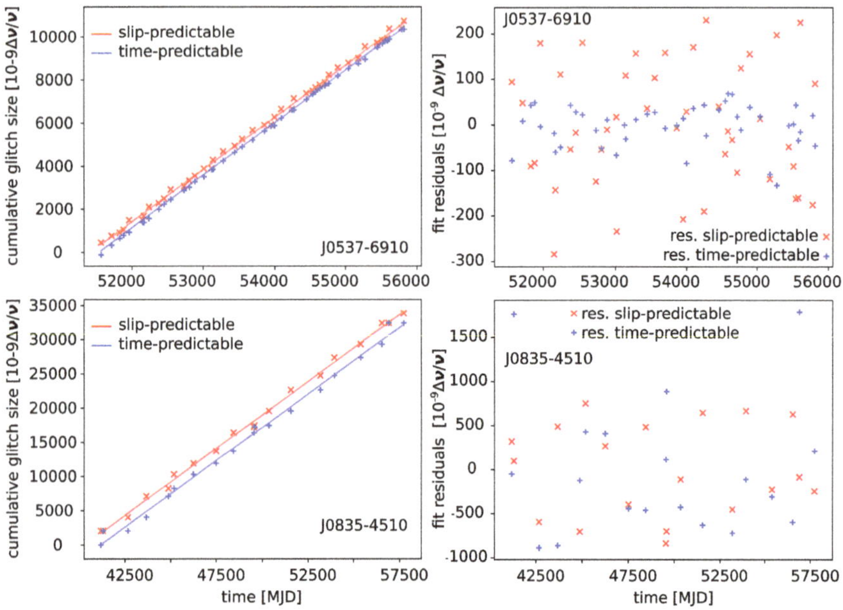

Fig. 3. The glitch activity of J0537-6910 and the Vela pulsar (J0835-6910). The upper and lower points of the cumulative glitch amplitude have been fitted with the standard linear regression procedure to test whether those two quasi-periodic pulsars display a tendency for the slip-predictable or time-predictable scenarios. The residuals of the fits are shown in the right panels. For J0537-6910 the residuals relative to the slip-predictable fit line are more dispersed than the ones relative to the time-predictable case, indicating some slight preference of the system for the latter.

from self-organized critical systems[4] (Melatos *et al.*, 2008). However, the statistical properties of the Vela pulsar and J0537-6910 are unique among glitching pulsars, as they typically have large glitches which occur fairly periodically. Their observed activity is shown in Fig. (3).

In particular, J0537-6910 is one of the most actively glitching pulsars we know of, probably because of its very high spin-down rate (see Ho *et al.* (2020) and Akbal *et al.* (2021) for a recent summary of its properties). Its glitching behaviour seems to be particularly favourable to glitch

[4]The concept of self-organized criticality is believed to provide a natural explanation of the statistics of earthquakes, including the Gutenberg-Richter law for the distribution of earthquake magnitudes (Bak *et al.*, 1994; Bak and Tang, 1989; Ito and Matsuzaki, 1990; Sornette and Sornette, 1989).

prediction due to the presence of some degree of correlation between the waiting time from one glitch to the next and the amplitude of the first glitch (Antonopoulou *et al.*, 2018; Ferdman *et al.*, 2018; Middleditch *et al.*, 2006). The analysis of its cumulative glitch amplitude in Fig. 3 also reveals an imprint of such a time-predictable behaviour.

Apart from the case of J0537-6910 and to a less extent J1801-2304, only a few statistically compelling correlations are found between the size of a glitch and the time to the preceding or the succeeding glitch (Melatos *et al.*, 2018). This may sound counterintuitive, if glitches are threshold-triggered events, as both the starquake and the vortex avalanches scenarios seems to indicate. However, this absence of obvious correlations can be reproduced in terms of a state-dependent Poisson process for the internal stress $\sigma(t)$, where the rate with which a glitch is triggered diverges as the internal stress reaches a threshold σ_c (Carlin and Melatos, 2019; Fulgenzi *et al.*, 2017). Alternatively, Fig. 2 suggests that the lack of obvious correlations may be due to a process that does not conserve neither σ^c nor σ^{low}. Within the vortex avalanche scenario the non-constancy of the upper and lower thresholds is naturally explained because the vortex configurations is expected to move from a local minimum to another in a complex landscape as sketched in Fig. 1: each starting minimum would be associated to a different σ^c, and every final position in the landscape to a certain σ^{low}. Unsurprisingly, the exact sequence of such values would depend on the past history of the system (also in view of the possible presence of hysteresis phenomena, Antonelli and Haskell 2020) and only very long observations could unveil some statistical properties of such landscape.

As the number of glitches detected in single pulsars grows enough for allowing systematic and reliable studies of the statistical properties of the glitching behaviour, it will be more and more important to ask why and how these properties change across the pulsar population in relation with the estimated pulsar age, magnetic field, temperature and spin-down rate.

3. Two-fluid Equations and Mutual Friction

Since the starquake model alone seems to fail to explain the observational pattern of pulsar glitches (see also, Chamel and Haensel (2008) and Haskell and Melatos (2015) on this point), much attention in the last decades has been devoted to understanding and refining fluid theories for the several thermodynamic and hydrodynamic degrees of freedom that are necessary to describe NS matter at the macroscopic scale. In fact, in the presence of superfluid phases, the set of thermodynamic variables needed to define a homogeneous equilibrium state must be enlarged to account for the possible presence of long lived metastable states with persistent currents. It follows that the system is a multi-fluid, namely a conducting medium where different (chemical or even abstract, like the entropy) species can flow with respect to each other (Carter 1989).

The basic multifluid formalism necessary to build pulsar glitch models is based on the seminal works of Carter and collaborators (see, e.g., Carter and Khalatnikov (1992) and Langlois *et al.* (1998)) and has already been reviewed by Andersson (2021), Chamel (2017b), and Haskell and Sedrakian (2017). An extended presentation of relativistic multifluids can be found in Andersson and Comer (2007), see also Gavassino and Antonelli (2020) for a comparison of different hydrodynamic languages used to describe the internal dynamics of superfluid NSs. Moreover, aspects of superfluidity and superconductivity in NSs and nuclear matter with a focus on pairing phenomena have been discussed by Sauls (1989) and Sedrakian and Clark (2019).

Since the neutrons in the crust are expected to form Cooper pairs in the $^1S_0$ channel of the neutron-neutron interaction potential (i.e., the superfluid order parameter in the crust is a scalar), the superfluid in the crust is expected to behave analogously to the laboratory superfluid Helium-4, whose hydrodynamics is well described by the HVBK hydrodynamics (Barenghi *et al.*, 2001; Sonin, 2016). On the other hand, according to the current understanding of nuclear interactions obtained from nucleon scattering data, the neutron pairing most favourable at densities above nuclear saturation is probably in the $^3P_2$ state rather than the $^1S_0$ pairing (Fujita and Tsuneto, 1972; Yasui *et al.*, 2019). In the following, we will neglect the complex issues that arise when the order parameter for neutron superfluidity is not a scalar (Brand and Pleiner, 1981), like the possible presence of domain walls (Yasui and Nitta, 2020), exotic vortices of half-integer circulation and spin textures (Leinson, 2020). Despite these difficulties, some formulations for the neutron superfluid and proton superconductor mixture in the core have been proposed as well (see, e.g., Andersson *et al.* (2006); Glampedakis *et al.* (2011); Gusakov and Dommes (2016); Mendell (1998); Rau and Wasserman (2020) and Sourie and Chamel (2021)).

Almost all glitch models to date are based on the assumption that (locally) it is possible to phenomenologically model neutron conduction in terms of two coupled components (or "chemical species"): a first fluid that should represent the neutron superfluid (usually labelled by an index $x = n$) and an effective charge-neutral fluid ($x = p$). This latter is a mixture of charged particles (or nuclear clusters if we are in the crust), and all the normal (i.e., non-superfluid) neutrons. All these species can be combined into the p-component as long as charge neutrality is satisfied over macroscopic regions (local charge imbalances are expected to be equilibrated by the electron fluid on very short timescales). Here, we adopt this simplified zero-temperature two-fluid effective description, where the generation of entropy and heat by friction is not taken into account.[5]

Fluid equations in the non-transfusive limit — We adopt the Newtonian formalism of Prix (2004) in the non-transfusive limit in which there are no chemical reactions involving the two species (see, e.g., Chamel and Carter (2006) and Gusakov and Andersson (2006) for other formulations of the hydrodynamic equations). This simplified description is the most commonly used setting when it comes to glitch modelling. The set of hydrodynamic equations consist of two ($x = n, p$) conservation equations,

$$\partial_t \rho_x + \nabla_i (\rho_x v_i^x) = 0 \,, \tag{10}$$

and of two momentum equations,

$$n_x \left(\partial_t + L_{v_x}\right) p_i^x + \rho_x \nabla_i \left(\tilde{\mu}_x - |v_x|^2/2\right) = F_i^x \,, \tag{11}$$

where $L_{v_x} p_i^x$ is the Lie derivative of the momentum per particle p_i^x along the kinematic velocity field v_i^x,

$$L_{v_x} p_i^x = v_x^k \partial_k p_i^x + p_j^x \partial_i v_x^j \,. \tag{12}$$

In (11), $\tilde{\mu}_x$ is the specific chemical potential of the relative component, and $\rho_x = m_x n_x$ is the mass density, where m_x is the baryon mass and n_x is the effective baryon number density. The effect of gravity or external fields can be included in the force density F_i^x.

Velocity, momenta and entrainment — The field v_x^i is sometimes referred to as kinematic velocity to stress the fact that it is the quantity that appears in the conservation Eq. (10). The momentum of the superfluid species p_i^n is, at the inter-vortex scale, proportional to the gradient of the phase of the superfluid order parameter, so that it is an irrotational field. Hence,

---

[5]An exception is the thermally driven glitch model, where the sudden vortex motion generates heat that affects the vortex creep rate and the related angular momentum exchange between the components (Larson and Link, 2002; Link and Epstein, 1996).

p_i^n is analogous to the "superfluid velocity" of the Tisza-Landau or HVBK models for superfluid Helium (see, Andersson and Comer (2011); Carter and Khalatnikov (1992), and Gavassino and Antonelli (2020), and references therein). At the macroscopic scale p_i^n is the averaged momentum in the fluid element (possibly containing several vortex lines), and its curl locally reflects the average arrangement of the quantum vortices.

To gain some intuition on the distinction between velocity and momentum we just have to recall their role. The velocity field can be defined starting from a well-defined conserved current for the particles labelled by x: this field is the product of a counting procedure, that is exactly the same that is used in simple fluids (just applied on a subset of the total ensemble of particles). The momentum, on the other hand, is a dynamical quantity. Assume a homogeneous system where there are no gradients and for which we have already defined the velocities $\mathbf{v}_n$ and $\mathbf{v}_p$ with the counting procedure. We can now make an experiment and apply a constant external force $\mathbf{F}_n^{ext}$ only to the neutrons: clearly, we expect $\mathbf{v}_n$ to change, but, thanks to the interaction between the two species, also $\mathbf{v}_p$ may change by some amount. If there were only neutrons, one would have $\mathbf{F}_n^{ext} = n_n \dot{\mathbf{p}}_n = \rho_n \dot{\mathbf{v}}_n$, but now

$$\mathbf{F}_n^{ext} = n_n \dot{\mathbf{p}}_n = \rho_{nn}^* \dot{\mathbf{v}}_n + \rho_{pn}^* \dot{\mathbf{v}}_p \,, \tag{13}$$

where ρ_{nn}^* and ρ_{pp}^* are some effective mass densities that depend on the details of the interaction and that were first introduced in the description of laboratory superfluid mixtures (Andreev and Bashkin, 1976; Khalatnikov, 1957).

The last piece of the puzzle that is missing is the following: why does it happen that $\mathbf{p}_n$, and not $\mathbf{v}_n$, is the quantity that is related to the gradient of the phase of the order parameter? At the formal level, this is a self-consistency requirement for this kind of hydrodynamic theory at the inter-vortex scale (Carter and Khalatnikov, 1992), see also Gavassino and Antonelli (2020) and references therein. The intuition behind this fact was first discussed by Anderson (1966) on the basis of the Josephson effect in the context of superfluid Helium.

In the end, the relation (13) between p_i^x and v_x^i turns out to be a convex combination of the two velocity fields, that can be written in terms of a dimensionless parameter ϵ_x (Prix, 2004),

$$p_x^i = m_x \left[(1-\epsilon_x) v_x^i + \epsilon_x v_y^i \right] , \quad \text{for} \quad x \neq y, \tag{14}$$

where $\rho_n \epsilon_n = \rho_p \epsilon_p$ to guarantee some fundamental properties of the hydrodynamic stress tensor T^{ik} to be discussed below (Carter *et al.*, 2006).

In NS cores, where both superfluid neutrons and the charged component move as fluids, the entrainment is due to nuclear interactions between the species that endows a neutron with a cloud of protons, and vice versa (Chamel and Allard, 2019; Chamel and Haensel, 2006; Gusakov and Haensel, 2005; Leinson, 2017).

On the other hand, in the inner crust the entrainment phenomenon is thought to be due to the Bragg scattering of the superfluid neutrons with the nuclear clusters in the lattice (Chamel, 2012). Here, ρ_n is not the total density of neutrons, but rather the density of the dripped neutrons that leaked outside the heavy nuclear clusters that constitute the solid lattice. Hence, ϵ_n also defines the mobility of those dripped neutrons and, ultimately, the effective density

$$\rho_n^* = \frac{\rho_n}{1 - \epsilon_n} , \tag{15}$$

that has to be ascribed to the part of neutrons that are free to move (Carter *et al.*, 2006). The quantity ρ_n^* also defines the effective moment of inertia of the neutrons in glitch models (Antonelli and Pizzochero, 2017; Chamel and Carter, 2006) and is chemically gauge independent (Carter *et al.*, 2006; Gavassino *et al.*, 2021), namely it does not depend on the ambiguity regarding how many neutrons should be counted as free in the inner crust. In the end, ρ_n^* can be considered to be the effective "superfluid density", the quantity that is analogous to the superfluid density in the Tisza–Landau model (Andersson and Comer, 2011; Carter *et al.*, 2006; Chamel and Carter, 2006; Prix, 2004). The precise calculation of ϵ_n in the inner crust is particularly delicate (Martin and Urban, 2016; Watanabe and Pethick, 2017), as its value can also be very dependent on the arrangement of the crustal lattice and the presence of disorder (Sauls *et al.*, 2020).

Mutual friction — In glitch models, the force F_i^x in (11) accounts for the so-called *mutual friction* (Hall and Vinen, 1956), which allows for momentum transfer between the two components. The adjective "mutual" emphasises the fact that, in the absence of external forces (i.e., $F_i^n + F_i^p = 0$), the total momentum density

$$j_i = n_n p_i^n + n_p p_i^p = \rho_n v_i^n + \rho_p v_i^p , \tag{16}$$

is conserved, in the sense that

$$\partial_t j^i + \nabla_k T^{ik} = 0 \qquad T^{ik} = \sum_x n_x v_x^i p_x^k + \Psi \delta^{ik} \,, \tag{17}$$

where Ψ is a thermodynamic potential that reduces to the usual pressure when the two species comove. The fact $T^{ik} = T^{ki}$ is guaranteed by the before-mentioned property $n_n \epsilon_n = n_p \epsilon_p$.

We now introduce the vortex-mediated mutual friction force. First, we have to understand why F_i^x is related to the presence of vortices. The idea is simple: given a vortex configuration, then also the phase of the order parameter is assigned. If the vortex configuration is frozen, the phase field can not change, nor its gradient. It follows that in order to change $\mathbf{p}_n$ we have to displace some vortices: the faster the displacement, the larger will be $\dot{\mathbf{p}}_n$. Hence, it is reasonable to expect that $\mathbf{F}^n$, which defines $\dot{\mathbf{p}}_n$ in the homogeneous system, can be written as a function of the local averaged vorticity and of the local average vortex velocity. In particular, we can expect $\dot{\mathbf{p}}_n$ (and so $\mathbf{F}^n$) to be proportional to the average "vortex current" in the fluid element. The problem is to define the concept of current for string-like objects, but it is possible to do so if the vortices are locally parallel (which does not necessarily mean that they have to be straight or form an array at the macroscopic scale). Under this hypothesis, the macroscopic vorticity

$$\boldsymbol{\omega} = \nabla \times \mathbf{p}_n / m_n = \kappa n_v \hat{\boldsymbol{\omega}} \,, \qquad \kappa n_v = |\boldsymbol{\omega}|, \tag{18}$$

reflects completely also the vortex arrangement in the local fluid element; κ is the quantized circulation around a single vortex, n_v is the density of vortices measured in a plane orthogonal to $\hat{\boldsymbol{\omega}}$, the local direction of the vortices. In other words, the hypothesis that the quantum vortices are locally straight and parallel is so strong that there is practically no loss of information in performing the average procedure over the fluid element to obtain the macroscopic momentum $\mathbf{p}_n$. If, on the other hand, vortices are locally arranged in a more complex, possibly tangled, way, then the single vector $\boldsymbol{\omega}$ cannot be used to describe the vortex arrangement. How to obtain the mutual friction in such a case is still an open problem, even though progress can be made (Andersson *et al.*, 2007; Haskell *et al.*, 2020; Mongiovì *et al.*, 2017) by using concepts of quantum turbulence imported from the study of vortex tangles in superfluid Helium (Barenghi *et al.*, 2001).

Now, we can derive the phenomenological form of $\mathbf{F}^n$, that must be a Galilean invariant. Since, we have at our disposal $\hat{\boldsymbol{\omega}}$ and the relative velocity $\mathbf{v}_{np} = \mathbf{v}_n - \mathbf{v}_p$, we can always project $\mathbf{F}^n$ on a (non-normalized) right-handed orthogonal basis:

$$\perp_{\hat{\omega}} \mathbf{v}_{np} = -\hat{\boldsymbol{\omega}} \times (\hat{\boldsymbol{\omega}} \times \mathbf{v}_{np}) , \quad \hat{\boldsymbol{\omega}} \times \mathbf{v}_{np} , \quad (\hat{\boldsymbol{\omega}} \cdot \mathbf{v}_{np})\hat{\boldsymbol{\omega}} , \tag{19}$$

where the operator $\perp_{\hat{\omega}}$ is the projector that kills the components parallel to $\hat{\boldsymbol{\omega}}$. Therefore, the force per unit volume $\mathbf{F}^n$ can be decomposed as

$$\mathbf{F}^n/(\rho_n \, |\boldsymbol{\omega}|) = \mathcal{B}_d \, \hat{\boldsymbol{\omega}} \times (\hat{\boldsymbol{\omega}} \times \mathbf{v}_{np}) + \mathcal{B}_c \, \hat{\boldsymbol{\omega}} \times \mathbf{v}_{np} + \mathcal{B}_k \, \hat{\boldsymbol{\omega}}(\hat{\boldsymbol{\omega}} \cdot \mathbf{v}_{np}), \tag{20}$$

where the factors ρ_n and $|\boldsymbol{\omega}|$ have been introduced to make the $\mathcal{B}$ coefficients dimensionless. If we want $\mathbf{F}^n$ to be a force that drives the system towards the zero-lag state $\mathbf{v}_{np} = 0$, then we have to require that the dissipative part is directed in the opposite direction to $\perp_{\hat{\omega}} \mathbf{v}_{np}$, so that $\mathcal{B}_d > 0$.

The $\mathcal{B}_k$ coefficient may have a role when the vortices develop Kelvin waves (helical displacements of a vortex's core), as discussed by Barenghi *et al.* (1983). For strictly straight vortices $\mathcal{B}_k = 0$ is usually neglected in glitch and Helium studies but it should be included in the case a certain amount of quantum turbulence is expected.[6] Relativistic extension of the geometric construction in (19) can be found in the seminal work of Langlois *et al.* (1998), see also Gavassino *et al.* (2021); Gusakov (2016).

Finally, the phenomenological $\mathcal{B}$ coefficients in (20) can be functions of the lag $\mathbf{v}_{np}$ and the vorticity, making the mutual friction nonlinear in the lag (Antonelli and Pizzochero, 2017; Celora *et al.*, 2020). The hard task is to find their expressions from the physics at the microscopic and mesoscopic scale.

3.1. *The relation between mutual friction and vortex dynamics*

The $\mathcal{B}$ coefficients in (20) are phenomenological macroscopic coefficients that have been introduced on the basis of a purely geometrical argument. We can shift the problem of their determination by assuming a certain phenomenological model for the dynamics of vortex lines. This will not solve all the problems, as vortex's dynamics will be governed by a new

[6]The phenomenological expression (20) is general enough that it can be supposed to be valid also if vortices are not parallel to each other in the fluid element. In particular, (20) also contains the Gorter–Mellink result for the mutual friction in the isotropic turbulent regime (see, Andersson *et al.* (2007)) as the limiting case $\mathcal{B}_c = 0$, $\mathcal{B}_d = \mathcal{B}_k \propto |\mathbf{v}_{np}|^2$. The difference with the non-turbulent case is that $\boldsymbol{\omega}$ alone is not sufficient to describe the vorticity in the fluid element, so more involved additional terms may be needed in (20).

set of phenomenological parameters (again, introduced geometrically), that should then be derived from the microscopic properties of the system at the scale of the vortex's core (roughly defined by the coherence length scale of the superfluid).

We can now make clear the relation between the average vortex velocity and the mutual friction outlined in the previous section (see, e.g., Sedrakian *et al.* (1999), Antonelli and Haskell (2020) and references therein for further details). For locally parallel vortex lines, we can introduce the average vortex velocity $\mathbf{v}_L$ by writing down the equation of transport for the macroscopic vorticity,

$$\partial_t \boldsymbol{\omega} + \nabla \times (\boldsymbol{\omega} \times \mathbf{v}_L) = 0\,. \tag{21}$$

In the above equation, $\mathbf{v}_L$ can be taken to be orthogonal to $\boldsymbol{\omega}$ with no loss of generality. In this way, the current of vortex lines in the fluid element is simply $n_v \mathbf{v}_L$. A geometrical definition of $\mathbf{v}_L$ in general relativity is discussed in Gavassino *et al.* (2020). Equation (21) must be consistent with the vorticity equation derived by taking the curl of (11) for $x = n$,

$$\partial_t \boldsymbol{\omega} + \nabla \times (\boldsymbol{\omega} \times \mathbf{v}_n) = \nabla \times (\rho_n^{-1}\, \mathbf{F}_n). \tag{22}$$

Equations (21) and (22) are consistent if

$$\mathbf{F}_n = -\rho_n \boldsymbol{\omega} \times (\mathbf{v}_L - \mathbf{v}_n) = -\rho_n |\boldsymbol{\omega}| \mathbf{f}_M\,, \tag{23}$$

that tells us that the friction between the components is zero when the vortices are advected in such a way that $\mathbf{v}_L = \mathbf{v}_n$. In the above equation we also defined the quantity $\mathbf{f}_M$, the *Magnus force* acting on a straight vortex segment.[7]

If the vortices stick to the normal component because of pinning, then $\mathbf{v}_L = \mathbf{v}_p$ and $\mathbf{F}_n = \rho_n \boldsymbol{\omega} \times \mathbf{v}_{np}$, namely $\mathcal{B}_c = 1$ and $\mathcal{B}_d = 0$, so that there is no dissipation in this limit. More generally, if we have a model that allow us to calculate the average vortex velocity in a fluid element $\mathbf{v}_L(\mathbf{v}_{np})$ for a given value of the lag, then we can substitute this expression back into (23) to obtain the expression of the $\mathcal{B}$ coefficients in (20).

Magnus and drag forces — In both Helium (Barenghi *et al.*, 1983) and NSs (Andersson *et al.*, 2006; Mendell, 1991), the simplest phenomenological model for vortex dynamics is in terms of balance of the Magnus force and

[7] The quantity $\mathbf{f}_M = \boldsymbol{\omega} \times (\mathbf{v}_L - \mathbf{v}_n)$ is just a rescaling of what in the literature is typically called Magnus force per unit length (of vortex line), namely $\rho_n \boldsymbol{\omega} \times (\mathbf{v}_L - \mathbf{v}_n)$. Note that our $\mathbf{f}_M$ is (apart from the overall sign) the same quantity appearing in (20).

a drag force (proportional to a dimensionless drag parameter $\mathcal{R}$) acting on a straight segment of vortex line,

$$\mathbf{f}_M - \mathcal{R} \perp_{\hat{\omega}} (\mathbf{v}_L - \mathbf{v}_p) = \hat{\omega} \times [\mathbf{v}_L - \mathbf{v}_n + \mathcal{R}\hat{\omega} \times (\mathbf{v}_L - \mathbf{v}_p)] = 0\,, \qquad (24)$$

that can be inverted in the plane orthogonal to $\hat{\omega}$ to find $\mathbf{v}_L$ as a function of $\mathbf{v}_n$ and $\mathbf{v}_p$. Plugging the result for $\mathbf{v}_L$ back into (23) allows to find

$$\mathcal{B}_c = \frac{\mathcal{R}^2}{1+\mathcal{R}^2} \quad \mathcal{B}_d = \frac{\mathcal{R}}{1+\mathcal{R}^2} \quad \mathcal{B}_k = 0. \qquad (25)$$

This tells us that, as long as $\mathcal{R}$ can be treated as a constant, the mutual friction is linear in the lag $\mathbf{v}_{np}$.

In the outer core, the presence of entrainment causes the magnetization of the neutron superfluid vortices, generated by the current of protons that are entrained by the local circulation associated with the vortex (Alpar *et al.*, 1984a; Andersson *et al.*, 2006). The scattering of the relativistic free electrons with these magnetised vortex cores is the main process that is taken into account in the estimates of $\mathcal{R}$.

In the inner crust, the drag depends also on the relative velocity between the superfluid vortices and the normal component. For low values of this relative velocity the drag is caused by phonon excitation in the crustal lattice (Jones, 1990), while for higher values of this parameter the drag force is caused by excitation of Kelvin waves on the superfluid vortices (Epstein and Baym, 1992; Jones, 1992). A recent review on the calculation of the drag parameter due to Kelvin waves excitation has been presented in Graber *et al.* (2018), while dynamical aspects related to the fact that vortices can bend are discussed in Link (2009), and Wlazlowski *et al.* (2016).

It is also worth mentioning that in the inner crust the complication induced by the *chemical gauge* problem (i.e., the problem of defining which neutrons can be considered to be free, in the sense that they are not bounded into the lattice clusters, see Carter *et al.* (2006)) demands a more general treatment than the one outlined above: in addition to the Magnus and the drag forces in (24) also other contributions should be included into the vortex equation of motion in order to ensure that the final hydrodynamic theory is well behaved under redefinitions of the effective density of free neutrons (Gavassino *et al.*, 2021).

Magnus, drag and pinning forces — If vortices are immersed in a non-homogeneous medium, then their interaction with the substrate must be taken into account. The vortex creep model of Alpar *et al.* (1984b) deals with this problem by assuming that the dependence of the jump rate from one pinning centre to another one is given by the typical Arrhenius formula, where the activation energy for the jump is corrected by the presence of the background superfluid flow. This makes $\mathbf{F}_n$ nonliner in the velocity lag (Gügercinoğlu and Alpar, 2017).

In fact, when an additional pinning force is included into Eq. (24) for the motion for the vortices (as done, e.g., in Haskell and Melatos (2016), Link (2009), and Sedrakian (1995)), the $\mathcal{B}$ coefficients are not constant anymore. For small $\mathbf{v}_{np}$ the parameter $\mathcal{B}_d$ is highly suppressed, while $\mathcal{B}_c \approx 1$. When the lag reaches a certain value (dependent on the details of the pinning landscape), the vortices start to move and $\mathcal{B}_d$ tends to grow. For very high values of the lag, the pinning interaction will be just a small correction to the Magnus and drag forces, so that we can expect that the linear regime in (25) is eventually recovered for very high values of the lag (remember, however, that the parameter $\mathcal{R}$ may not be constant in the first place).

Pinning can also induce hysteresis in the pinning-depinning transition as the lag is first increased and then decreased in a cyclic transformation, making the behaviour of the parameters $\mathcal{B}_c$ and $\mathcal{B}_d$ history-dependent (Antonelli and Haskell, 2020). This may provide a natural explanation for the behaviour of most pulsars, which is neither time-predictable nor slip-predictable (as sketched in Fig. 2).

3.2. *Circular models*

In general, the fluid Eqs. (10) and (11) can be solved only numerically (Howitt *et al.*, 2016; Peralta *et al.*, 2005). Thus, in order to obtain a glitch model that is more easily tractable, it is necessary to simplify the problem by restricting ourselves to a specific subset of all possible fluid motions. In this section, we assume a circular (not necessarily uniform) velocity field and derive the equations of motion for the angular velocity of the two components: this constitutes the backbone of many glitch models.

Let us work with the usual cylindrical coordinates (x, φ, z), where x is the distance from the rotation axis of the star (directed along z). The spherical radius is $r = \sqrt{x^2 + z^2}$. In the absence of precession and neglecting meridional flows,[8] the two species can only rotate as

$$\begin{aligned} \mathbf{v}_p &= x\,\Omega_p(x,z,t)\,\mathbf{e}_\varphi, \\ \mathbf{v}_n &= x\,\left(\Omega_p(x,z,t) + \Omega_{np}(x,z,t)\right)\,\mathbf{e}_\varphi. \end{aligned} \tag{26}$$

[8]In General Relativity, this would be the circular spacetime approximation (Antonelli *et al.*, 2018; Gavassino *et al.*, 2020; Sourie and Chamel, 2021; Sourie *et al.*, 2016).

It is also useful to define an auxiliary angular velocity variable directly from the momentum $\mathbf{p}_n$,

$$\Omega_v(x,z,t) = p_n^\varphi/x = \Omega_p + (1-\epsilon_n)\Omega_{np} \,. \tag{27}$$

Therefore, the macroscopic vorticity (18) reads

$$\boldsymbol{\omega} \;=\; -x\partial_z\Omega_v\,\mathbf{e}_x \;+\; (2\Omega_v + x\partial_x\Omega_v)\,\mathbf{e}_z \,, \tag{28}$$

that reduces to the standard results $\boldsymbol{\omega} = 2\Omega_v\,\mathbf{e}_z$ and $n_v = 2\Omega_v/\kappa$ if the superfluid is rigidly rotating at the macroscopic scale.

We can use (28) and (20) to obtain (we are working with locally parallel vortices, so $\mathcal{B}_k = 0$)

$$\mathbf{F}_n = \rho_n\, x\, \Omega_{np}\Big(\mathcal{B}_c\,[x\partial_z\Omega_v\mathbf{e}_z + (2\Omega_v + x\partial_x\Omega_v)\,\mathbf{e}_x] - \mathcal{B}_d\,|\boldsymbol{\omega}|\,\mathbf{e}_\varphi\Big) \,. \tag{29}$$

We are interested in specifying the Eq. (11) directed along the azimuthal φ direction for both the normal and superfluid components. We are not interested in the other two directions, namely the components of (11) along $\mathbf{e}_x$ and $\mathbf{e}_z$, because they do not contain any time derivative: their role is to define the instantaneous configuration of the rotating NS, whose deviation from the spherical hydrostatic equilibrium must be constantly adjusted as Ω_n and Ω_p vary in time. In the slowly rotating limit (Andersson and Comer, 2001), it is not a bad approximation to neglect the components of (11) along $\mathbf{e}_x$ and $\mathbf{e}_z$ and assume some given stratification of the star, i.e., $\rho_{n,p} = \rho_{n,p}(t,x,z) \approx \rho_{n,p}(r)$. Hence, we will restrict ourselves to the dynamics along $\mathbf{e}_\varphi$, that is particularly simple since the Lie derivative and the gradient term (because of the axial symmetry hypothesis) do not have any component directed along $\mathbf{e}_\varphi$. A direct substitution into (11) gives:

$$x\rho_n(\partial_t\Omega_n - \epsilon_n\partial_t\Omega_{np}) = x\rho_n\partial_t\,\Omega_v \;=\; F_n^\varphi \,, \tag{30}$$

$$x\rho_p(\partial_t\Omega_p + \epsilon_p\partial_t\Omega_{np}) \;=\; -F_n^\varphi + F_{Bp}^\varphi \,, \tag{31}$$

$$F_n^\varphi = -\rho_n x \mathcal{B}_d |\boldsymbol{\omega}| \Omega_{np} \,, \tag{32}$$

$$|\boldsymbol{\omega}| = \sqrt{(2\Omega_v + x\partial_x\Omega_v)^2 + (x\partial_z\Omega_v)^2} \,, \tag{33}$$

where we have also added an extra external force F_{Bp}^φ to model the braking effect of dipole radiation on the normal component (an external torque due to gravitational wave emission would act on both components). Note that, despite $\mathcal{B}_c \neq 0$, this parameter only appears in the two equations that define the stellar structure. If we sum the two dynamical equations in (30) and (31) we have ($\rho = \rho_n + \rho_p$)

$$x(\rho\,\partial_t\Omega_p + x\rho_n\partial_t\Omega_{np}) = F^{\varphi}_{Bp}\,. \tag{34}$$

that, in the limit $F^{\varphi}_{Bp} = 0$ is the φ component of (17). The above equation may be used in place of, say, (31).

Another widely used assumption in glitch modelling is that the normal component rotates rigidly. This assumption is beyond the fluid Eq. (11), as it is motivated on some piece of physics that is not implemented in the simple dynamical equations we are using here. Hence, when doing this we have to ensure the consistency of the fluid equations: if we simply put $\Omega_p = \Omega_p(t)$ into the above equation, then inconsistencies may arise. Assuming rigid motion for the normal component forces us to substitute one of the local equations (say (34) for convenience) with an integral one,

$$\int d^3x\, x^2\,(\rho\,\partial_t\Omega_p + \rho_n\,\partial_t\Omega_{np}) = \int d^3x\, x\, f^{\varphi}_{Bp}\,. \tag{35}$$

It is only at this point that we can really restrict the motion of the charged component to be perfectly rigid: assuming $\Omega_p = \Omega_p(t)$ in the above equations allows us to bring it outside the integral sign to obtain

$$I\dot{\Omega}_p + \int d^3x\, x^2\rho_n\partial_t\Omega_{np} = -I|\dot{\Omega}_\infty|\,, \tag{36}$$

where I is the total moment of inertia of the star for the given stratification $\rho_n(x,z,t) + \rho_p(x,z,t) \approx \rho(r)$ and $|\dot{\Omega}_\infty|$ is a phenomenological parameter (possibly a function of time and/or the angular velocities) that sets the strength of the total braking torque,

$$|\dot{\Omega}_\infty| = \frac{1}{I}\left|\int d^3x\, x\, f^{\varphi}_{Bp}\right|\,. \tag{37}$$

The other equation that we need can be taken to be (30),

$$\partial_t\,\Omega_v(t,x,z) = -\mathcal{B}_d\,|\boldsymbol{\omega}|\,\Omega_{np} = -\mathcal{B}_d\,|\boldsymbol{\omega}|\,\frac{\Omega_{vp}}{1-\epsilon_n}\,, \tag{38}$$

where

$$\Omega_{vp}(x,z,t) = \Omega_v - \Omega_p = (1-\epsilon_n)\,\Omega_{np}\,, \tag{39}$$

is the lag between the superfluid angular velocity[9] Ω_v and the velocity of the normal component.

If we want to use Ω_v in place of Ω_n as primary variable, then (36) can be equivalently written as

$$I\dot{\Omega}_p + \int d^3x\, x^2\, \rho_n^*\, \partial_t \Omega_{vp} = -I|\dot{\Omega}_\infty|\,, \tag{40}$$

where ρ_n^* is the entrainment corrected superfluid density introduced in (15). In this way the equations are formulated in terms of the chemically gauge invariant quantities Ω_p, Ω_v, ρ_n^* and ρ.

Axially symmetric dynamical equations — To summarize, for a given stratification of the NS defined by $\rho_{n,p} \approx \rho_{n,p}(r)$ and $\epsilon_n \approx \epsilon_n(r)$, the axially symmetric dynamical equations are (36) and (38). Due to the form of the vorticity in (33), the two equations can be conveniently written by selecting as primary variables $\Omega_p(t)$ and $\Omega_v(t,x,z)$. By introducing an average operation over the local moment of inertia defined as

$$dI_v = d^3x\, x^2\, \rho_n^*\,; \quad I_v = \int dI_v\,; \quad \langle\, f\,\rangle = I_v^{-1}\int dI_v\, f(t,x,z)\,, \tag{41}$$

the system defined by (40) and (38) is equivalent to

$$I\partial_t\, \Omega_p(t) + I_v\, \partial_t\langle\, \Omega_{vp}\,\rangle = -I|\dot{\Omega}_\infty|\,, \tag{42}$$

$$\partial_t \Omega_v(t,x,z) = -\mathcal{B}_d(r,\Omega_{vp})\,|\boldsymbol{\omega}|\,\frac{\Omega_{vp}(t,x,z)}{1-\epsilon_n(r)}\,, \tag{43}$$

where $\Omega_{vp}(t,x,z) = \Omega_v(t,x,z) - \Omega_p(t)$, $\langle\Omega_{vp}\rangle = \langle\Omega_v\rangle - \Omega_p$ and $|\boldsymbol{\omega}|$ is given in (33). Since the stratification is given, the drag parameter $\mathcal{B}_d$ can be assumed to be fixed and almost spherically stratified, i.e., $\mathcal{B}_d \approx \mathcal{B}_d(r)$, or to contain also an extra dependence on the lag Ω_{vp}, see the discussion below (20).

Notice that to derive (42) and (43) we have used only the momentum Eq. (11) because the conservation Eq. (10) are automatically satisfied when assuming axial symmetry and no meridional flows.

---

[9]We recall that the variable Ω_v is defined by the vortex configuration and is, basically, the *superfluid velocity* of Tisza and Landau (not to be confused with the *velocity of the species that is superfluid*, Ω_n). Hence, Ω_v is not an exotic variable, but, rather, it is the standard choice for laboratory superfluids. It has also the nice property of being chemical-gauge independent (Carter *et al.*, 2006; Gavassino *et al.*, 2021).

Finally, (42) and (43) have been derived in the limit of slack vortices, in which the effect of bending a quantum vortex is neglected (Antonelli *et al.*, 2018). If the vortex tension (Andersson *et al.*, 2007; Haskell and Melatos, 2016; Link, 2014) is taken into account (as it is sometimes done in the HVBK equations, Barenghi *et al.* (2001)), then an additional term containing derivatives along z and x should appear in the left hand side of (43). Opposite to the slack limit, we have the extreme scenario in which Ω_v is cylindrical (i.e., it has no z dependence, so that $\boldsymbol{\omega}$ is everywhere parallel to the rotation axis), which is formally recovered if the vortex tension is infinite. Such an unrealistic scenario is, however, still useful from the practical point of view as it allows to further simplify the equations: several glitch models are actually *cylindrical*, meaning that they are built on this simplifying assumption (e.g., Alpar *et al.* (1984b); Antonelli and Pizzochero (2017); Graber *et al.* (2018); Haskell *et al.* (2018); Haskell *et al.* (2012); Khomenko and Haskell (2018) and Pizzochero (2011)).

Models with rigid components — The phenomenological model of Baym and collaborators defined by Eqs. (1) and (2) lacks direct connection with the local properties of NS matter. However, it is possible to link its parameters with the quantities appearing in (42) and (43). The rigid model can be recovered by averaging (43) over the superfluid region by using (41),

$$\langle \partial_t \Omega_v \rangle = \partial_t \langle \Omega_v \rangle = -\left\langle \mathcal{B}_d \, |\boldsymbol{\omega}| \, \frac{\Omega_{vp}}{1-\epsilon_n} \right\rangle . \tag{44}$$

Clearly, the approximate reduction to a rigid model by performing the average of the equations will work well if stratification and spatial correlations between the different quantities are weak, that is

$$\left\langle \mathcal{B}_d \, |\boldsymbol{\omega}| \, \frac{\Omega_{vp}}{1-\epsilon_n} \right\rangle \approx \left\langle \frac{\mathcal{B}_d \, |\boldsymbol{\omega}|}{1-\epsilon_n} \right\rangle \langle \Omega_{vp} \rangle \approx 2 \langle \Omega_v \rangle \left\langle \frac{\mathcal{B}_d}{1-\epsilon_n} \right\rangle \langle \Omega_{vp} \rangle . \tag{45}$$

The factor $\langle \Omega_p \rangle \approx \Omega_p$ is almost constant, but the above expression can still be highly nonlinear if $\mathcal{B}_d$ depends on the local lag. The linear model of Baym and collaborators can only be found if $\mathcal{B}_d$ is locally constant over the timescale of interest. In this case, a direct comparison of (42) and (44) with (1) and (2) allows us to make the following identifications:

$$\begin{aligned} & I_s = I_v \,; && I_o = I - I_v \,; \\ & \Omega_s = \langle \Omega_v \rangle \,; && \Omega_o = \Omega_p \,; \\ & \frac{1}{\tau} = \frac{I - I_v}{I \, \tau_{exp}} \approx 2\Omega_p \left\langle \frac{\mathcal{B}_d}{1-\epsilon_n} \right\rangle . && \end{aligned} \tag{46}$$

For example, this tells us how a fit of the timescale τ_{exp} from observations can be interpreted in terms of the physical input at smaller scales. General relativistic corrections should be also taken into account, which affect both the expressions for the moments of inertia and for τ (Antonelli *et al.*, 2018; Gavassino *et al.*, 2020; Sourie *et al.*, 2016).

A similar reduction to a rigid model can be performed by favouring Ω_n in place of Ω_v. In this case, extra internal torques due to the entrainment coupling appear in the final equations (Chamel and Carter, 2006; Sidery *et al.*, 2010).

Finally, the superfluid region (i.e., the domain of Ω_v, that extends in the inner crust and outer core) can also be split into an arbitrary number of different, non-overlapping, sub-regions to obtain the multi-component generalization of the model of Baym and collaborators (see, e.g., Montoli *et al.* (2020a) for the formal derivation of the three-components analogue of Baym's model).

4. Dynamical Phases and the Vela's 2016 Glitch Precursor

It is interesting to make some general statements about the dynamics defined by (42) and (43). Those equations do not allow to track the stresses in the solid components and to directly model starquakes and changes of the moments of inertia I and I_v should be implemented by hand, as in Baym's model of Section 2.1. Hence, we assume here the Anderson and Itoh (1975) paradigm, in which the moments of inertia I and I_v can be taken to be constant.

The angular momentum balance Eqs. (42) and (43) have two different roles. The momentum balance (42) allows to identify some general dynamical phases, but the ability of a real pulsar to switch from one phase to another depends on the details of friction (i.e., on the details of the $\mathcal{B}_d(\Omega_{vp})$ function) or, more generally, on the statistical properties of the landscape in Fig. 1. In other words, the possibility to explore certain phases critically depends on how the angular momentum can be exchanged between the components and transported between different regions during the pulsar's life.

Despite we still do not have a complete and satisfactory understanding of these issues, it is still instructive to consider each dynamical phase as a separate (theoretical) possibility. The following analysis is valid for a system where the superfluid component can rotate non-uniformly (this system has an infinite number of degrees of freedom, so that each phase is highly

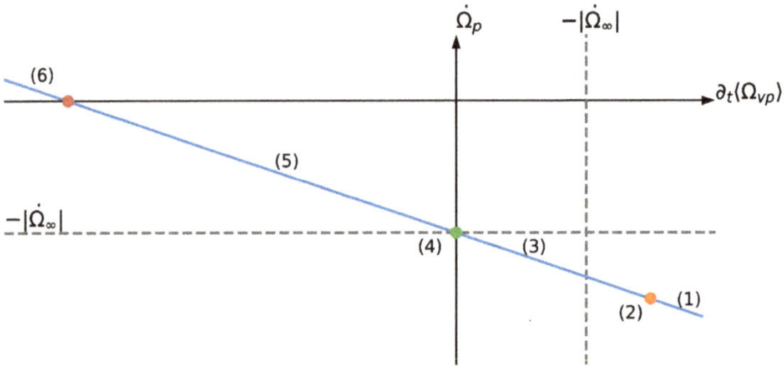

Fig. 4. The evolution of the total angular momentum (42) can be regarded as a constraint (the blue line) in the $(\partial_t\langle\Omega_{vp}\rangle, \partial_t\Omega_p)$ plane, whatever the number of internal components (formally infinite if the superfluid can rotate non-uniformly) or the details of the mutual friction. The numbers indicate the different dynamical phases described in the main text. The slope of the curve depends on the ratio I_v/I.

degenerate and can be realized in a number of ways), but the same reasoning can be applied to simpler models with two or more rigid components, that have a finite number of degrees of freedom and are defined by ordinary differential equations.

We plot the different phases defined by the angular momentum balance (42) in Fig. 4. Going from negative to positive values of $\partial_t\Omega_p$ we have the following phases:

(1) *Inward vortex creep* — From (42), we see that $\partial_t\Omega_p < -I/(I-I_v)|\dot{\Omega}_\infty|$ can be realized only if the internal torque contributes to slowing down the observable component (the internal torque has the same sign as the external one). Since $\mathcal{B}_d$ is a positive function (or functional) of the local lag, as demanded by the second law of thermodynamics (Gavassino *et al.*, 2021; Gusakov, 2016), we should have that this phase can be realized only if there is a sufficiently extended region in which the lag is negative (in fact, $\text{sign}\Omega_{vp} = \text{sign}\Omega_{np}$). This also implies that the vortex velocity is, on average, directed inward.
(2) *Average pinning* — In this dynamical phase, the superfluid does not spin-down. Mathematically, we have $\partial_t\langle\Omega_v\rangle = 0$. The lag between the two components builds up as $\partial_t\langle\Omega_{vp}\rangle = -\partial_t\Omega_p$. Equation (40) implies that $\partial_t\Omega_p = -I/(I-I_v)|\dot{\Omega}_\infty|$. This phase can be achieved with perfect pinning (the superfluid is *completely decoupled* from the normal component) or by a balance of inward and outward creep (unlikely unless the actual lag has very big gradients and is partially reversed).

(3) *Moderate outward creep* — The average lag satisfies $0 < \partial_t \langle \Omega_{vp} \rangle < -\partial_t \Omega_p$, meaning that the superfluid is only partially decoupled. This state is likely to occur during the long term post-glitch relaxation: the reduced post-glitch lag slowly increases, i.e., $\partial_t \langle \Omega_{vp} \rangle > 0$. Consequently, the superfluid gradually recouples with the observable component (the system is expected to reach the steady state if no other glitches occur in the meanwhile).

(4) *Steady state* — Here the coupling of the superfluid with the normal component is stronger than in the previous phase, and such that the lag is on average constant in time, namely $\partial_t \langle \Omega_{vp} \rangle = 0$. Despite both components spin-down at the same rate $\partial_t \Omega_p = \partial_t \langle \Omega_v \rangle = -|\dot{\Omega}_\infty|$, the local lag Ω_{vp} does not have to do so: for example, this situation may be realized also if a sequence of packets of unpinned vortices creep outward.

(5) *Speed up* — This phase can be defined as $-|\dot{\Omega}_\infty| < \partial_t \Omega_p < 0$, meaning that the observable component is spinning down but the average lag is decreasing in such a way that $-I/I_v |\dot{\Omega}_\infty| < \partial_t \langle \Omega_{vp} \rangle < 0$: the normal component is receiving some extra angular momentum from the superfluid but at such a low rate that it is still spinning down. Such a phase has been probably been observed in the Crab, and it has been speculated that it may be the result of a small increase in the Crab pulsar's internal temperature (Vivekanand, 2017) that has the effect of increasing the outward vortex creep rate (Alpar *et al.*, 1984b).

(6) *Spin-up* — In this phase the observable component spins up, $\partial_t \Omega_p > 0$ and the lag is decreasing with a high rate $\partial_t \langle \Omega_{vp} \rangle < -I/I_v |\dot{\Omega}_\infty|$. In real glitches the inequality is actually believed to be stronger, $\partial_t \langle \Omega_{vp} \rangle \ll -I/I_v |\dot{\Omega}_\infty|$ but a slow spin-up event has been observed in the Crab pulsar following a fast, unresolved, glitch (Shaw *et al.*, 2018). The particular value $\partial_t \Omega_p = 0$ can be realized only if the internal torque due to the superfluid exactly balances the external one. This can happen instantaneously when the observable component reaches the maximum amplitude in a glitch. In the vicinity of this event the lag can be both positive or negative (depending on the fact that an overshoot takes place or not, see also Section 6).

This analysis is independent on the details of the mutual friction, but it depends on the fact that the moments of inertia are assumed to be constant.

The Vela precursor — Ashton *et al.* (2019) observed a feature in the Vela pulsar's phase residuals around the Vela's 2016 glitch recorded by Palfreyman *et al.* (2018). This feature can be interpreted as a spin-down (an antiglitch), occurring just before the main spin-up event and possibly linked to the triggering of the glitch. Assuming that such observed behaviour can be explained only in terms of angular momentum exchange between the superfluid and the normal component, it is interesting to try to understand to which dynamical phase this spin-down "precursor" may belong to.

According to the fit performed by Ashton and collaborators, this sudden decrease of the Vela's rotational frequency ($\nu \approx 11\,\mathrm{Hz}$) has an amplitude $\Delta\nu \sim -5 \times 10^{-6}\,\mathrm{Hz}$ and it is likely to occur over a time span of $\Delta t \sim 100\mathrm{s}$ or shorter, implying $\dot{\Omega}_p \approx 2\pi\Delta\nu/\Delta t \sim -10^{-7}\,\mathrm{rad/s^2}$. Since a spin-down of the normal component would increase the lag, they also speculated that this precursor may trigger the subsequent glitch by causing a critical lag between crustal superfluid and the crust.

We should compare the above numbers with the secular spin-down parameter of the Vela, $|\dot{\Omega}|_\infty \approx 10^{-10}\mathrm{rad/s^2}$: by looking at Fig. 4 we immediately have that the Vela should be in either phase 1, 2 or 3 during such event. Let us assume that phase 2 is realized, which means a complete decoupling of the superfluid (or that in different parts of the superfluid the creep is balanced between outward creep and inward creep). In this case, the observable component could spin-down at a faster rate as the external torque acts on a reduced moment of inertia, especially if $I_v \approx I$, which is the extreme situation where the NS is almost entirely superfluid. Assuming phase 2, we have $-\dot{\Omega}_p/|\dot{\Omega}|_\infty = I/(I - I_v) \approx 10^3$, so that the the blue line in Fig. 4 would be practically vertical. This scarcity of normal matter in a NS (i.e., $I - I_v \sim 10^{-3} I$) is not predicted by any reasonable model of NS structure. Hence, we may assume that the Vela was in phase 1 (phase 3 is excluded since already phase 2 is not a viable option). However, this is also not a realistic possibility: vortices can undergo inward creep (so that the mutual friction torque acts in the same direction of the external torque) if the lag is momentarily reversed, which is impossible as this event immediately precedes a standard spin-up glitch, where the lag has to be positive.

In conclusion, such a huge anti-glitch precursor cannot be explained in terms of the standard cartoon of pulsar glitches without invoking the occurrence of the *inward-creep* phase just before the main spin-up event. Since we just used the general Eq. (42) for the dynamics of the total angular momentum, which is just an angular momentum balance and is insensitive to all the details of the internal dynamics, assuming complex models for the mutual friction or adding additional (fluid or rigid) components does not change this conclusion.

Possible explanations of the precursor — Gügercinoğlu and Alpar (2020) speculated that the precursor can be a consequence of the formation of high vortex density regions surrounded by vortex depletion regions called *traps* (Cheng *et al.*, 1988). In this case, the displacement of vortices is a byproduct of a starquake (that eventually is also the cause of the glitch but does not significantly modify the moments of inertia). According to this picture, the dislocations of the lattice induced by the quake provide stronger pinning sites for the vortex lines. These new sites can attract vortices over microscopic distances, hence the formation of traps. Since the velocity and direction of the displaced vortices depend on the concentration of dislocations, it is still not clear how the creation of traps can induce the average inward creep needed to fully justify the precursor.

A tentative explanation in terms of stochastic fluctuations of the internal torque has also been proposed by Ashton *et al.* (2019) and Carlin and Melatos (2020). In the case the continuous transfer of angular momentum from the superfluid to the normal component is particularly noisy, then the precursor may just be the effect of an abnormal fluctuation in the internal torque. Such a fluctuation increases in a short time the average lag between the components, possibly pushing it above the depinning threshold in some regions, hence triggering the subsequent glitch. It follows that understanding if such internal torque fluctuations — that, in turn, are fluctuations in the state of motion of many vortices — can actually take place in the real system is key to falsify this possibility. Moreover, the statistical occurrence of such large fluctuations should be closely related to the observed statistical occurrence of glitches.

Future observations of well-resolved glitches have, thus, the potential to challenge the current understanding and, possibly, to inspire new physical paradigms for glitch modelling. However, there is also the additional complication that what we directly observe is not the normal component but rather a signal generated in the magnetosphere of the pulsar, that has its own dynamics. Since the antiglitch cannot be attained by invoking neither the internal torque nor the external one (or both, as in phase 1), a more conservative interpretation is that the precursor does not represent a a change in Ω_p. On the contrary, it is likely to reflect a change pertaining the emission region of the Vela (Montoli *et al.*, 2020a). The recent analysis of Ashton *et al.* (2020) revealed a flickering magnetospheric activity of the Vela around the 2016 glitch epoch. If confirmed, this would provide a further clue that the magnetosphere is indeed a complex dynamical region (see also Bransgrove *et al.* (2020), that simulated the magnetospheric perturbations induced by a crustquake).

5. Constraints on Neutron Star Structure

In this section, we present some methods that can be used to extract information on a glitching NS through its observed glitch behaviour. As often occurs in astrophysics, extrapolation of information is an indirect process, whose result depends on some state of the art microscopic input parameters and on the ability to build models that reproduce the relevant piece of physics. In the following, we describe two static constraints on the pinning forces and on entrainment, coming, respectively, from the observation of glitches of large amplitude and very active pulsars. Then, we discuss the information that can be extracted from the dynamical modelling of the spin-up phase.

5.1. *Static constraint from the maximum glitch amplitude*

Observations of glitches of large frequency jump $\Delta\nu$ can be used to understand if the microscopically calculated pinning forces are enough to explain it. Within the assumption of circular motion described in the previous section, from (42) we have that (ignoring the external torque that acts on much longer timescales)

$$2\pi\Delta\nu = -\frac{I_v}{I}\langle\Delta\Omega_{vp}\rangle\,, \tag{47}$$

where $\Delta\Omega_{vp}$ is the difference between the lag after and before the glitch, locally. The exact difference $\Delta\Omega_{vp}$ is impossible to observe or estimate and it also depends on the past history of the pulsar. However, we can give a theoretical upper limit Ω^*_{vp} to the local lag Ω_{vp}: defining as f_P the maximum Magnus force per unit length that pinning can sustain, from $|\mathbf{f}_M| = f_P$ with $\mathbf{v}_L = \mathbf{v}_p$ we obtain

$$\Omega_{vp}(t,x,z) < \Omega^*_{vp}(x,z) = \frac{f_P(r)(1-\epsilon_n(r))}{\kappa\,\rho_n(r)\,x}\,. \tag{48}$$

At this point, an upper limit $\Delta\Omega_{\max}$ on the glitch amplitude can be obtained by assuming that before the glitch the angular momentum reservoir was full (its maximum capacitance being defined by the pinning forces), while it is fully emptied after the glitch. It follows that

$$2\pi\Delta\nu < \Delta\Omega_{\max} = \frac{I_v}{I}\langle\Omega^*_{vp}\rangle\,, \tag{49}$$

namely (we recall that ν is the frequency, not the angular velocity)

$$2\pi\Delta\nu < \Delta\Omega_{\max} = \frac{\pi^2}{I\kappa}\int dr\, r^3 f_P(r)\,. \tag{50}$$

Interestingly, the above result does not depend neither on the entrainment parameter ϵ_n nor the density $\rho_n(r)$, but it only depends on the radial profile $f_P(r)$ and on the total moment of inertia I, which in turn depends on the total density profile $\rho(r)$ only. Clearly, the region where $f_P(r) > 0$, namely the region that contributes to the integral in (50), should be contained within the region defined by $\rho_n(r) > 0$. Moreover, the constraint (50) has been derived in the limit of slack vortices, but the exact same result is valid also for the extreme scenario of straight vortices discussed below (43).

When general relativistic corrections in the slow rotation approximation (Andersson and Comer, 2001; Hartle, 1967) are taken into account, the constraint (50) reads (Antonelli *et al.*, 2018)

$$\Delta\Omega_{\max} = \frac{\pi^2}{I\kappa}\int_0^{R_d} dr\, r^3\, e^{\Lambda(r)}\, \frac{\mathcal{E}(r) + P(r)}{m_n n(r) c^2} f_P(r)\,, \tag{51}$$

where Λ is the metric function which describes length dilatation, $\mathcal{E}$ is the total mass-energy density and P is the pressure and n the total baryon density (the equation-of-state (EoS), is $E = E(n)$ and $P = P(n)$). Here, the radial profiles are obtained by integrating the TOV equations, in the sense that the corrections to the stellar structure of the second order in Ω_p can be safely neglected for slowly rotating two-component NSs (Andersson and Comer, 2001).

Again, the result in (51) depends only on the pinning force profile, the mass of the star and the EoS. Thus, once the microphysical parameters like the EoS and the pinning forces are fixed, it provides a way to constrain the mass of a pulsar via the observation of a large glitch. This is represented in Fig. 5; not surprisingly, lighter pulsars can, in principle, undergo larger glitches. If we observe a large glitch of amplitude $\Delta\nu$ from a pulsar, all the input which predict a theoretical glitch smaller than the observed one are ruled out, at least for that object. By fixing the EoS and f_P, this translates into an upper limit on the mass of a particular glitching object. To date, the pulsar that displayed the glitch of largest amplitude is J0205+6449

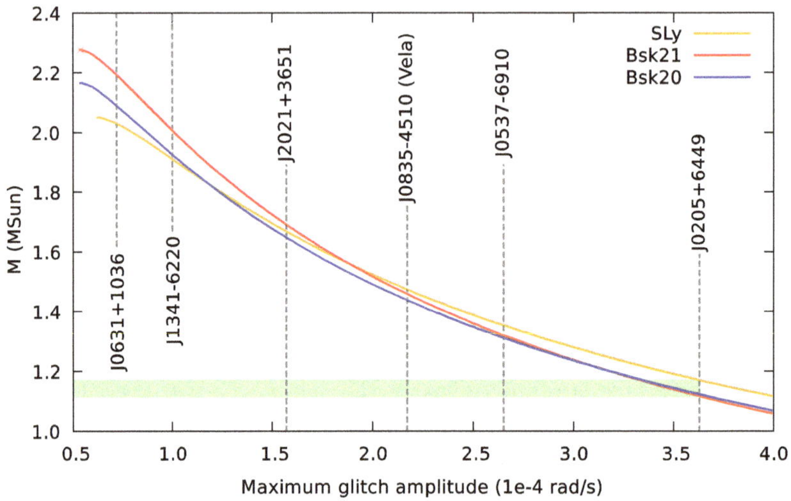

Fig. 5. The maximum glitch amplitude $\Delta\Omega_{\max}$ in (51) calculated as a function of mass for different unified EoSs: SLy4 (Douchin and Haensel, 2001), BSk20 and BSk21 (Goriely *et al.*, 2010). The largest glitch displayed by some pulsars is also shown: the record-holder is J0205+6449 (Livingstone *et al.*, 2009), that should have a mass lower than 1.1–1.2$M_\odot$, depending on the EoS employed. The pinning forces employed are the ones derived by Seveso *et al.* (2016) for the inner crust, see also, Antonelli *et al.* (2018).

(Livingstone *et al.*, 2009), that, therefore, should have a mass lower than 1.1–1.2 $M_\odot$ when the pinning forces of Seveso *et al.* (2016) are used in the inner crust. Since, the theoretical minimum mass of a NS estimated from supernovae explosions is ~1.17 $M_\odot$ (Suwa *et al.*, 2018), this means that the input used to calculate $\Delta\Omega_{\max}$ in Fig. 5 is only marginally acceptable. Hence, the largest glitch of J0205+6449 tells us that it may be necessary to extend the pinning region into the outer core, especially if even larger glitches will be detected in the future or new and more refined microscopic calculations will provide estimates of pinning forces in the crust that are lower than the ones used here.

5.2. *Static constraint from the activity parameter*

Other type of information can be obtained by considering the so-called glitch activity, that represents the mean angular acceleration due to glitches (Fuentes *et al.*, 2017; Lyne *et al.*, 2000). For a pulsar that has undergone $N_{\rm gl}$ glitches of amplitude $\Delta\Omega_i$ at times t_i with size $\Delta\Omega_i$ in a time span T_{obs}, the absolute activity can be defined as

$$\mathcal{A}_a = \frac{1}{T_{obs}} \sum_{i=1}^{N_{gl}} \Delta\Omega_i \,. \tag{52}$$

This definition works well under the assumption that the statistical properties of the stochastic sequence $(t_i, \Delta\Omega_i)$ do not depend on the window of observation, namely if the activities relative to sub-intervals are practically constant.

If T_{obs} is not know and we have only the information $(t_i, \Delta\Omega_i)$ available, a quick way to estimate $\mathcal{A}_a$ would be

$$\mathcal{A}_a \approx \frac{1}{t_{N_{\rm gl}} - t_1} \left[\sum_{i=2}^{N_{gl}-1} \Delta\Omega_i + \frac{\Delta\Omega_1 + \Delta\Omega_{N_{gl}}}{2} \right] , \tag{53}$$

which corrects the bias induced by the fact that in the interval from t_1 to $t_{N_{\rm gl}}$ starts and ends with the detection of a glitch.

Due to small-number statistics, this may be difficult to assess in practice, apart for few obvious cases like the Vela and J0537-6910, where the cumulative glitch series has a clear linear trend. For instance, too few glitches have been detected in some pulsars, so that it may not be safe to conclude that the inferred value of $\mathcal{A}_a$ corresponds to the value that would be extrapolated if the observed sequence of glitches were longer.[10]

In any case, the absolute activity is typically inferred by performing an ordinary linear regression on the cumulative glitch sequence, as in Fig. 3, even though this may lead to an underestimation of the associated uncertainty for cumulative data (Beck, 1974; Mandel, 1957; Montoli *et al.*, 2021).

Constraint on the moment of inertia of the pinning region — It is possible to infer some constraints on the moment of inertia fraction for the angular momentum reservoir by employing the activity parameter (Link *et al.*, 1999). Let us revise this argument, that is as robust as the one presented in the previous subsection, in the sense that it does not depend on the

[10]Moreover, also having a high number of glitches N_{gl} detected in a pulsar does not necessarily guarantee that the standard linear fit procedure provides a good estimate of $\mathcal{A}_a$. This can be intuitively understood by considering a sequence in which the first (or last) glitch has an amplitude much higher than the sum of all the others. A linear fit to the points of the cumulative amplitude will return a value of $\mathcal{A}_a$ that is much smaller than the one simply calculated via (52). Again, the problem is that in this extreme example the activities calculated in subintervals are not scattered around the value provided by (52).

poorly-known details of the mutual friction in (43), but only on the general statement (42). We formally divide Ω_p into the contributions due to the smooth and slow evolution observed during the waiting-times (R) and the fast one due to glitches (G), namely

$$\dot{\Omega}_p = \dot{\Omega}_p^R + \dot{\Omega}_p^G \,. \tag{54}$$

Now, we introduce an average over a long time interval of extension T,

$$[\![f]\!] = T^{-1} \int_T dt\, f(t, ...) \,, \tag{55}$$

so that, by definition,

$$[\![\dot{\Omega}_p^G]\!] = \mathcal{A}_a \,. \tag{56}$$

Consider now the total angular momentum balance (42), and assume I_v constant in time (possible readjustments of the stellar structure are neglected). We have

$$[\![I_v \langle \partial_t \Omega_{vp} \rangle]\!] + I\mathcal{A}_a + I [\![\dot{\Omega}_p^R]\!] = -I |\dot{\Omega}_\infty| \,. \tag{57}$$

The first term is always bounded due to the finiteness of the angular momentum reservoir,

$$[\![I_v \langle \partial_t \Omega_{vp} \rangle]\!] = \frac{I_v}{T} \Big(\langle \Omega_{vp} \rangle_{t=T} - \langle \Omega_{vp} \rangle_{t=0} \Big) < \frac{I_v}{T} \langle \Omega_{vp}^* \rangle = \frac{I}{T} \Delta\Omega_{\max} \,, \tag{58}$$

where we have used the average in (55). In this way, if T is long enough to guarantee $\Delta\Omega_{\max}/T \ll \mathcal{A}_a < |\dot{\Omega}_\infty|$, Eq. (57) can be approximated as

$$\mathcal{A}_a + [\![\dot{\Omega}_p^R]\!] = -|\dot{\Omega}_\infty| \,. \tag{59}$$

We can use this expression to find a theoretical upper bound on the observed activity. To do so we need to set a robust limiting value to $\dot{\Omega}_p^R$. As discussed in Section 4, we can invoke the extreme scenario in which the vortices are pinned between glitches ($\dot{\Omega}_v^R = 0$). This idealized situation is probably never realized in the real pulsar because of both thermal creep (Alpar *et al.*, 1984b) and disorder in the pinning landscape (Antonelli and Haskell, 2020), so we can safely conclude that

$$[\![\dot{\Omega}_p^R]\!] < -\frac{I\,|\dot{\Omega}_\infty|}{I - I_v} . \tag{60}$$

Finally, plugging (59) into the inequality above gives the sought upper bound on the observed activity,

$$\frac{\mathcal{A}_a}{|\dot{\Omega}_\infty|} < \frac{I_v}{I - I_v} . \tag{61}$$

The above constraint reduces to the one derived by Link *et al.* (1999) for $\epsilon_n = 0$. The entrainment corrections to the moments of inertia were first derived by Chamel and Carter (2006): their result, albeit derived by using a different — but equivalent — Newtonian formalism, is consistent with (61). Relativistic corrections to I_v can also be included (Antonelli *et al.*, 2018; Montoli *et al.*, 2021).

The constraint (61) is usually derived by starting from a model with two rigid components rotating with uniform angular velocities Ω_p and Ω_n. Within this framework and in the limit of small I_v, the constraint turns out to be (Andersson *et al.*, 2012; Chamel, 2013; Chamel and Carter, 2006)

$$\frac{\mathcal{A}_a}{|\dot{\Omega}_\infty|} < \frac{I_n}{(1 - \langle\epsilon_n\rangle)\, I} ; \quad I_n = \lim_{\epsilon_n \to 0} I_v , \tag{62}$$

where $\langle\epsilon_n\rangle$ is the averaged entrainment in the superfluid domain, see also Sidery *et al.* (2010).

If we assume the pinned superfluid region to be limited in the crust of the star, and we fix the microphysical parameters, namely the EoS and the entrainment parameter, then the moments of inertia I_v and I in (61) depend only on the stellar mass. Therefore, similarly to the case for maximum glitch amplitude, also here we obtain a constraint on the mass of the pulsar (see Fig. 6). Apart from the EoS, and in both cases the constraint depends on a particular microphysical parameter: the maximum pinning force for the constraint from the largest glitch and the entrainment for the activity constraint. Thus, the two different methods would allow for two tests on completely different microphysical parameters, if the real mass of the star were known.

The entrainment parameter for the super-flow in the crust of a NS has been obtained in Chamel (2012), by calculating the effects of Bragg scattering due to the presence of the crustal lattice. The main result is a negative entrainment parameter $\epsilon_n \sim -10$ in the crust of the star, which

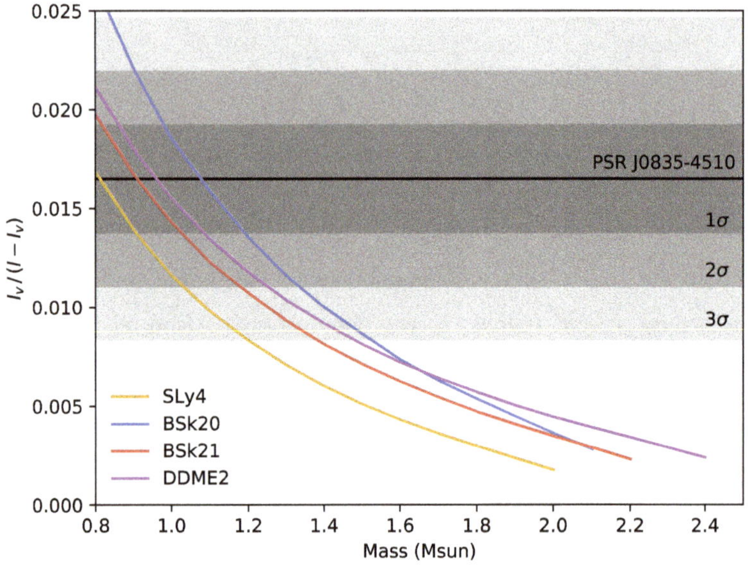

Fig. 6. The curve defining the activity constraint (61) for some EoSs. Here, the expression of I_v is the general relativistic one in the slow rotation approximation (Antonelli *et al.*, 2018). The superfluid reservoir is limited to the inner and the entrainment is that of Chamel (2012). The ratio $\mathcal{A}_a/|\dot{\Omega}_\infty|$ is plotted for the Vela pulsar, along with the uncertainty as obtained via the bootstrap procedure described in Montoli *et al.* (2021).

implies a severely reduced superfluid density ρ_n^* and a smaller I_v, making the requirement in (61) more difficult to be met.In fact, studies on the activity parameter for the Vela yield an unrealistically low constraint on the mass of the pulsar for many EoSs, if a crustal reservoir and the entrainment calculated in Chamel (2012) are assumed (see, Andersson *et al.*, 2012; Chamel, 2013; Delsate *et al.*, 2016). Several ways have been proposed to overcome this impasse, including the use of stiff EoSs (Piekarewicz *et al.*, 2014), a Bayesian analysis of the EoS uncertainty (Carreau *et al.*, 2019) or an extension of the neutron superfluid participating in the glitch beyond the crust-core transition (Ho *et al.*, 2015; Montoli *et al.*, 2020b).

Furthermore, the exact values of the entrainment parameter in the inner crust are still matter of debate. Apart from calculation based on the effects of band structure on the effective neutron superfluid density in the crust (Carter *et al.*, 2005; Chamel, 2012), other alternative estimates have been proposed, which include hydrodynamical approaches (Martin and Urban, 2016) or Bragg scattering in a disordered crustal lattice

(Sauls *et al.*, 2020), both of which yield milder entrainment effects in the crust. According to the analysis of Watanabe and Pethick (2017), the effects of band structure — that should lead to strong entrainment and to a smaller value of I_v — are suppressed when the pairing gap is of order or greater than the strength of the crustal lattice potential — see the discussion in Chamel (2017a). In all those cases, the reduction of the neutron superfluid density would be less than estimated on the basis of band theory calculations and, consequently, it may be premature to rule out models of glitches based on superfluidity in the crust alone.

Finally, different ways of estimating the activity parameter yield different values for the uncertainty of this parameter. For example, dropping the homoscedasticity hypothesis for the cumulative glitch amplitude data that define the glitch activity gives rise to larger uncertainties (see Fig. 6 and Montoli *et al.* (2021)).

5.3. *Stronger constraints*

As soon as we tentatively extend the pinning region partially in the inner core, the moment of inertia I_v grows so much that the curves in Fig. 6 are shifted upwards in such a way that the mass of the Vela is consistent with any value smaller than the maximum one allowed by the EoS. However, observations of both the maximum glitch amplitude and the glitch activity may be used in tandem to put stronger — but less robust — constraints on the pulsar's mass (Montoli *et al.*, 2020b). In fact, while the constraints coming from the maximum glitch amplitude and the activity are derived on the basis of the angular momentum balance only, we now have to model the internal dynamics, which amounts to select a particular mutual friction prescription.

Starting from corotation of the two components, forward integration in time of the dynamical Eqs. (42) and (43) allows to follow the evolution of the lag $\Omega_{vp}(x, z, t)$ and of the associated quantity $\Delta\Omega_m(t)$ defined as

$$\Delta\Omega_m(t) = \frac{I_v}{I}\langle \Omega_{vp} \rangle < \Delta\Omega_{\max} , \tag{63}$$

that is nothing but a theoretical upper bound on the amplitude of a glitch triggered at time t (assuming that at $t = 0$ the star was in a state of corotation). In other words, $\Delta\Omega_m(t)$ sets a theoretical upper limit for the glitch amplitude at time t, given that a large glitch occurred at $t = 0$.

Again, since this is just an upper bound, it is not really important if glitches in real pulsars can empty the whole reservoir: this is just an assumption to construct the theoretical upper bound $\Delta\Omega_m(t)$.

The information coming from the glitch activity is used to roughly estimate the time t needed to replenish the angular momentum reservoir to a level that is sufficient for the largest glitch amplitude $\Delta\Omega_{\rm obs}$ observed in the pulsar under study, $t_{act} \sim \Delta\Omega_{\rm obs}/\mathcal{A}_a$. Asking that $\Delta\Omega_m(t_{act}) > \Delta\Omega_{\rm obs}$ provides the sough constraint on the pulsar's mass, once the microphysical input and the mutual friction prescriptions are fixed.

An alternative and interesting possibility is to adapt the activity constraint in (61) by taking into account the thermal history of a pulsar. Ho *et al.* (2015) used (62) to constrain the mass of a sample of active pulsars by considering a superfluid reservoir that extends into the outer core, accounting for the temperature dependence of superfluidity. They inferred the temperature of pulsars from observations to constrain how much the superfluid reservoir extends into the core. This can be done by choosing a model of critical temperature of neutron (singlet-state) superfluidity as a function of baryon number density (Lombardo and Schulze, 2001) and simulating the internal temperature of a pulsar. If the age of the pulsar is known — or, at least, its characteristic age $P/(2\dot{P})$ — this gives an estimate of where the internal temperature of the cooling core is lower than the critical temperature for superfluidity. Hence, this approach makes it possible to obtain mass upper bounds by combining pulsar glitch data and the temperature dependence of superfluidity.[11]

Differently from (63), this method does not assume anything about the mutual friction, yet some extra information is needed. Instead of simulating the internal rotational dynamics, the difficulty is to simulate the cooling, that depends on several ingredients, most importantly the superfluid gaps (Page *et al.*, 2006). Moreover, the characteristic age is not always an accurate indication of the true age of a pulsar.

[11]The cooling and rotational evolution in a NS are coupled. For example, the late-time thermal evolution is affected by the spin-down due to the deposition of heat due to the action of mutual friction: besides the neutrino and photon emissions that are energy sinks, rotational energy is converted into heat by friction due to differential rotation of the superfluid (Larson and Link, 1999; Page, 1997). Conversely, the internal temperature also regulates the thermal creep (Alpar *et al.*, 1984b) and possibly, localized heat deposition may increase the creep rate to initiate a glitch (Larson and Link, 2002; Link and Epstein, 1996).

In the end, current mass estimates of glitching pulsar like the ones presented in Ho *et al.* (2015) and Pizzochero *et al.* (2017) should be taken more as an indication of how much is the influence of the assumed microscopic parameters on the final results.

6. Dynamical Constraints from Spin-up Observations

We already discussed in Section 2.1, how observations of the slow relaxation after a glitch is the main evidence for the presence of superfluid components in NSs. Also, the short spin-up timescale associated to the 2000 and 2004 glitches in the Vela pulsar — of the order of less than a minute (Dodson *et al.*, 2007, 2002) — and the particular slow-rise detected in the Crab pulsar (Shaw *et al.*, 2018) can be interpreted as hints for the fact that the mutual friction is not simply linear, but that, on the contrary, its strength is modulated as the superfluid evolves across the various dynamical phases (Celora *et al.*, 2020; Gügercinoğlu and Alpar, 2019; Haskell *et al.*, 2018). Despite this is the general picture at the basis of the vortex-mediated friction, we anticipate that the mutual friction is typically taken to be linear — as in the system (65) below — as long as the model is supposed to be applied only to a single dynamical phase (see, e.g., Haskell *et al.*, 2012; Graber *et al.*, 2018; Sourie and Chamel, 2020; Pizzochero *et al.*, 2020).

The recent detection of a glitch in the Vela pulsar in December 2016 (Palfreyman *et al.*, 2018) has opened a window on the elusive spin-up phase. The peculiarity of this detection lies in the fact that this is the first pulse-to-pulse observation of a glitch in the act, allowing for unprecedented timing resolution of the spin-up and the following phases (see Section 4). In fact, analysis of the phase residuals revealed that the initial fast spin-up occurs over a timescale $\tau_{exp} \lesssim 10\,\text{s}$, as well as the presence of an *overshoot* in the data in the first seconds of the glitch (Ashton *et al.*, 2019; Pizzochero *et al.*, 2020), a transient dynamical phase sometimes observed in glitch simulations that account for the non-uniformity of the superfluid velocity lag and stellar stratification (Antonelli and Pizzochero, 2017; Graber *et al.*, 2018; Haskell *et al.*, 2012; Larson and Link, 2002).

It is possible to use the minimal Baym's model in (5) to fit the data of Palfreyman *et al.* (2018) and extract a coupling timescale that, in turn, may be used to infer the mutual friction strength $\mathcal{B}_d$ during the spin-up. Taking the spin-up timescale to be $\tau \approx \tau_{exp} \sim 10\,\text{s}$ and assuming that the glitch is due to the superfluid in the crust — the superfluid in the core is assumed to be corotating with the normal (Alpar *et al.*, 1984a), so that $I_v/I \sim 1\%$ – from (46) we conclude that[12]

$$\begin{aligned} \langle \mathcal{B}_d \rangle &\sim 10^{-2} \quad \text{if } \langle \epsilon_n \rangle \sim 10\,, \\ \langle \mathcal{B}_d \rangle &\sim 10^{-3} \quad \text{if } \langle \epsilon_n \rangle \ll 1\,, \end{aligned} \tag{64}$$

in the inner crust crust, where we recall that $2\Omega_p \sim 100\,\text{rad/s}$ for the Vela. The two cases refer, roughly, to the strong (Chamel, 2012, 2017a) and weak (Martin and Urban, 2016; Sauls *et al.*, 2020) entrainment cases, respectively.

The rough estimate in (64) depends on the fact that we reduced a fluid model to a much simpler simple model with only two rigid components. Clearly, such a simplification cannot be valid if, for example, the superfluid in the core is coupled to the normal component on a timescale that is comparable or longer to the one of the superfluid in the crust. Moreover, it is not possible to reproduce the overshoot feature since the response of the system in (5) can only fit an exponential rise.

6.1. *Models with three rigid components*

The minimal glitch model which allows for an overshoot requires at least three rigid components. This can be formally obtained from the two-component fluid Eqs (42) and (43) by splitting the whole superfluid domain into two non-overlapping regions (Montoli *et al.*, 2020a). Taking the average over these two regions — see the discussion below (46) — gives us the three-component analogue of the Baym's model,

$$\begin{aligned} &x_p\dot{\Omega}_p + x_1\dot{\Omega}_1 + x_2\dot{\Omega}_2 = -|\dot{\Omega}_\infty|\,, \\ &\dot{\Omega}_1 = -b_1\,(\Omega_1 - \Omega_p), \\ &\dot{\Omega}_2 = -b_2\,(\Omega_2 - \Omega_p), \end{aligned} \tag{65}$$

where p indicates the normal (observable) component, $i = 1,2$ labels the two superfluid regions, $x_i = I_i/I$ is the moment of inertia fraction of the region i and b_i is the inverse of the coupling timescale between the i region and the normal component. The normal fraction $x_p = 1 - x_1 - x_2$ is not an independent parameter of the model. The system (65) can be solved

[12]The estimate $\mathcal{B}_d \sim 10^{-5}$ given in Eq. (2) of Ashton *et al.* (2019) is based on the assumption of a single superfluid component that is limited to the inner crust ($I_v \sim 0.01I$). Since they do not explicitly include entrainment, it should be reproduced by our $\langle \epsilon_n \rangle \ll 1$ case. The two orders of magnitude discrepancy is due to a missing I_v/I factor.

analytically. Following the same notation used in (5), the angular velocity residue with respect to the steady-state spin-down law reads

$$\Delta\Omega_p(t) = \Delta\Omega_p^{\infty}\left[1 - \omega\, e^{-t\lambda_+} - (1-\omega)\, e^{-t\lambda_-}\right], \tag{66}$$

where ω, $\Delta\Omega_p^{\infty}$ and $\lambda_\pm$ are functions of the parameters $x_{1,2}$, $b_{1,2}$ and $|\dot{\Omega}_\infty|$, as well as of the initial conditions for the two superfluids and the normal components. In particular, $\Delta\Omega_p^{\infty}$ is the angular velocity residue of the star at large t, after both exponential transients have fully relaxed.

To extract information on the parameters $x_{1,2}$ and $b_{1,2}$, the formula (66) must be supplemented with the exact dependence of its four parameters ω, $\Delta\Omega_p^{\infty}$ and $\lambda_\pm$ on the physical parameters $x_{1,2}$, $b_{1,2}$ and the initial conditions. For example, the two relaxation timescales $1/\lambda_\pm$ in (66) are functions of all the four parameters $b_{1,2}$ and $x_{1,2}$,

$$\lambda_\pm = \frac{1}{2}\Bigg[\beta_1(1-x_2) + \beta_2(1-x_1) \pm \sqrt{[\beta_1(1-x_2) + \beta_2(1-x_1)]^2 - 4\beta_1\beta_2 x_p}\Bigg], \quad \beta_i = b_i/x_p\,. \tag{67}$$

The explicit formulae for ω and $\Delta\Omega_p^{\infty}$, as well as the link of the averaged physical parameters $b_{1,2}$ and $x_{1,2}$ with the local microscopic input (including entrainment), are given in Montoli *et al.* (2020a).

Physical interpretation of the parameters — Exactly as the Baym's model, also the system (65) is agnostic about the location of the superfluid components, so that the moment of inertia fractions $x_{1,2}$ of the two superfluid components do not necessarily correspond to crust and core superfluid. For example, the two superfluid components may be both located in the core of the star (Sourie and Chamel, 2020). The hope is to fit (66) from observations of a well-resolved spin-up, and then to infer the most likely values for $x_{1,2}$ and $b_{1,2}$, so to constrain the superfluid regions and the (averaged) friction parameters.

In the original Baym's model defined in (1) and (2) the entrainment coupling was not explicit, but hidden into the expressions of the phenomenological parameters in terms of the local ones. We can reduce the full fluid problem to the set of phenomenological ordinary differential equations in (65) by following the same reasoning that led to (46). Now, instead of introducing a single average over the whole superfluid region, we formally split it into two non-overlapping domains (labelled by $i = 1, 2,$),

$$dI_v = d^3x\, x^2\, \rho_n^*; \quad I_i = \int_i dI_v; \quad \langle f \rangle = I_i^{-1} \int_i dI_v\, f\,. \tag{68}$$

In this way, the inverse coupling timescales and the moments of inertia fractions can be expressed as (compare with Eq. 46)

$$\begin{aligned} &x_i = I_i/I\,; && x_p = (I - I_1 - I_2)/I\,; \\ &\Omega_1 = \langle \Omega_v \rangle_1\,; && \Omega_2 = \langle \Omega_v \rangle_2\,; \\ &b_i \approx 2\Omega_p \left\langle \frac{\mathcal{B}_d}{1-\epsilon_n} \right\rangle_i && \approx 2\Omega_p \frac{\langle \mathcal{B}_d \rangle_i}{\langle 1-\epsilon_n \rangle_i}\,. \end{aligned} \tag{69}$$

Again, the presence of entrainment affects the interpretation of the macroscopic parameters x_i and b_i in terms of the local input.

It is worth to remark that the idea behind the reduction to a model with rigid components is that the correlations in (45) are negligible. Hence, it may be a good strategy to identify the two regions in such a way that the resulting parameters $b_{1,2}$ are as different as possible, so that the superfluid tend to respond in a as uniform as possible way within a region. Because of this, it is reasonable to assume that the two superfluid components represent regions with different physical properties (like the superfluid in the crust and the one in the inner core, but we will see that this is not the only possibility).

Finally, the identities in (69) also tell us that the description of the dynamical phases in Section 4 is still valid, since

$$I_v \, \langle \partial_t \Omega_v \rangle = I(x_1 \dot{\Omega}_1 + x_2 \dot{\Omega}_1)\,; \quad I_v = I(x_1 + x_2)\,. \tag{70}$$

In fact, the averaging procedure used to reduce a fluid model to a model with two or more rigid components reproduces perfectly the the total angular momentum balance (42).

6.2. *Overshoot condition*

The maximum value $\Delta\Omega_{\text{over}}$ of the angular velocity jump during a glitch can be obtained by finding the maximum value that can be attained in (66). There are two possible situations: the maximum of $\Delta\Omega_p(t)$ is reached asymptotically — namely $\Delta\Omega_{\text{over}} = \Delta\Omega_p^\infty$ at large times — or there exists a finite time for which $\Delta\Omega_{\text{over}} > \Delta\Omega_p^\infty$. In the latter case, the solution $\Delta\Omega_p(t)$ describes an overshoot and this happens if $\omega > 1$. However, the parameter ω is, in general, a rather complex function of the initial lags and the physical parameters, so that this overshoot condition is not of simple physical interpretation.

It is possible to simplify the problem by restring our attention to a particular subset of solutions. The simplest overshoot condition can be found when the unknown initial conditions rely on some physical assumptions on the role of the two components (Pizzochero *et al.*, 2020). One of the two superfluid components, say 1, is assumed to be a "passive" one that does not change its creep rate — meaning that b_1 is constant before and after the triggering of the glitch. Hence, it is reasonable to set its initial lag to the steady-state value, $\Omega_1 - \Omega_p = |\dot{\Omega}_\infty|/b_1$ at $t = 0$. On the other hand, the component 2 can be assumed to be the "active" one, namely the component that undergoes a transition from a pinned to an unpinned state. Since this is the component that is pinned before the glitch, it is also the component that provides the extra angular momentum for the observed spin-up.

In this setting, it is possible to show that the mathematical condition for an overshoot is just $b_1 < b_2$: the post-glitch timescale $1/b_2$ of the active component (that in the pre-glitch state was only loosely coupled to the rest of the star because of pinning) must be smaller than the coupling timescale $1/b_1$ of the passive component (Pizzochero *et al.*, 2020). From the physical point of view, the overshoot occurs when the superfluid region that stores the angular momentum for the glitch (the active component) can transfer its excess of angular momentum to the normal component faster than the typical timescale the rest of the superfluid reacts with. This idea is further clarified by considering the limit $b_1 \ll b_2$: the passive component maintains its state of motion, being practically decoupled, while the active superfluid imparts its momentum to the normal component only, spinning it up. At this point, the lag between the passive superfluid and the normal component is smaller than its positive steady-state value or even possibly reversed. The post-overshoot relaxation, during which the lag $\Omega_1 - \Omega_p$ will tend to approach again the steady-state value, is modulated by the slower response of the passive superfluid.

This kind of behaviour is also clear from the simulations of the Vela 2016 glitch performed by Graber *et al.* (2018), where the passive and active components are assumed to be the superfluid in the core and the one in the inner crust, respectively.

Sourie and Chamel (2020) applied the above picture to study the possible impact of the pinning to the flux-tubes in the outer core, where the magnetic field is expected to be predominantly toroidal and pinning to fluxoids may be the most effective (Gügercinoglu and Alpar, 2014). In their setting the active component is the superfluid in the outer core, while the passive one is the superfluid in both the inner core and crust. In this way, they have been able to translate the simplified overshoot condition $b_1 < b_2$ into a statement on the number of fluxoids that are, on average, pinned to a vortex. When the number of pinned fluxoids exceeds a certain threshold, the parameter b_2 is suppressed (pinning tends to suppress mutual friction) and no overshoot is possible.

6.3. *Constraint on the moment of inertia of the passive component*

By using a three-component model of the kind in (65) and its particular solution given in Pizzochero *et al.* (2020), Sourie and Chamel (2020) proposed a way to constrain the moment of inertia fraction x_1 of the passive component from the observation of an overshoot in the angular velocity residue. This constraint can also be derived directly from the full solution to the problem including entrainment corrections (Montoli *et al.*, 2020b).

We have seen that an overshoot is likely to occur when $b_1 < b_2$. By expanding the value of $\Delta\Omega_{\rm over}$ in the limit $b_1 \ll b_2$, it is possible to show that

$$x_1 > 1 - \Delta\Omega_p^\infty / \Delta\Omega_{\rm over}\,, \quad \text{if} \quad b_1 \ll b_2\,. \tag{71}$$

Hence, the a measurement of both the asymptotic amplitude $\Delta\Omega_p^\infty$ and of the maximum amplitude at the peak of the overshoot $\Delta\Omega_{\rm over}$ provides a lower bound to the superfluid fraction x_1. For example, observing $\Delta\Omega_{\rm over} \approx 1.1 \times \Delta\Omega_p^\infty$ would immediately imply $x_1 > 10\%$, ruling out the possibility that the passive component is the inner crust alone.

Both $\Delta\Omega_p^\infty$ and $\Delta\Omega_{\rm over}$ may be obtained by fitting the data of a spin-up observation: we try to test the constraint (71) with the data of the Vela 2016 glitch in Fig. 7. The median of the probability distribution of x_1 is larger than that of $1 - \Delta\Omega_p^\infty/\Delta\Omega_{\rm over}$. In the covariance plot most of the points fall in the region below the diagonal, thus respecting the constraint in (71). Since (71) is valid in the limit $b_1 \ll b_2$ (but the analysis in Fig. 7 is not), this also is a clue that it is likely that in the 2016 Vela glitch the two coupling parameters $b_{1,2}$ have values different by at least an order of magnitude. In fact, posteriors for $b_{1,2}$ (as evaluated in Montoli *et al.*, 2020a) indicate that the 16th-84th percentile intervals are $b_1 \sim 0.004 - 0.009\,\mathrm{s}^{-1}$ and $b_2 \sim 0.08 - 24.64\,\mathrm{s}^{-1}$.

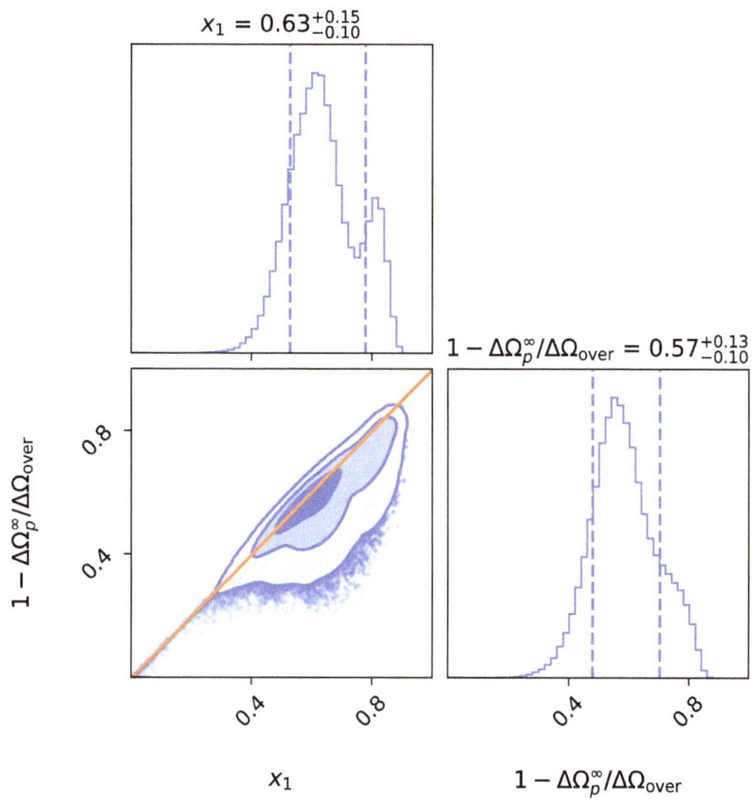

Fig. 7. Corner plot for the posterior distribution of the quantities x_1 and $1 - \Delta\Omega_p^\infty/\Delta\Omega_{\rm over}$ fitted from the Vela 2016 spin-up (Montoli *et al.*, 2020a). The orange diagonal represents the constraint (71) of Sourie and Chamel (2020): most of the posterior points satisfy the constraint.

7. Conclusions

We discussed some of the fundamental ideas behind the current understanding and modelling of pulsar glitches. Since the phenomenological minimal model of Baym *et al.* (1969) was developed, we now have now a better understanding of how the hydrodynamic modelling of NS interiors works. While phenomenological models are useful to fit observations, information on the physics of the interior of a NS can be extracted only if we can reliably link the parameters of the fits to a more fundamental physical description of the system.

As a pedagogical exercise, we derived — starting from a local fluid description that is more directly linked to the microphysics of the system — the axially symmetric equations that constitute the backbone of several glitch models. In this way, the phenomenological parameters of Baym's model (or its natural three-component generalization) can be expressed in terms of the local properties of dense nuclear matter.

We have revised the current state of the art in constraining some aspects of NS structure from timing observations of pulsar glitches. We discussed two robust constraints coming, respectively, from the observation of glitches of large amplitude (Antonelli *et al.*, 2018) and from the average glitch activity over an extended period of time (Andersson *et al.*, 2013; Carreau *et al.*, 2019; Chamel, 2013; Delsate *et al.*, 2016; Link *et al.*, 1999; Montoli *et al.*, 2021; Piekarewicz *et al.*, 2014). Observation of large glitch amplitudes (the record holder is J0205+6449) can be used to test microscopic estimates of pinning forces, while the high activity of some pulsars (the Vela being the most active) provides a way to test the entrainment coupling. The two tests are completely independent (apart from the fact that both rely on the choice of an EoS) and seem to indicate that part of the superfluid in the outer core should participate in storing angular momentum: the upper limit on the inferred mass coming from the constraints is, for some objects, close or smaller than the theoretical minimum NS mass of $1.17\,M_\odot$ inferred from supernovae explosion simulations (Suwa *et al.*, 2018).

This is interesting, as pinning of vortices to flux tubes is actually a possible mechanism to store angular momentum, as reviewed by Alpar (2017). Those two static constraints may always be refined, but at the expense of solving the hydrodynamic equations in the NS interiors, or invoking some simplified model that accounts for the angular momentum exchange between the components (Montoli *et al.*, 2020b).

Then, we moved to consider a class of dynamical studies (as opposed to the two static constraints from glitch amplitudes and activities). In particular, the possibility of observe all the phases of a glitch in the act (Palfreyman *et al.*, 2018) justifies the efforts towards a better understanding and less phenomenological modelling of the processes that mediate the angular momentum exchange in a glitch. For instance, our ability to extract information on NS's structure and superfluid properties from glitch observations is directly linked to our ability to develop and solve reliable hydrodynamic models for the internal torques between the components of a mature NS.

In turn, this boils down to understand the vortex-mediated mutual friction between the superfluid components and the normal (observable) one.

One of the most important points is that the standard linear model of mutual friction (Andersson *et al.*, 2006; Epstein and Baym, 1992; Mendell, 1991), is modified by the presence of impurities (i.e., pinning, Alpar *et al.* (1984b); Bulgac *et al.* (2013); Seveso *et al.* (2016) and Wlazlowski *et al.* (2016)), quantum turbulence (Andersson *et al.*, 2007; Haskell *et al.*, 2020), long-range mutual interaction between vortices (Howitt *et al.*, 2020) and memory effects including hysteresis (i.e., kinetic processes that are out of equilibrium, see Antonelli and Haskell (2020)). All these issues regarding the vortex-mediated mutual friction still need further investigation and understanding. This is not surprising, since some aspects of vortex (or flux-tube) creep and pinning are still not fully understood even in the context of terrestrial superconductors (Blatter *et al.*, 1994; Reichhardt and Nori, 1999), where the observed complex dynamics and memory phenomena arise from the fine balance between competing processes that are included in the equation of motion for the vortices (or flux-tubes).

Hopefully, as more resources will be devoted to the continuous monitoring of promising active pulsars — like Crab (Basu *et al.*, 2020; Shaw *et al.*, 2018), the Vela (Palfreyman *et al.*, 2018) and J0537-6910 (Ho *et al.*, 2020) — a whole new wealth of data will allow to put more stringent constraints on NS structure with the aid of next-generations glitch models.

Planned multi-element telescopes — such as the Square Kilometer Array (Stappers *et al.*, 2018) — can also be extremely useful for high cadence pulsar monitoring, especially if multiple subarrays are available (Basu *et al.*, 2020). When coupled with automated glitch detection pipelines, this will considerably increase the number and quality of detected glitches across the pulsar population, ushering in a new era of discovery in a field that has entered into its 50s (Bell Burnell, 2017).

Acknowledgements

Partial support comes from PHAROS, COST Action CA16214, and from INFN, the Italian Institute for Nuclear Physics. Marco Antonelli acknowledges support from the Polish National Science Centre grant SONATA BIS 2015/18/E/ST9/00577, P.I.: B. Haskell.

References

Akbal, O., Alpar, M. A., Buchner, S., and Pines, D. (2017). Nonlinear interglitch dynamics, the braking index of the Vela pulsar and the time to the next glitch, *Monthly Noices of the Royal Astronomical Society*, **469**, pp. 4183–4192.

Akbal, O., Gügercinoğlu, E., and Alpar, M. A. (2021). Glitch analysis and braking index determination of the unique source PSR J0537-6910, arXiv e-prints, arXiv:2106.03341.

Alpar, M. A. (1977). Pinning and threading of quantized vortices in the pulsar crust superfluid, *Astrophysical Journal*, **213**, pp. 527–530.

Alpar, M. A. (2017). Flux-Vortex pinning and neutron star evolution, *Journal of Astrophysics and Astronomy*, **38**(3), p. 44.

Alpar, M. A., Chau, H. F., Cheng, K. S., and Pines, D. (1994). Postglitch relaxation of the Crab pulsar: Evidence for crust cracking, *Astrophysical Journal Letters*, **427**, pp. L29–L31.

Alpar, M. A., Langer, S. A., and Sauls, J. A. (1984a). Rapid postglitch spin-up of the superfluid core in pulsars, *Astrophysical Journal*, **282**, pp. 533–541.

Alpar, M. A., Pines, D., Anderson, P. W., and Shaham, J. (1984b). Vortex creep and the internal temperature of neutron stars. I - General theory, *Astrophysical Journal*, **276**, pp. 325–334.

Anderson, P. W. (1966).Considerations on the flow of superfluid helium, *Reviews of Modern Physics*, **38**, pp. 298–310.

Anderson, P. W. and Itoh, N. (1975). Pulsar glitches and restlessness as a hard superfluidity phenomenon, *Nature*, **256**(5512), p. 25.

Anderson, P. W. and Kim, Y. B. (1964). Hard superconductivity: Theory of the motion of abrikosov flux lines, *Review of Modern Physics*, **36**, pp. 39–43.

Andersson, N. (2021). A superfluid perspective on neutron star dynamics, *Universe*, **7**(1), pp. 17.

Andersson, N. and Comer, G. L. (2001). Slowly rotating general relativistic superfluid neutron stars, *Classical and Quantum Gravity*, **18**, pp. 969–1002.

Andersson, N. and Comer, G. L. (2007). Relativistic fluid dynamics: Physics for many different scales, *Living Reviews in Relativity*, **10**, pp. 1–83.

Andersson, N. and Comer, G. L. (2011). Entropy entrainment and dissipation in finite temperature superfluids, *International Journal of Modern Physics D*, **20**(7), pp. 1215–1233.

Andersson, N., Comer, G. L., and Prix, R. (2003). Are pulsar glitches triggered by a superfluid two-stream instability? *Physics Review Letter*, **90**, p. 091101.

Andersson, N., Glampedakis, K., Ho, W. C. G., and Espinoza, C. M. (2012). Pulsar glitches: The crust is not enough, *Physical Review Letters*, **109**(24), p. 241103.

Andersson, N., Krüger, C., Comer, G. L., and Samuelsson, L. (2013). A minimal model for finite temperature superfluid dynamics, *Classical and Quantum Gravity*, **30**(23), p. 235025.

Andersson, N., Sidery, T., and Comer, G. L. (2006). Mutual friction in superfluid neutron stars, *Monthly Notices of the Royal Astronomical Society*, **368**(1), pp. 162–170.

Andersson, N., Sidery, T., and Comer, G. L. (2007). Superfluid neutron star turbulence, *Monthly Notices of the Royal Astronomical Society*, **381**(2), p. 747.

Andreev, A. F. and Bashkin, E. P. (1976). Three-velocity hydrodynamics of superfluid solutions, *Soviet Journal of Experimental and Theoretical Physics*, **42**, p. 164.

Antonelli, M. and Haskell, B. (2020). Superfluid vortex-mediated mutual friction in non-homogeneous neutron star interiors, *Monthly Notices of the Royal Astronomical Society*, **499**(3), pp. 3690–3705.

Antonelli, M., Montoli, A., and Pizzochero, P. M. (2018). Effects of general relativity on glitch amplitudes and pulsar mass upper bounds, *Monthly Notices of the Royal Astronomical Society*, **475**(4), pp. 5403–5416.

Antonelli, M. and Pizzochero, P. M. (2017). Axially symmetric equations for differential pulsar rotation with superfluid entrainment, *Monthly Notices of the Royal Astronomical Society*, **464**, p. 721–733.

Antonopoulou, D., Espinoza, C. M., Kuiper, L., and Andersson, N. (2018). Pulsar spin-down: the glitch-dominated rotation of PSR J0537-6910, *Monthly Notices of the Royal Astronomical Society*, **473**(2), pp. 1644–1655.

Ashton, G., Lasky, P. D., Graber, V., and Palfreyman, J. (2019). Rotational evolution of the Vela pulsar during the 2016 glitch, *Nature Astronomy*, **3**, pp. 1143–1148.

Ashton, G., Lasky, P. D., Nathan, R., and Palfreyman, J. (2020). Flickering of the Vela pulsar during its 2016 glitch, *arXiv e-prints*, arXiv:2011.07927.

Ashton, G., Prix, R., and Jones, D. I. (2017). Statistical characterization of pulsar glitches and their potential impact on searches for continuous gravitational waves, *Physical Review D*, **96**, 6, pp. 063004.

Bak, P., Christensen, K., and Olami, Z. (1994). Self-organized criticality: Consequences for Statistics and predictability of earthquakes, *Washington DC American Geophysical Union Geophysical Monograph Series*, **83**, pp. 69–74.

Bak, P. and Tang, C. (1989). Earthquakes as a self-organized critical phenomenon, *Journal of Geophysics Research*, **94**, B11, pp. 15,635–15,637.

Barenghi, C. F., Donnelly, R. J., and Vinen, W. F. (1983). Friction on quantized vortices in helium II. A review, *Journal of Low Temperature Physics*, **52**(3-4), pp. 189–247.

Barenghi, C. F., Donnelly, R. J., and Vinen, W. F. (eds.) (2001). *Quantized vortex dynamics and superfluid turbulence*, *Lecture Notes in Physics*, Vol. 571, Springer Berlin Heidelberg, Berlin, Heidelberg.

Basu, A., Joshi, B. C., Krishnakumar, M. A., Bhattacharya, D., Nandi, R., Bandhopadhay, D., Char, P., and Manoharan, P. K. (2020). Observed glitches in eight young pulsars, *Monthly Notices of the Royal Astronomical Society*, **491**(3), pp. 3182–3191.

Baym, G., Epstein, R. I., and Link, B. (1992). Dynamics of vortices in neutron stars, *Physica B Condensed Matter*, **178**(1–4), pp. 1–12.

Baym, G., Pethick, C., and Pines, D. (1969). Superfluidity in Neutron Stars, *Nature*, **224**(5220), pp. 673–674.

Baym, G., Pethick, C. J., Pines, D., and Ruderman, M. (1969). Spin up in neutron stars:the future of the vela pulsar, *Nature*, **224**(5222), pp. 872–874.

Beck, J. V. (1974). Parameter estimation with cumulative errors, *Technometrics*, **16**(1), pp. 85–92.

Bell Burnell, J. (2017). The past, present and future of pulsars, *Nature Astronomy*, **1**, pp. 831–834.

Blatter, G., Feigel'man, M. V., Geshkenbein, V. B., Larkin, A. I., and Vinokur, V. M. (1994). Vortices in high-temperature superconductors, *Review Modern Physics*, **66**, pp. 1125–1388.

Brand, H. and Pleiner, H. (1981). Broken symmetries and hydrodynamics of superfluid $^3p_2$-neutron-star matter, *Physics Review D*, **24**, pp. 3048–3057.

Bransgrove, A., Beloborodov, A. M., and Levin, Y. (2020). A Quake Quenching the Vela Pulsar, *The Astrophysics Journal*, **897**(2), p. 173.

Bulgac, A., Forbes, M. M., and Sharma, R. (2013). Strength of the vortex-pinning interaction from real-time dynamics, *Physics Review Letter*, **110**(24), p. 241102.

Carlin, J. B. and Melatos, A. (2019). Generating quasi-periodic pulsar glitches using a state-dependent Poisson process, *Monthly Notices of the Royal Astronomical Society*, **483**, 4, pp. 4742–4750.

Carlin, J. B. and Melatos, A. (2020). Long-term statistics of pulsar glitches triggered by a Brownian stress accumulation process, *Monthly Notices of the Royal Astronomical Society*, **494**(3), pp. 3383–3391.

Carlin, J. B. and Melatos, A. (2021). Long-term statistics of pulsar glitches due to history-dependent avalanches, *The Astrophysics Journal*, **917**(1), p. 1.

Carreau, T., Gulminelli, F., and Margueron, J. (2019). General predictions for the neutron star crustal moment of inertia, *Physical Review C*, **100**(5), p. 055803.

Carter, B. (1989). Covariant theory of conductivity in ideal fluid or solid media, Vol. 1385, Chap. 1, Berlin Springer Verlag, pp. 1–64.

Carter, B., Chamel, N., and Haensel, P. (2005). Entrainment coefficient and effective mass for conduction neutrons in neutron star crust: Simple microscopic models, *Nuclear Physics A*, **748**, pp. 675–697.

Carter, B., Chamel, N., and Haensel, P. (2006). Entrainment coefficient and effective mass for conduction neutrons in neutron star crust:. macroscopic treatment, *International Journal of Modern Physics D*, **15**, pp. 777–803.

Carter, B. and Khalatnikov, I. (1992). Momentum, vorticity, and helicity in covariant superfluid dynamics, *Annals of Physics*, **219**(2), pp. 243–265.

Celora, T., Khomenko, V., Antonelli, M., and Haskell, B. (2020). The effect of non-linear mutual friction on pulsar glitch sizes and rise times, *Monthly Notices of the Royal Astronomical Society*, **496**(4), pp. 5564–5574.

Chamel, N. (2012). Neutron conduction in the inner crust of a neutron star in the framework of the band theory of solids, *Physical Review C*, **85**(3), p. 035801.

Chamel, N. (2013). Crustal entrainment and pulsar glitches, *Physical Review Letter*, **110**(1), p. 011101.

Chamel, N. (2017a). Entrainment in superfluid neutron-star crusts: hydrodynamic description and microscopic origin, *Journal of Low Temperature Physics*, **189**(5–6), pp. 328–360.

Chamel, N. (2017b). Superfluidity and superconductivity in neutron stars, *Journal of Astrophysics and Astronomy*, **38**, p. 43.

Chamel, N. and Allard, V. (2019). Entrainment effects in neutron-proton mixtures within the nuclear energy-density functional theory: Low-temperature limit, *Physical Review C*, **100**(6), p. 065801.

Chamel, N. and Carter, B. (2006). Effect of entrainment on stress and pulsar glitches in stratified neutron star crust, *Monthly Notices of the Royal Astronomical Society*, **368**(2), pp. 796–808.

Chamel, N. and Haensel, P. (2006). Entrainment parameters in a cold superfluid neutron star core, *Physical Review C*, **73**(4), p. 045802.

Chamel, N. and Haensel, P. (2008). Physics of neutron star crusts, *Living Reviews in Relativity*, **11**(1), p. 10.

Cheng, K. S., Pines, D., Alpar, M. A., and Shaham, J. (1988). Spontaneous superfluid unpinning and the inhomogeneous distribution of vortex lines in neutron stars, *The Astrophysical Journal*, **330**, pp. 835–846.

Chugunov, A. I. and Horowitz, C. J. (2010). Breaking stress of neutron star crust, *Monthly Notices of the Royal Astronomical Society*, **407**(1), pp. L54–L58.

Crawford, F. and Demiański, M. (2003). A Comparison of Measured Crab and Vela Glitch Healing Parameters with Predictions of Neutron Star Models, *The Astrophysical Journal*, **595**, pp. 1052–1057.

Delsate, T., Chamel, N., Gürlebeck, N., Fantina, A. F., Pearson, J. M., and Ducoin, C. (2016). Giant pulsar glitches and the inertia of neutron star crusts, *Physical Review D*, **94**(2), p. 023008.

Dodson, R., Lewis, D., and McCulloch, P. (2007). Two decades of pulsar timing of Vela, *Astrophysics and Space Science*, **308**, pp. 585–589.

Dodson, R. G., McCulloch, P. M., and Lewis, D. R. (2002). High time resolution observations of the january 2000 glitch in the vela pulsar, *Astrophysical Journal Letters*, **564**, pp. L85–L88.

Donati, P. and Pizzochero, P. M. (2004). Fully consistent semi-classical treatment of vortex-nucleus interaction in rotating neutron stars, *Nuclear Physics A*, **742**, pp. 363–379.

Douchin, F. and Haensel, P. (2001). A unified equation of state of dense matter and neutron star structure, *Astronomy Astrophysics*, **380**, pp. 151–167.

Epstein, R. I. and Baym, G. (1992). Vortex drag and the spin-up time scale for pulsar glitches, *The Astrophysical Journal*, **387**, pp. 276–287.

Espinoza, C. M., Lyne, A. G., Stappers, B. W., and Kramer, M. (2011). A study of 315 glitches in the rotation of 102 pulsars, *Monthly Notices of the Royal Astronomical Society*, **414**, pp. 1679–1704.

Ferdman, R. D., Archibald, R. F., Gourgouliatos, K. N., and Kaspi, V. M. (2018). The Glitches and rotational history of the highly energetic young pulsar PSR J0537-6910, *The Astrophysical Journal*, **852**(2), p. 123.

Fuentes, J. R., Espinoza, C. M., and Reisenegger, A. (2019). Glitch time series and size distributions in eight prolific pulsars, *Astronomy Astrophysics*, **630**(A115).

Fuentes, J. R., Espinoza, C. M., Reisenegger, A., Stappers, B. W., Shaw, B., and Lyne, A. G. (2017). The glitch activity of neutron stars, *ArXiv e-prints*.

Fujita, T. and Tsuneto, T. (1972). The ginzburg-landau equation for 3P2 pairing: superfluidity in neutron stars, *Progress of Theoretical Physics*, **48**(3), pp. 766–782.

Fulgenzi, W., Melatos, A., and Hughes, B. D. (2017). Radio pulsar glitches as a state-dependent poisson process, *Monthly Notices of the Royal Astronomical Society*, **470**, 4, pp. 4307–4329.

Gavassino, L. and Antonelli, M. (2020). Thermodynamics of uncharged relativistic multifluids, *Classical and Quantum Gravity*, **37**(2), p. 025014.

Gavassino, L., Antonelli, M., and Haskell, B. (2021). Superfluid dynamics in neutron star crusts: the iordanskii force and chemical gauge covariance, *Universe*, **7**(2), p. 28.

Gavassino, L., Antonelli, M., Pizzochero, P. M., and Haskell, B. (2020). A universal formula for the relativistic correction to the mutual friction coupling time-scale in neutron stars, *Monthly Notices of the Royal Astronomical Society*, **494**(3), pp. 3562–3580.

Giliberti, E., Cambiotti, G., Antonelli, M., and Pizzochero, P. M. (2020). Modelling strains and stresses in continuously stratified rotating neutron stars, *Monthly Notices of the Royal Astronomical Society*, **491**(1), pp. 1064–1078.

Glampedakis, K. and Andersson, N. (2009). Hydrodynamical trigger mechanism for pulsar glitches, *Physical Review Letter*, **102**(14), p. 141101.

Glampedakis, K., Andersson, N., and Samuelsson, L. (2011). Magnetohydrodynamics of superfluid and superconducting neutron star cores, *Monthly Notices of the Royal Astronomical Society*, **410**, pp. 805–829.

Goriely, S., Chamel, N., and Pearson, J. M. (2010). Further explorations of Skyrme-Hartree-Fock-Bogoliubov mass formulas. XII. Stiffness and stability of neutron-star matter, *Physical Review C*, **82**(3), p. 035804.

Graber, V., Andersson, N., and Hogg, M. (2017). Neutron stars in the laboratory, *International Journal of Modern Physics D*, **26**(8), pp. 1730015–347.

Graber, V., Cumming, A., and Andersson, N. (2018). Glitch rises as a test for rapid superfluid coupling in neutron stars, *The Astrophysical Journal*, **865**(1), p. 23.

Gügercinoglu, E. and Alpar, M. A. (2014). Vortex creep against toroidal flux lines, crustal entrainment, and pulsar glitches, *The Astrophysical Journal*, **788**(1), p. L11.

Gügercinoğlu, E. and Alpar, M. A. (2020). The 2016 Vela glitch: a key to neutron star internal structure and dynamics, *Monthly Notices of the Royal Astronomical Society*, **496**(2), pp. 2506–2515.

Gügercinoğlu, E. and Alpar, M. A. (2017). Neutron star dynamics under time-dependent external torques, *Monthly Notices of the Royal Astronomical Society*, **471**(4), pp. 4827–4831.

Gügercinoğlu, E. and Alpar, M. A. (2019). The largest Crab glitch and the vortex creep model, *Monthly Notices of the Royal Astronomical Society*, **488**(2), pp. 2275–2282.

Gusakov, M. E. (2016). Relativistic formulation of the hall-vinen-bekarevich-khalatnikov superfluid hydrodynamics, *Physical Review D*, **93**, p. 064033.

Gusakov, M. E. and Andersson, N. (2006). Temperature-dependent pulsations of superfluid neutron stars, *Monthly Notices of the Royal Astronomical Society*, **372**(4), pp. 1776–1790.

Gusakov, M. E. and Dommes, V. A. (2016). Relativistic dynamics of superfluid-superconducting mixtures in the presence of topological defects and an electromagnetic field with application to neutron stars, *Physical Review D*, **94**(8), p. 083006.

Gusakov, M. E. and Haensel, P. (2005). The entrainment matrix of a superfluid neutron proton mixture at a finite temperature, *Nuclear Physics A*, **761**(3–4), pp. 333–348.

Haber, A. and Schmitt, A. (2017). Critical magnetic fields in a superconductor coupled to a superfluid, *Physical Review D*, **95**, p. 116016.

Hall, H. E. and Vinen, W. F. (1956). The rotation of liquid helium II. I. experiments on the propagation of second sound in uniformly rotating helium II, *Proceedings of the Royal Society A: Mathematical, Physical and Engineering Sciences*, **238**(1213), pp. 204–214.

Hartle, J. B. (1967). Slowly rotating relativistic stars. I. equations of structure, *The Astrophysical Journal*, **150**, p. 1005.

Haskell, B., Antonopoulou, D., and Barenghi, C. (2020). Turbulent, pinned superfluids in neutron stars and pulsar glitch recoveries, *Monthly Notices of the Royal Astronomical Society*, **499**(1), pp. 161–170.

Haskell, B., Khomenko, V., Antonelli, M., and Antonopoulou, D. (2018). Crust or core? Insights from the slow rise of large glitches in the Crab pulsar, *Monthly Notices of the Royal Astronomical Society*, **481**(1), pp. L146–L150.

Haskell, B. and Melatos, A. (2015). Models of pulsar glitches, *International Journal of Modern Physics D*, **24**(03), p. 1530008.

Haskell, B. and Melatos, A. (2016). Pinned vortex hopping in a neutron star crust, *Monthly Notices of the Royal Astronomical Society*, **461**(2), pp. 2200–2211.

Haskell, B., Pizzochero, P. M., and Sidery, T. (2012). Modelling pulsar glitches with realistic pinning forces: A hydrodynamical approach, *Monthly Notices of the Royal Astronomical Society*, **420**(1), pp. 658–671.

Haskell, B. and Sedrakian, A. (2017). Superfluidity and superconductivity in neutron stars, *ArXiv e-prints* .

Ho, W. C. G., Espinoza, C. M., Antonopoulou, D., and Andersson, N. (2015). Pinning down the superfluid and measuring masses using pulsar glitches, *Science advances*, **1**(9), p. e1500578.

Ho, W. C. G., Espinoza, C. M., Arzoumanian, Z., Enoto, T., Tamba, T., Antonopoulou, D., Bejger, M., Guillot, S., Haskell, B., and Ray, P. S. (2020). Return of the big glitcher: NICER timing and glitches of PSR J0537-6910, *Monthly Notices of the Royal Astronomical Society*, **498**(4), pp. 4605–4614.

Howitt, G., Haskell, B., and Melatos, A. (2016). Hydrodynamic simulations of pulsar glitch recovery, *Monthly Notices of the Royal Astronomical Society*, **460**(2), pp. 1201–1213.

Howitt, G., Melatos, A., and Delaigle, A. (2018). Nonparametric estimation of the size and waiting time distributions of pulsar glitches, *The Astrophysical Journal*, **867**(1), p. 60.

Howitt, G., Melatos, A., and Haskell, B. (2020). Simulating pulsar glitches: an N-body solver for superfluid vortex motion in two dimensions, *Monthly Notices of the Royal Astronomical Society*, **498**(1), pp. 320–331.

Ito, K. and Matsuzaki, M. (1990). Earthquakes as self-organized critical phenomena, *Journal of Geophysics Research*, **95**(B5), pp. 6853–6860.

Itoh, N. (1983). Prediction of pulsar glitch frequency based on the hard superfluid model, *Progress of Theoretical Physics*, **69**(1), pp. 338–340.

Jensen, H. (1998). *Self-Organized Criticality*, Cambridge University Press.

Jones, P. B. (1990). Rotation of the neutron-drip superfluid in pulsars: the resistive force, *Monthly notics of the Royal Astronomical Society*, **243**, pp. 257–262.

Jones, P. B. (1992). Rotation of the neutron-drip superfluid in pulsars: the Kelvin phonon contribution to dissipation, *Monthly Notices of the Royal Astronomical Society*, **257**, pp. 501–506.

Jones, P. B. (2003). Nature of fault planes in solid neutron star matter, *The Astrophysical Journal*, **595**(1), pp. 342–345.

Khalatnikov, I. M. (1957). Hydrodynamics of solutions of two superfluid liquids, *Soviet Journal of Experimental and Theoretical Physics*, **5**(4), p. 542.

Khomenko, V., Antonelli, M., and Haskell, B. (2019). Hydrodynamical instabilities in the superfluid interior of neutron stars with background flows between the components, *Physical Review D*, **100**(12), p. 123002.

Khomenko, V. and Haskell, B. (2018). Modelling pulsar glitches: the hydrodynamics of superfluid vortex avalanches in neutron stars, *Publications of the Astronomical Society of Australia*, **35**, p. e020.

Langlois, D., Sedrakian, D. M., and Carter, B. (1998). Differential rotation of relativistic superfluid in neutron stars, *Monthly Notices of the Royal Astronomical Society*, **297**, pp. 1189–1201.

Larson, M. B. and Link, B. (1999). Superfluid friction and late-time thermal evolution of neutron stars, *The Astrophysical Journal*, **521**(1), pp. 271–280.

Larson, M. B. and Link, B. (2002). Simulations of glitches in isolated pulsars, *Monthly Notices of the Royal Astronomical Society*, **333**(3), pp. 613–622.

Leinson, L. B. (2017). Non-linear approach to the entrainment matrix of superfluid nucleon mixture at zero temperature, *Monthly Notices of the Royal Astronomical Society*, **470**(3), pp. 3374–3387.

Leinson, L. B. (2020). Vortex lattice in rotating neutron spin-triplet superfluid, *Monthly Notices of the Royal Astronomical Society*, **498**(1), pp. 304–309.

Link, B. (2009). Dynamics of quantum vorticity in a random potential, *Physical Review Letter*, **102**(13), p. 131101.

Link, B. (2014). Thermally activated post-glitch response of the neutron star inner crust and core. I. theory, *The Astrophysical Journal*, **789**, p. 141.

Link, B. and Epstein, R. I. (1996). Thermally driven neutron star glitches, *The Astrophysical Journal*, **457**, p. 844.

Link, B., Epstein, R. I., and Lattimer, J. M. (1999). Pulsar constraints on neutron star structure and equation of state, *Physical Review Letter*, **83**, pp. 3362–3365.

Livingstone, M. A., Ransom, S. M., Camilo, F., Kaspi, V. M., Lyne, A. G., Kramer, M., and Stairs, I. H. (2009). X-ray and radio timing of the pulsar in 3C 58, *The Astrophysical Journal*, **706**(2), pp. 1163–1173.

Lombardo, U. and Schulze, H.-J. (2001). *Superfluidity in Neutron Star Matter*, Vol. 578, p. 30.

Lyne, A. G., Shemar, S. L., and Smith, F. G. (2000). Statistical studies of pulsar glitches, *Monthly Notices of the Royal Astronomical Society*, **315**, pp. 534–542.

Manchester, R. N. (2018). Pulsar glitches and their impact on neutron-star astrophysics, *arXiv e-prints*, arXiv:1801.04332.

Mandel, J. (1957). Fitting a straight line to certain types of cumulative data, *Journal of the American Statistical Association*, **52**(280), pp. 552–566.

Martin, N. and Urban, M. (2016). Superfluid hydrodynamics in the inner crust of neutron stars, *Physical Review C*, **94**(6), p. 065801.

Melatos, A., Howitt, G., and Fulgenzi, W. (2018). Size-waiting-time correlations in pulsar glitches, *The Astrophysical Journal*, **863**(2), p. 196.

Melatos, A. and Peralta, C. (2007). Superfluid turbulence and pulsar glitch statistics, *Astrophysical Journal, Letters*, **662**, pp. L99–L102.

Melatos, A., Peralta, C., and Wyithe, J. S. B. (2008). Avalanche dynamics of radio pulsar glitches, *The Astrophysical Journal*, **672**(2), pp. 1103–1118.

Mendell, G. (1991). Superfluid hydrodynamics in rotating neutron stars. II. dissipative effects, *The Astrophysical Journal*, **380**, p. 530.

Mendell, G. (1998). Magnetohydrodynamics in superconducting-superfluid neutron stars, *Monthly Notices of the Royal Astronomical Society*, **296**(4), pp. 903–912.

Middleditch, J., Marshall, F. E., Wang, Q. D., Gotthelf, E. V., and Zhang, W. (2006). Predicting the Starquakes in PSR J0537-6910, *The Astrophysical Journal*, **652**(2), pp. 1531–1546.

Mongiovì, M. S., Russo, F. G., and Sciacca, M. (2017). A mathematical description of glitches in neutron stars, *Monthly Notices of the Royal Astronomical Society*, **469**(2), pp. 2141–2150.

Montoli, A., Antonelli, M., Haskell, B., and Pizzochero, P. (2021). Statistical estimates of the pulsar glitch activity, *Universe*, **7**(1), p. 8.

Montoli, A., Antonelli, M., Magistrelli, F., and Pizzochero, P. (2020a). A Bayesian estimate of the superfluid moments of inertia from the 2016 glitch in the Vela pulsar, *arXiv e-prints*, arXiv:2005.01594.

Montoli, A., Antonelli, M., and Pizzochero, P. M. (2020b). The role of mass, equation of state, and superfluid reservoir in large pulsar glitches, *Monthly Notices of the Royal Astronomical Society*, **492**(4), pp. 4837–4846.

Morley, P. D. and Garcia-Pelayo, R. (1993). Scaling law for pulsar glitches, *Europhysics Letters*, **23**(3), pp. 185–189.

Page, D. (1997). Fast cooling of neutron stars: superfluidity versus heating and accreted envelope, *Astrophysical Journal Letter*, **479**(1), pp. L43–L46.

Page, D., Geppert, U., and Weber, F. (2006). The cooling of compact stars, *Nuclear Physics A*, **777**, pp. 497–530.

Page, D., Prakash, M., Lattimer, J. M., and Steiner, A. W. (2011). Rapid cooling of the neutron star in cassiopeia A triggered by neutron superfluidity in dense matter, *Physical Review Letter*, **106**(8), p. 081101.

Palfreyman, J., Dickey, J. M., Hotan, A., Ellingsen, S., and van Straten, W. (2018). Alteration of the magnetosphere of the Vela pulsar during a glitch, *Nature*, **556**(7700), pp. 219–222.

Peralta, C., Melatos, A., Giacobello, M., and Ooi, A. (2005). Global three-dimensional flow of a neutron superfluid in a spherical shell in a neutron star, *The Astrophysical Jurnal*, **635**(2), pp. 1224–1232.

Piekarewicz, J., Fattoyev, F. J., and Horowitz, C. J. (2014). Pulsar glitches: The crust may be enough, *Physical Review C*, **90**(1), p. 015803.

Pines, D. (1999). Pulsar Glitches: To what extent do these probe Crustal Superfluidity, Core-Crust Coupling, and the Equation of State of Dense Neutron Matter? in Z. Arzoumanian, F. Van der Hooft, and E. P. J. van den Heuvel (eds.), *Pulsar Timing, General Relativity and the Internal Structure of Neutron Stars*, p. 199.

Pines, D. and Alpar, M. A. (1985). Superfluidity in neutron stars, *Nature*, **316**(6023), pp. 27–32.

Pines, D., Shaham, J., and Ruderman, M. A. (1974). *Neutron Star Structure from Pulsar Observations,* in C. J. Hansen (ed.), *Physics of Dense Matter*, Vol. 53, p. 189.

Pizzochero, P. M. (2011). Angular momentum transfer in vela-Like pulsar glitches, *The Astrophysical Journal*, **743**(1), p. L20.

Pizzochero, P. M., Antonelli, M., Haskell, B., and Seveso, S. (2017). Constraints on pulsar masses from the maximum observed glitch, *Nature Astronomy*, **1**, p. 0134.

Pizzochero, P. M., Montoli, A., and Antonelli, M. (2020). Core and crust contributions in overshooting glitches: The Vela pulsar 2016 glitch, *Astronomy Astrophysics*, **636**, p. A101.

Posselt, B. and Pavlov, G. G. (2018). Upper limits on the rapid cooling of the central compact object in cas A, *The Astrophysical Journal*, **864**(2), p. 135.

Prix, R. (2004). Variational description of multifluid hydrodynamics: Uncharged fluids, *Physical Review D*, **69**, p. 4.

Radhakrishnan, V. and Manchester, R. N. (1969). Detection of a change of state in the pulsar PSR 0833-45, *Nature*, **222**(5190), pp. 228–229.

Rau, P. B. and Wasserman, I. (2020). Relativistic finite temperature multifluid hydrodynamics in a neutron star from a variational principle, *Physical Review D*, **102**, p. 063011.

Reichhardt, C. and Nori, F. (1999). Phase locking, devil's staircases, farey trees, and arnold tongues in driven vortex lattices with periodic pinning, *Physical Review Letter*, **82**(2), pp. 414–417.

Reichley, P. E. and Downs, G. S. (1969). Observed decrease in the periods of pulsar PSR 0833-45, *Nature*, **222**(5190), pp. 229–230.

Ruderman, M. (1969). Neutron starquakes and pulsar periods, *Nature*, **223**(5206), pp. 597–598.

Ruderman, M. (1991). Neutron star crustal plate tectonics. I. magnetic dipole evolution in millisecond pulsars and low-mass x-ray binaries, *The Astrophysical Journal*, **366**, p. 261.

Ruderman, M., Zhu, T., and Chen, K. (1998). Neutron star magnetic field evolution, crust movement, and glitches, *The Astrophysical Journal*, **492**(1), pp. 267–280.

Ruderman, M. A. (1976). Crust-breaking by neutron superfluids and the VELA pulsar glitches, *The Astrophysical Journal*. **203**, p. 213.

Sauls, J. A. (1989). Superfluidity in the interiors of neutron stars, in *Timing neutron stars*, Springer, pp. 457–490.

Sauls, J. A., Chamel, N., and Alpar, M. A. (2020). Superfluidity in disordered neutron stars crusts, *arXiv e-prints*, arXiv:2001.09959.

Sedrakian, A. and Clark, J. W. (2019). Superfluidity in nuclear systems and neutron stars, *European Physical Journal A*, **55**(9), p. 167.

Sedrakian, A., Wasserman, I., and Cordes, J. M. (1999). Precession of isolated neutron stars. I. effects of imperfect pinning, *The Astrophysical Journal*, **524**(1), pp. 341–360.

Sedrakian, A. D. (1995). Vortex repinning in neutron star crusts, *Monthly Notices of the Royal Astronomical Society*, **277**(1), pp. 225–234.

Seveso, S., Pizzochero, P. M., Grill, F., and Haskell, B. (2016). Mesoscopic pinning forces in neutron star crusts, *Monthly Notices of the Royal Astronomical Society*, **455**, pp. 3952–3967.

Shapiro, S. L. and Teukolsky, S. A. (1983). *Black Holes, White Dwarfs, and Neutron Stars: The Physics of Compact Objects*, 1st edn., Wiley-VCH.

Shaw, B., Lyne, A. G., Stappers, B. W., Weltevrede, P., Bassa, C. G., Lien, A. Y., Mickaliger, M. B., Breton, R. P., Jordan, C. A., Keith, M. J., and Krimm, H. A. (2018). The largest glitch observed in the Crab pulsar, *Monthly Notices of the Royal Astronomical Society*, **478**(3), pp. 3832–3840.

Shemar, S. L. and Lyne, A. G. (1996). Observations of pulsar glitches, *Monthly Notices of the Royal Astronomical Society*, **282**, pp. 677–690.

Shimazaki, K. and Nakata, T. (1980). Time-predictable recurrence model for large earthquakes, *Geophysical Research Letters*, **7**(4), pp. 279–282.

Shternin, P. S., Yakovlev, D. G., Heinke, C. O., Ho, W. C. G., and Patnaude, D. J. (2011). Cooling neutron star in the Cassiopeia A supernova remnant: evidence for superfluidity in the core, *Monthly Notices of the Royal Astronomical Society*, **412**(1), pp. L108–L112.

Sidery, T., Passamonti, A., and Andersson, N. (2010). The dynamics of pulsar glitches: contrasting phenomenology with numerical evolutions, *Monthly Notices of the Royal Astronomical Society*, **405**(2), p. 1061.

Smoluchowski, R. and Welch, D. O. (1970). Progressive deformation of the crust of pulsars, *Physical Review Letter*, **24**, pp. 1191–1192.

Sonin, E. B. (2016). *Dynamics of Quantised Vortices in Superfluids*, Cambridge University Press.

Sornette, A. and Sornette, D. (1989). Self-organized criticality and earthquakes, *Europhysics Letters*, **9**, p. 197.

Sornette, D. (2002). Predictability of catastrophic events: Material rupture, earthquakes, turbulence, financial crashes, and human birth, *Proceedings of the National Academy of Science*, **99**, pp. 2522–2529.

Sourie, A. and Chamel, N. (2020). Vortex pinning in the superfluid core of neutron stars and the rise of pulsar glitches, *Monthly Notices of the Royal Astronomical Society*, **493**(1), pp. L98–L102.

Sourie, A. and Chamel, N. (2021). Vortex pinning in the superfluid core of relativistic neutron stars, *Monthly Notices of the Royal Astronomical Society*, **503**(1), pp. 1407–1417.

Sourie, A., Oertel, M., and Novak, J. (2016). Numerical models for stationary superfluid neutron stars in general relativity with realistic equations of state, *Physical Review D*, **93**(8), p. 083004.

Stappers, B. W., Keane, E. F., Kramer, M., Possenti, A., and Stairs, I. H. (2018). The prospects of pulsar timing with new-generation radio telescopes and the Square Kilometre Array, *Philosophical Transactions of the Royal Society of London A*, **376**(2120), p. 20170293.

Suwa, Y., Yoshida, T., Shibata, M., Umeda, H., and Takahashi, K. (2018). On the minimum mass of neutron stars, *Monthly Notices of the Royal Astronomical Society*, **481**(3), pp. 3305–3312.

Vivekanand, M. (2017). A non-glitch speed-up event in the crab pulsar, *Astronomy Astrophysics*, **597**, p. L9.

Warszawski, L. and Melatos, A. (2011). Gross-Pitaevskii model of pulsar glitches, *Monthly Notices of the Royal Astronomical Society*, **415**(2), pp. 1611–1630.

Warszawski, L., Melatos, A., and Berloff, N. G. (2012). Unpinning triggers for superfluid vortex avalanches, *Physical Review B*, **85**, p. 104503.

Watanabe, G. and Pethick, C. J. (2017). Superfluid density of neutrons in the inner crust of neutron stars: New life for pulsar glitch models, *Physical Review Letter*, **119**, p. 062701.

Watkins, N. W., Pruessner, G., Chapman, S. C., Crosby, N. B., and Jensen, H. J. (2016). 25 Years of self-organized criticality: concepts and controversies, *Space Science Reviews*, **198**(1–4), pp. 3–44.

Weltevrede, P., Johnston, S., and Espinoza, C. M. (2011). The glitch-induced identity changes of PSR J1119-6127, *Monthly Notices of the Royal Astronomical Society*, **411**(3), pp. 1917–1934.

Wijngaarden, M. J. P., Ho, W. C. G., Chang, P., Heinke, C. O., Page, D., Beznogov, M., and Patnaude, D. J. (2019). Diffusive nuclear burning in cooling simulations and application to new temperature data of the Cassiopeia A neutron star, *Monthly Notices of the Royal Astronomical Society*, **484**(1), pp. 974–988.

Wlazlowski, G., Sekizawa, K., Magierski, P., Bulgac, A., and Forbes, M. M. N. (2016). Vortex pinning and dynamics in the neutron star crust, *Physical Review Letters*, **117**, p. 23.

Wood, T. S., Graber, V., and Newton, W. G. (2020). Superconducting phases in a two-component microscale model of neutron star cores, *arXiv e-prints*, arXiv:2011.02873.

Yasui, S., Chatterjee, C., Kobayashi, M., and Nitta, M. (2019). Reexamining Ginzburg-Landau theory for neutron ${}^{P}3_2$ superfluidity in neutron stars, *Physical Review C*, **100**(2), p. 025204.

Yasui, S. and Nitta, M. (2020). Domain walls in neutron ${}^3p_2$ superfluids in neutron stars, *Physical Review C*, **101**, p. 015207.

https://doi.org/10.1142/9789811220944_0008

Chapter 8

Effects of a Strong Phase Transition on Supernova Explosions, Compact Stars and Their Mergers

Andreas Bauswein*, David B. Blaschke† and Tobias Fischer†

*GSI Helmholtzzentrum für Schwerionenforschung GmbH
Darmstadt, Germany

†Institute of Theoretical Physics
University of Wroclaw, Poland

Abstract: We outline a theoretical approach supporting strong phase transitions from normal nuclear matter to the deconfined quark–gluon plasma, in the equation of state (EOS) for compact star matter. Implications of this hypothesis are discussed for astrophysical applications. Special emphasis is devoted to potentially detectable signatures, which can be directly related with the occurrence of a sufficiently strong phase transition. Therefore, simulations of core-collapse supernovae and binary compact star mergers are considered, including the subsequent emission of gravitational waves and, in the case of supernova, in addition the neutrinos play the role of messengers.

Keywords: phase transition; quark–gluon plasma; compact star.

1. Introduction

The state of matter at the interior of core-collapse supernovae and binary neutron star mergers can reach extreme conditions, in terms of temperatures up to several 10^{11} K [1], baryon densities in excess of normal nuclear saturation density and large isospin asymmetry. The latter is

[1] 1.16×10^{10} K $\simeq$ 1 MeV.

given by the proton fraction (in general charge fraction) Y_p, where $Y_p = 0$ for neutron matter (no net charges are present) and $Y_p = 0.5$ for isospin symmetric matter. The equation of state (EOS) for simulations of supernovae and binary neutron star mergers must hence cover such an extended three-dimensional domain, where presently first-principle nuclear matter calculations are not available. Instead, model EOS are being developed for supernova studies or simulations of neutron star mergers (for a comprehensive review about the EOS in supernova studies, c.f., Fischer *et al.*, 2017). These combine several domains with different degrees of freedom, e.g., heavy nuclei at low temperatures, inhomogeneous nuclear matter composed of light and heavy nuclei together with unbound nucleons, and homogeneous matter at high temperatures and densities. In this chapter we reflect on the role of the EOS in core-collapse supernovae and during binary neutron star mergers, with particular focus on the recent finding about the presence of a phase transition from normal hadronic matter to the quark–gluon plasma.

In the case of core-collapse supernova studies, accurate neutrino transport is essential for the prediction of the neutrino signal from these explosive events, as was observed from SN1987A (c.f., Bionta *et al.*, 1987; Hirata *et al.*, 1988). This observation is until now a benchmark. It confirmed that neutrinos from the next galactic event will become *the* observable signal, from which we will be able to learn not only details about these stellar explosions but also about the state of matter at the supernova interior. In addition to neutrinos, also gravitational waves are emitted from the interior of core-collapse supernovae. The *classical* emission is associated with signal that originates from the core bounce of moderately up to rapidly rotating stellar cores collapse (c.f., Dimmelmeier *et al.*, 2007, 2008; Kotake *et al.*, 2006; Ott *et al.*, 2004, 2006; Scheidegger *et al.*, 2008), which, however, seems difficult to be detectable. In general, the SN GW signal does not obey any template character. For example, it depends sensitively on the details of the radiation hydrodynamics models used to provide GW signals form simulations and on the EOS, as was demonstrated by Scheidegger *et al.* (2010). An alternative idea for the detection prospects of gravitational wave from the long-term post-bounce evolution has been proposed by Murphy *et al.* (2009). The foundation of this is the excitation of a few dominant fluid modes, which, in turn, give rise to the emission of GWs. This idea has been further explored and a connection could be established between the neutrino and GW emission characteristics using sophisticated general relativistic three-dimensional SN simulations by Kuroda *et al.* (2017). Nevertheless, the lack of any SN GW measurement, for any of the observing runs of aLIGO/VIRGO, demonstrate the complication of the problem and point to the need of improvements, both, of the GW detectors in the kHZ range as well as on the data analysis.

Gravitational waves are also key observable from binary neutron star mergers. Since the EOS of neutron star matter determines the stellar structure and the dynamics of a neutron star merger, gravitational waves carry information about the properties of high-density matter. This in particular includes the prospect to learn about the presence of deconfined quark matter in compact stars from the observations of gravitational waves produced by neutron star mergers.

While galactic supernova explosions and neutron star mergers provide the ultimate sites for testing the location and characteristics of the deconfinement transition in the QCD phase diagram at high baryon densities and low temperatures, inaccessible to HIC experiments, they are very rare events. Since SN1987A occurred we have not observed a supernova neutrino signal and since GW170817, there was no binary neutron star merger detected which would allow the extraction of the neutron star compactness from the measurement of its tidal deformability. Up to now, we have no detection of the postmerger gravitational wave signal. Therefore, compact stars which are observed as pulsars in different astrophysical environments, remain excellent tools to calibrate our knowledge of the dense matter EOS albeit at zero temperature.

We note that the *ab-initio* simulations of lattice QCD cannot make at present any predictions for the thermodynamics in the QCD phase diagram at finite and even high baryon density, due to the sign problem; see Muroya *et al.* (2003) for an introductory review on this issue. Present techniques of lattice quantum field theory for evaluating the QCD partition function are applicable only for low baryochemical potentials $\mu_B/T \lesssim 2$. It is found that the QCD transition is a chiral crossover with a pseudo-critical temperature $T_c(0) = 156.5 \pm 1.5$ MeV at $\mu_B = 0$ (Bazavov *et al.*, 2019). There is no indication for a critical endpoint (CEP) for $\mu_B/T \lesssim 2$ and $T > 0.9\, T_c(0)$ (Bazavov *et al.*, 2017). From recent lattice QCD simulations at very low quark masses and their extrapolation to the chiral limit it has been concluded that the chiral phase transition temperature is $T_c^{(0)} = 132^{+3}_{-6}$ MeV. For the temperature of the CEP (if that exists at all) has to hold $T_{\mathrm{CEP}} < T_c^{(0)}$ (Ding *et al.*, 2019).

At low baryon chemical potentials and high temperatures the QCD EOS measured in lattice QCD simulations is well confirmed by a Bayesian analysis of high-statistics data obtained by heavy-ion collision (HIC) experiments at the Relativistic Heavy-Ion Collider and at the Large Hadron Collider (Pratt *et al.*, 2015). It is the hope that in systematic beam energy scan programs of HIC experiments at lower collision energies ($\sqrt{s_{NN}} = 3 \ldots 11$ GeV being the relevant energy range in the nucleon-nucleon center of mass system of the collision), similar analyses of the high-statistics data would reveal indications of the CEP as a landmark in the QCD phase diagram. This hope, however, has not been fulfilled yet. For a recent review, see An *et al.* (2022). The problem has been clearly elucidated in the work by Senger (2021), where he shows that when the temperature of the chemical freezeout (dilepton spectrum) should fulfill the above lattice QCD constraint for T_{CEP}, the collision energy has to be below $\sqrt{s_{NN}} = 6(3)$ GeV. In order to reach the mixed phase of a first-order phase transition with the dynamical trajectories of HIC simulations in the QCD phase diagram, a collision energy of at least $\sqrt{s_{NN}} = 5$ GeV is required (Arsene *et al.*, 2007). Following this line of argument, even if a CEP exists, its temperature is too low to be reached in HIC that could probe the QCD phase transition.

Therefore, in order to investigate the hadron-to-quark matter phase transition at low temperatures, one has to rely on observations of compact stars in their multi-messenger astrophysical context: as proto-neutron stars (PNS) in supernova explosions, as pulsars in isolation and in binaries as well as in their final stages of their life, in the inspiral, merger and postmerger phases of binary neutron star mergers.

In this contribution, we review several, recent aspects of the role that a strong QCD phase transition can play in these three. We discuss the birth of compact stars in core-collapse supernova events as well as the astrophysics of binary neutron star merger events.

2. Hybrid Neutron Star Formation in Core-Collapse Supernova Explosions

PNS are the central, compact objects of core-collapse supernovae. They are the transitional state of the initial collapse of the stellar core of massive stars and the remnant neutron star. The stellar core collapse is triggered due to the photo-disintegration of heavy, iron-group nuclei and due to electron captures on protons bound in these nuclei. Consequently, the density and the temperature increase continuously during the collapsing stellar core, which simultaneously deleptonises. When nuclear saturation density is reached in the center, the short range repulsive nuclear force balances gravity such that the core bounces back with the formation of a hydrodynamics shock wave. The latter propagates quickly out of the stellar core and stalls at a radius of around 100–200 km due to the continuous photodisintegration of infalling heavy nuclei from above and the launch of the ν_e-burst associated with the shock propagation across the neutrino sphere of last scattering. As a result, the dynamic shock turns into an accretion front.

The supernova explosions are related with the revival of the stalled bounce shock wave and the subsequent ejection of the stellar mantle that surrounds the PNS. It has long been considered via the liberation of energy from the PNS interior to a thin layer of accumulated material at the PNS surface (for reviews about core collapse supernovae, see Janka *et al.*, 2007; Mirizzi *et al.*, 2016). Several scenarios have been explored in the literature. The magneto-rotational mechanism of LeBlanc and Wilson (1970) (for recent works, c.f., Mösta *et al.*, 2014; Mösta *et al.*, 2015; Takiwaki *et al.*, 2009; Winteler *et al.*, 2012), considers the required energy through the winding up of the magentic field stress in the presence of initially high magnetic field and rapidly rotating stellar cores. Another scenario is the neutrino-heating mechanism of Bethe and Wilson (1985). With angle and energy dependent neutrino transport methods once can demonstrate the development of a gain region, below which neutrinos dominantly decouple from matter and hence establish net cooling, while above neutrinos are still dominantly trapped and hence net heating is established. This mechanism has been demonstrated to lead to supernova explosions for a variety of massive progenitor stars (cf. Bruenn *et al.*, 2013; Lentz *et al.*, 2015; Melson *et al.*, 2015; Mller *et al.*, 2012; Takiwaki *et al.*, 2012). However, in the framework of multi-dimensional simulations accurate Boltzmann neutrino transport cannot be employed due to the current computational limitations. Instead, approximate neutrino transport schemes are commonly used, whose range of applicability is being debated currently (cf., Nagakura *et al.*, 2018).

2.1. *Simulations of core-collapse supernovae with hadron-quark matter hybrid EOS*

Besides the aforementioned scenarios for the onset of the supernova explosion, another mechanism was discovered by Sagert *et al.* (2009), which is due to the occurance of a phase transition at high densities from hadronic matter to the quark–gluon plasma. The latter was being treated within the simple but powerful bag model. During the phase transition a large amount of latent heat is released in a highly dynamical fashion, which in turn triggers the onset of the supernova explosion even in spherically symmetric simulations. Moreover, it leaves an observable millisecond burst in the neutrino signal, as was first demonstrated by Dasgupta *et al.* (2010) for the currently operating Super-Kamiokande neutrino detector. These milestones demonstrate the sensitivity of EOS uncertainties related to our understanding of core-collapse supernovae and underline the need of a more elaborate understanding of the EOS at high densities including better constraints in particular.

For example, there is one problem concerning the hybrid EOS used in Sagert *et al.* (2009), namely that it does not fulfill the constraint which arises from the observations of pulsars with masses as high as 2 $M_\odot$, demonstrated, e.g., by Antoniadis *et al.* (2013), Fonseca *et al.* (2016), Cromartie *et al.* (2020) and Fonseca *et al.* (2021). This well-known problem of bag-model like hybrid EOS has been overcome by the inclusion of linear repulsive vector interaction, such as the VBAG model of Klähn and Fischer (2015), similar to effective nuclear mean-field models. Higher-order repulsive interactions were considered in Benić *et al.* (2015) and Kaltenborn *et al.* (2017), in light of the existence of a third family of compact stars as well as the so-called twin phenomenon where two families of compact stars exist with nearly the same mass but different radii. The corresponding RDF 1.2 hybrid EOS is shown in Fig. 1, for β-equilibrium and zero temperature conditions in Fig. 1(a), together with the mass–radius relation where the horizontal blue band marks the present maximum neutron star constraint of $2.08 \pm 0.07\,M_\odot$ from the measurement of the pulsar PSR J0740 + 6620 (Antoniadis *et al.*, 2013; Fonseca *et al.*, 2021). The vertical lines in Fig. 1 indicate the onset of the phase transition (solid) and reaching the pure quark matter phase (dash-dotted), which is due to Maxwell's condition for the first-order phase transition construction. In between is the region of thermodynamic instability. The need of data in between the onset of quark matter and reaching the pure quark matter phase, for hydrodynamics simulations, results in a hadron-quark mixed phase where the pressure gradient is flat in the zero-temperature case (see Fig. 1(a)). It is interesting to note that at finite entropy, examined at the example of $s = 3\ k_{\rm B}$, which is a representative value for the PNS interior, the phase boundaries are located at lower densities and the phase transition region widens (see Fig. 1(b)).

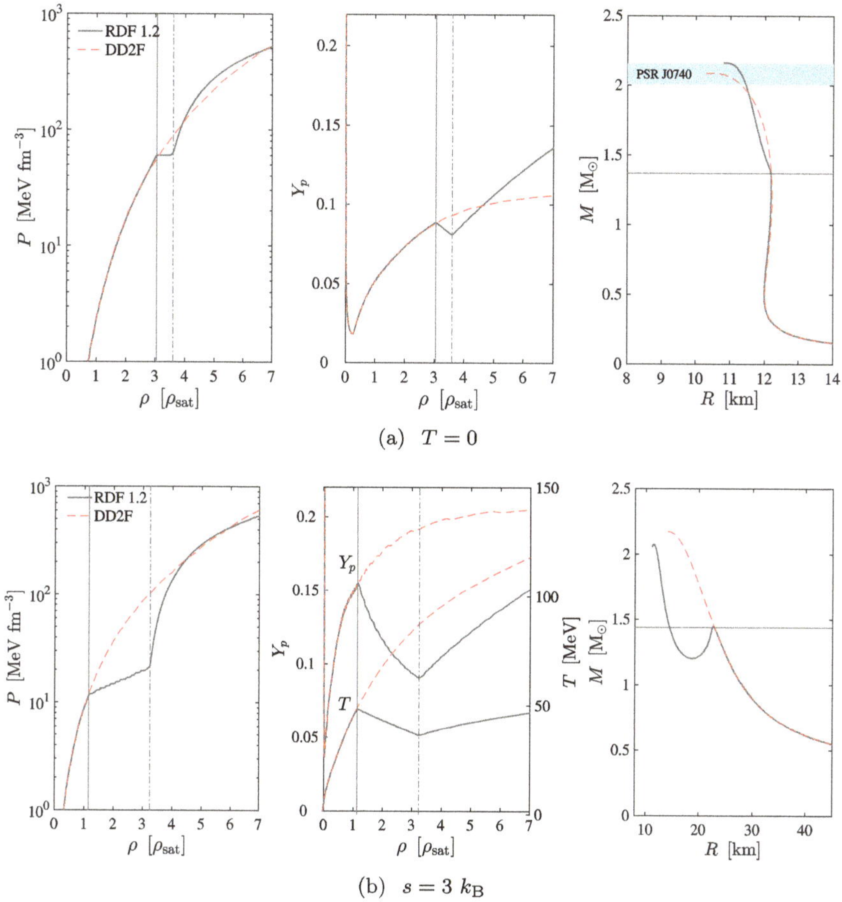

Fig. 1. EOS assuming β-equilibrium, showing the total pressure, P, the charge fraction, Y_p and the temperature T, as a function of the baryon density, ρ, as well as the corresponding mass–radius, $(M - R)$ relations, comparing the RDF 1.2 hadron-quark hybrid model (solid grey lines) and the DD2F reference hadronic model (red dashed lines) Figure taken from, Fischer (2021).

This has important consequences for the stability of PNS, in connection with cold hybrid stars, for which the onset mass increases but the maximum mass decreases. (a detailed discussion about the subject can be found in Hempel *et al.*, 2016).

To support the temperature dependence of the phase boundaries, $\rho_{\rm onset}$ and $\rho_{\rm final}$, Fig. 2 shows qualitative results for the DD2F-RDF 1.2 hadron-quark model EOS, together with curves of constant entropy per particle of

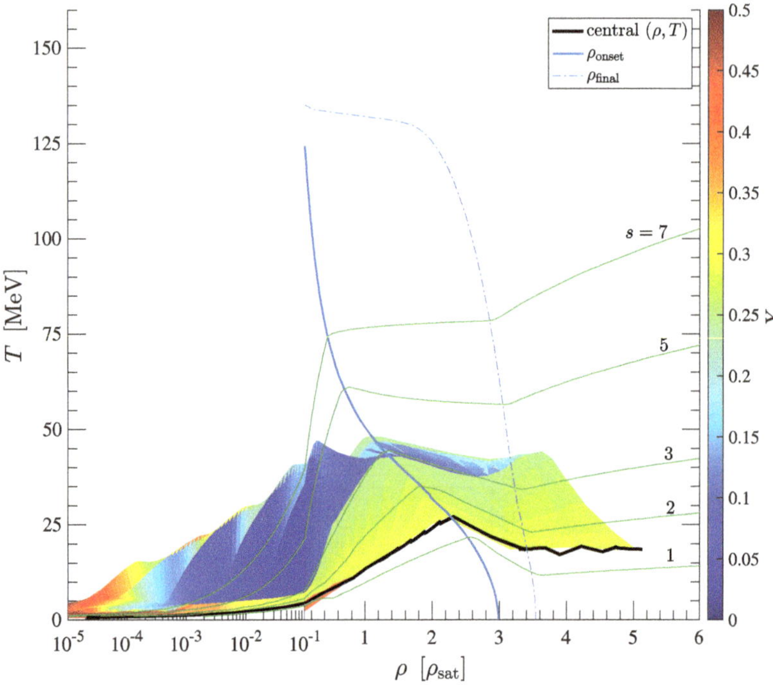

Fig. 2. Supernova temperature–density region occupied in simulations that feature a hadron-quark phase transition, where the color-coding is due to the electron abundance. Shown are also the hadron-quark matter phase boundaries for the onset of quark matter, $\rho_{\rm onset}$, and for reaching the pure quark-matter phase, $\rho_{\rm final}$, for the DD2F-RDF 1.2 hybrid EOS, as well as curves of constant entropy (green solid lines) together with the central supernova trajectory (thick solid black line). Figure taken from Fischer (2021).

$s = 1 - 7\ k_{\rm B}$. There is another important aspect related to the finite temperature dependence on the phase transition construction, the EOS softens substantially in the hadron-quark coexistence region (see Fig. 1(b)). During the post-bounce evolution, the central PNS grows not only in mass but also the central density increases continuously, due to the continuous mass accretion. Once the central density reaches $\rho_{\rm onset}$ the PNS becomes unstable. The initial contraction proceeds into an adiabatic collapse, which halts when a sufficient amount of the PNS interior reaches the pure quark matter phase, where the EOS stiffens again. In the RDF EOS class this is additionally supported due to the vector interactions. As a consequence, an additional strong hydrodynamics shock wave is formed at the phase boundaries, which propagates quickly to increasingly larger radii (see the thick

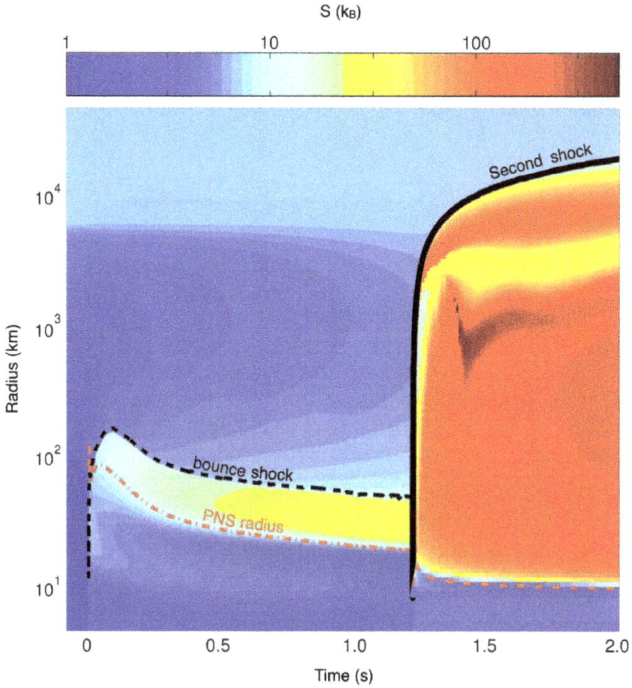

Fig. 3. Post-bounce evolution of the supernova launched from a 50 $M_\odot$ progenitor where the DD2F-RDF 1.2 hybrid EOS has been employed with the phase transition at about 1.2 s, where the entropy per particle being the color-code. Shown are the PNS radius, the bounce shock and the second shock, which forms as a direct consequence of the first-order hadron-quark phase transition. Figure taken from Fischer *et al.* (2018).

solid black line in Fig. 3). This second shock wave takes over the bounce shock, which triggers the onset of the supernova explosion. Afterwards, on the timescale of several hundreds of milliseconds, the central PNS, now featuring a massive core composed of deconfined quark matter (see Fig. 4), settles into a quasi-stationary state, i.e., the central density saturates at about five times the saturation density (see Fig. 4).

The classical picture of core-collapse supernova phenomenology, originally proposed by Woosley *et al.* (2002), considers the high-mass progenitor contributions to the failed explosion branch, which yield the formation of black holes instead. It becomes evident that this picture might be incomplete when taking into account a supernova explosion scenario, driven by the hadron-quark phase transition. However, this cannot account for the bulk part of all massive star explosions. It affects mainly the high-mass end

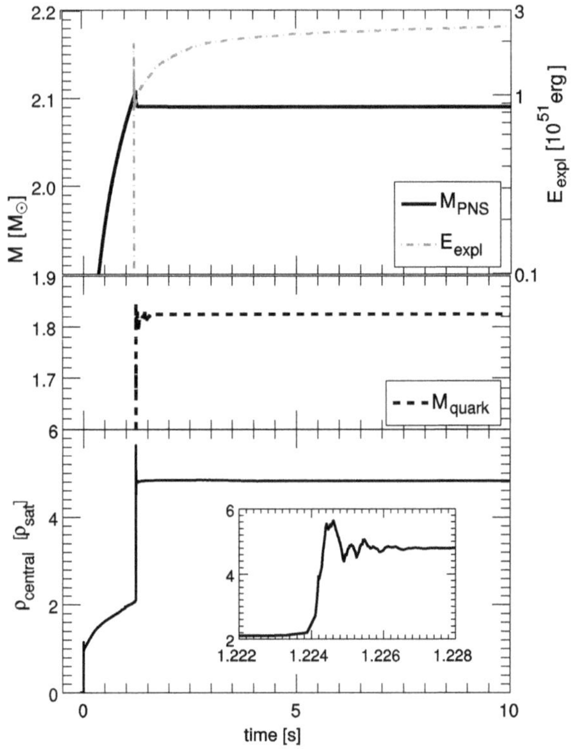

Fig. 4. Post-bounce supernova evolution of a 50 $M_\odot$ progenitor star that takes into account the hadron-quark matter phase transition, which occurs in this model after at about 1.2 s. The figure shows the explosion energy, $E_{\rm expl}$, the mass enclosed inside the PNS, $M_{\rm PNS}$, as well as the quark-matter core, $M_{\rm quark}$, and the central density, $\rho_{\rm central}$. Figure taken from Fischer *et al.* (2018).

of massive stars, due to the rather high central densities required for the phase transition to occur, which are not reached for low- and intermediate-mass progenitors on the order of $\leq 25\ M_\odot$. This picture of core-collapse supernova phenomenology has been proposed recently by Kuroda *et al.* (2021). It also takes into account that the class of core-collapse supernova explosions of very massive progenitors, in the mass range of $M > 30\ M_\odot$, driven by the hadron-quark phase transition can also only be small fraction. Otherwise there would be a conflict with stellar population synthesis analyses, concerning the number of black holes in this mass range. Furthermore, the role of metallicity has yet incompletely been studied. From Fischer (2021) is becomes clear that the situation might be completely different at (extremely) low metallicity, usually connected with the early evolution of the universe.

2.2. *Observational signals of the hadron-quark phase transition*

One question remains, namely, is there an observable signature from the supernova explosions driven by the hadron-quark first-order phase transition? The present simulations based on multi-purpose hybrid EOS, applicable for core-collapse supernova studies, reveal that there are potentially observable signatures in both, the neutrino and the gravitational wave signals. These cannot only directly related with the formation of a quark matter phase.

2.2.1. *Neutrino emission*

The emission of a burst-like signals is related with the dynamics of the formation and propagation of the second shock wave. This millisecond neutrino burst is released when the second shock wave, which forms a a direct consequence of the hadron-quark phase transition, propagates across the neutrino spheres of last inelastic scattering. This neutrino burst is detectable for a galactic event, as illustrated in Fig. 5 examined at the Super-Kamiokande detector. This burst is absent in *standard* neutrino-driven core-collapse supernova explosions (c.f., Wu *et al.*, 2015). Hence, if observed for a future galactic event, it represents a smoking gun signature for the appearance of quark matter. If not observed, one might be able to constrain the conditions, in terms of baryon density and temperature, up to which the hadron-quark phase transition can be excluded. Similar results have been obtained previously by Sagert *et al.* (2009) and Fischer *et al.* (2011), however, employing a simplistic bag-model EOS for quark matter which is incompatible with the constrain deduced from the existence of massive pulsars of $\sim$2 $M_{\odot}$.

2.2.2. *Emission of gravitational waves*

In general, core-collapse supernovae GWs stem from various different phases of the evolution. During the early post-bounce phase, high-amplitude GWs are emitted from convective motion, which lasts ups to several tens of milliseconds. After that, during the proceeding post-bounce mass accretion phase prior to the explosion onset, the GW emission is determined from a

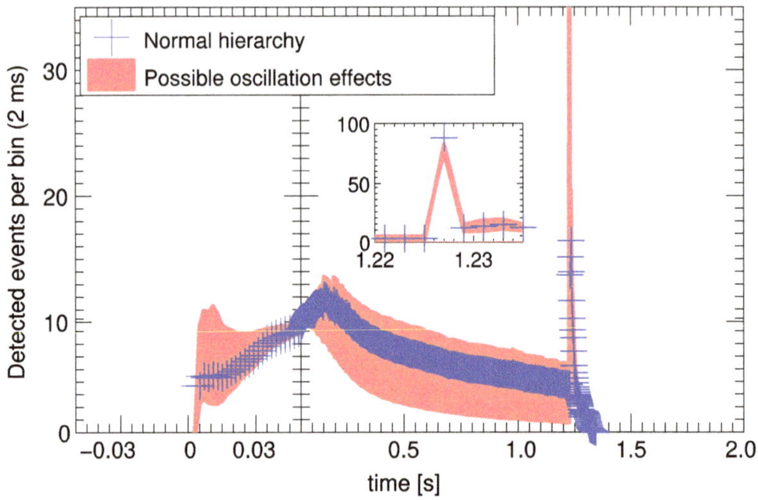

Fig. 5. Neutrino signal from a core-collapse supernova explosion launched from a 50 $M_\odot$ progenitor that takes into account the hadron-quark phase transition using the DD2F-RDF 1.2 hybrid EOS, evaluated at the Super-Kamiokande water Cherenkov detector at a fiducal distance of 10 kpc from the Earth. Figure taken from Fischer *et al.* (2018).

variety of aspects, e.g., neutrino-driven convection, the presence of hydrodynamics instabilities such as the standing accretion shock instability. One feature seen in all present core-collapse SN simulations is the continuously increasing GW frequency with time (c.f., Kuroda *et al.*, 2016; Mezzacappa *et al.* , 2020; Müller *et al.*, 2013; Murphy *et al.*, 2009; Shibagaki *et al.*, 2021; Vartanyan *et al.*, 2019, and references therein). This feature is considered to be associated with PNS oscillations as well as convection at the PNS interior at high density, which, in turn, excites the dominant post-bounce GW f-mode (Morozova *et al.*, 2018; Sotani and Takiwaki, 2020; Torres-Forne *et al.*, 2019). It is also present in the axially-symmetric SN simulations of Zha *et al.* (2020) and Kuroda *et al.* (2021) employing a hadron-quark hybrid EOS, however, prior to the phase transition when the post-bounce evolution is determined by the hadronic phase.

Figure 6 shows the post-bounce evolution of the GW amplitude (top panel) for an axially-symmetric SN simulation of Kuroda *et al.* (2021), launched from a 50 $M_\odot$ progenitor employing the hadron-quark hybrid EOS of Fischer *et al.* (2018). Shown is only the non-vanishing component in axial symmetry observed along the equatorial plane, i.e., perpendicular to the symmetric axis. Note that since the SN simulations employ an

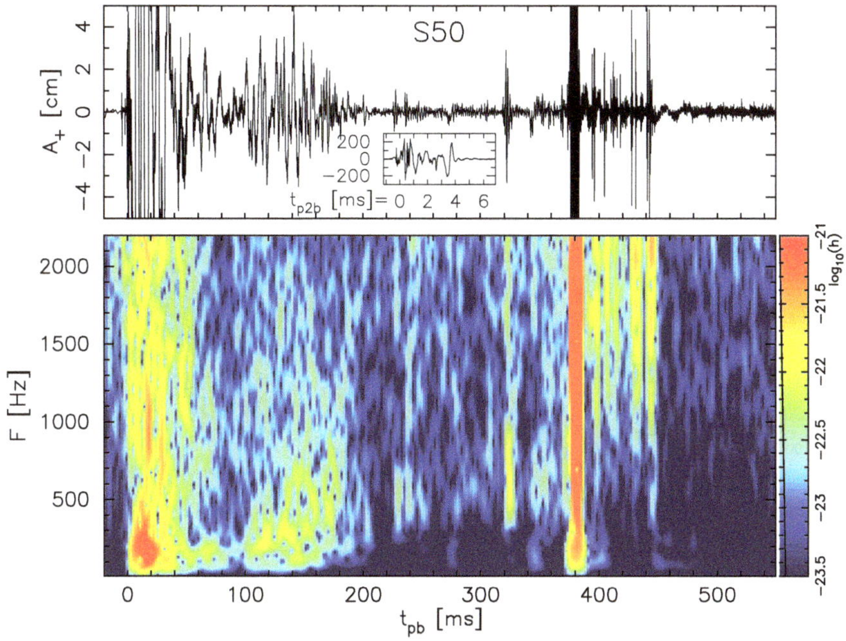

Fig. 6. Gravitational waveform A_+ (top panels) and spectrogram (bottom panels) for an axially-symmetric core-collapse SN simulation with hadron-quark hybrid EOS, launched from a 50 $M_\odot$ progenitor. The inlay at the top panel shows a zoom in of A_+ corresponding to the evolution shortly after the phase transition. Figure taken from Kuroda *et al.* (2021).

effective treatment of General Relativity, the GW amplitude A_+, more precisely the GW strain, is obtained from the standard quadrupole formula (for details, see Kuroda *et al.*, 2014; Shibata and Sekiguchi, 2003, and references therein). The bottom panel of Fig. 6 shows the spectrogram of the GW emission, which is obtained using a short time Fourier analysis. It shows the aforementioned continuously rising dominant f-mode frequency, starting at around 50–100 Hz rising to several hundreds of Hz. The sudden rise of the GW amplitude and frequency, at around 400 ms post bounce, is the occurence of the hadron-to-quark matter phase transition and the associated PNS collapse, second bounce and formation of the second shock wave. It lasts for only few milliseconds. After that, the PNS oscillations continue for several tens of milliseconds, which are responsible for the GW emission between about 400 and 450 ms, shown in Fig. 6.

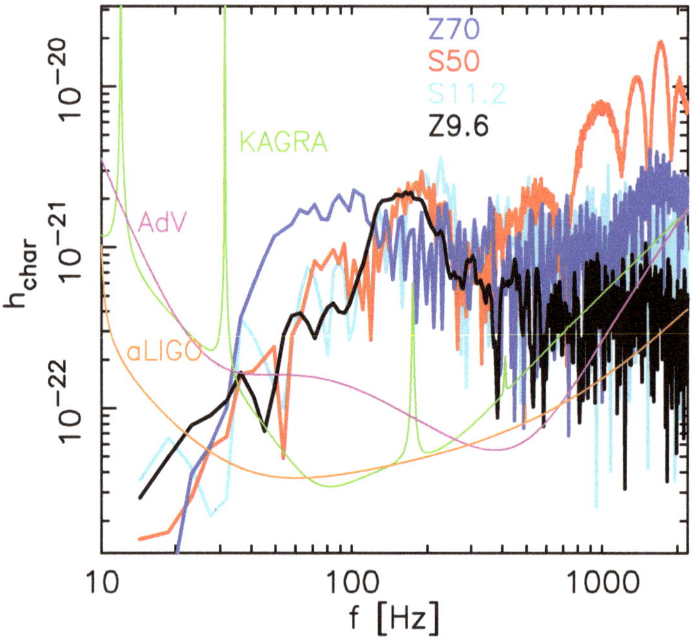

Fig. 7. Gravitational wave signal, spectral amplitudes with respect to the frequency, for a sample of axially-symmetric core-collapse SN simulations launched from different progenitors: 9.6 $M_{\odot}$ (Z9.6), 11.2 $M_{\odot}$ (S11.2), 50 $M_{\odot}$ (S50) and 70 $M_{\odot}$ (Z70). All models are of solar metallicity except Z9.6 and Z70, which correspond to the class of extremely metal poor stars. Figure taken from Kuroda *et al.* (2021).

In order to address the question about the detectability of such GW signal, in particular the possibility to distinguish it from SN that are not associated with the hadron-quark phase transition, Kuroda *et al.* (2021) compared the GW emission with those of canonical, neutrino-driven SN explosions as well as one failed SN model in which case a black hole forms instead. The latter is launched from a low-metalicity 70 $M_{\odot}$ progenitor (Z70). In comparison to this one, as well as two other neutrino-driven SN models launched from a low-metallicity 9.6 $M_{\odot}$ (Z9.6) and a solar-metallicity 11.2 $M_{\odot}$ progenitor (S11.2), the high-frequency kHz GW signal is enhanced by nearly one order of magnitude for the 50 $M_{\odot}$ SN simulation with hadron-to-quark matter phase transition (see Fig. 7). The origin of these high amplitudes is PNS oscillations and PNS convection after the phase transition took place. It is absent for Z70 because a black hole forms about 1 ms after the phase transition. The future observation of such a high GW amplitude component at kHz frequencies, which is likely to be absent in all other SN, represents a smoking gun signature for the presence of a hadron-to-quark matter phase transition *and* a resulting stable remnant compact star with quark matter at the interior. This might be used as constraint for the presence of quark matter at high baryon densities, complementary to lattice QCD and experimental HIC programs.

One caveat in predicting the GW emission from axially symmetric simulations might arise in the overestimation of convective motion due to the presence of a fixed symmetry axis. This aspect tends to overestimate the GW amplitude. Another, yet incompletely understood feature is the presence of a quiescent phase of GW emission during the post-bounce evolution (present between 200 and 350 ms post bounce in Fig. 6 prior to the phase transition induced sudden rise of the amplitude). It has been speculated whether it corresponds to the period when the shock wave gradually recedes. Only with simulations based on three spatial dimensional setup will enable us to resolve this issue.

2.2.3. *Nucleosynthesis yields of heavy elements*

The final phase of a supernova evolution after the supernova explosion onset, is characterized by the deleptonization of the nascent PNS, i.e., the diffusion of the trapped neutrinos. It lasts for several tens of seconds and is associated with the ejection of a low-mass outflow known as the neutrino-driven wind, which has long been studied as a possible site for the nucleosynthesis of heavy elements in the universe (c.f., Cowan *et al.*, 2021; Thielemann *et al.*, 2011, and references therein). It depends sensitively on the nucleosynthesis conditions of the neutrino-driven wind, i.e., entropy per particle and the neutron excess, as well as on their evolution (Woosley *et al.*, 1994). Recent investigations have confirmed that canonical, neutrino-driven core-collapse SN explosions cannot yield a strong r-process nucleosynthesis component (c.f., Fischer *et al.*, 2020a; Martínez-Pinedo *et al.*, 2014). The reason for this are the generally too low entropies per particle, on the order of 50–100 $k_{\rm B}$, and insufficient neutron excess or even generally proton-rich conditions if PNS convection is considered (Mirizzi *et al.*, 2016; Roberts *et al.*, 2012). No elements heavier than molybdenum, with atomic number $Z = 42$, can be produced. The key input physics in the SN simulations is three-flavor Boltzmann neutrino transport, for the accurate prediction of the neutrino fluxes and spectra as well as their evolution, which determine nearly entirely these aforementioned nucleosynthesis conditions (Qian and Woosley, 1996). Contrary, studies that are lacking neutrino transport have to assume the nucleosynthesis conditions.

Furthermore, recent multi-dimensional SN simulations point to the presence of late-time mass accretion, i.e., after the successful SN explosion onset, which lasts for seconds (c.f., Bollig *et al.*, 2021; Burrows and Vartanyan, 2021; Müller *et al.*, 2017; Nagakura *et al.*, 2021, and references therein). The exception might be the class of neutrino-driven explosions of low-mass stellar progenitor in the mass of 8–9 $M_{\odot}$ (Radice *et al.*, 2017; Stockinger *et al.*, 2020) as well as those which originate from binary stellar evolution leading to stripped envelope SN explosions (Müller *et al.*, 2019). If large down flows of matter persist after the SN explosion onset, then the neutrino-driven wind is likely to be absent and the nucleosynthesis results of these model might be very different from those of the neutrino-driven wind (Witt *et al.*, 2021).

The conditions found in SN explosions that are triggered through the hadron-to-quark matter phase transition by Fischer *et al.* (2020b) are very different from those of neutrino-driven SN explosions. One finds:

(a) an early, low-entropy and neutron-rich component with electron fraction of $Y_e \simeq 0.2 - 0.25$, which is ejected dynamically together with the expanding explosion shock wave,
(b) an intermediate component characterized by rising entropy and Y_e, and
(c) a late-time component.

The latter is the high-entropy neutrino-driven wind, featuring an entropy per baryon on the order of several 100 $k_{\rm B}$, however, being subject of large uncertainties as pointed out before. The integrated nucleosynthesis results of three SN explosions driven by the QCD phase transition are shown in Fig. 8, for three different progenitors of 35, 40 and 50 $M_{\odot}$, showing a robust strong r-process component with the production of elements not only corresponding to the third r-process peak at mass number $A \simeq 195$ but even transactinides such as plutonium. However, the abundances of the rare-Earth elements as well as those of the third r-process peak are suppressed by several orders of magnitude in comparison to the second r-process peak at nuclear mass number $A \simeq 125$. Hence, this class of massive star explosions cannot account for the bulk part of r-process nucleosynthesis in the galaxy. Instead, it is as a rare r-process site, as was the considered in Farouqi *et al.* (2021), together with the class of magneto-rotationally driven core-collapse SN (LeBlanc and Wilson, 1970; Mösta *et al.*, 2014; Mösta *et al.*, 2015; Winteler *et al.*, 2012). The latter requires a high magnetic field strength of the progenitor in combination with rapid rotation, which is likely to be realized only for a subset of all core-collapse SN.

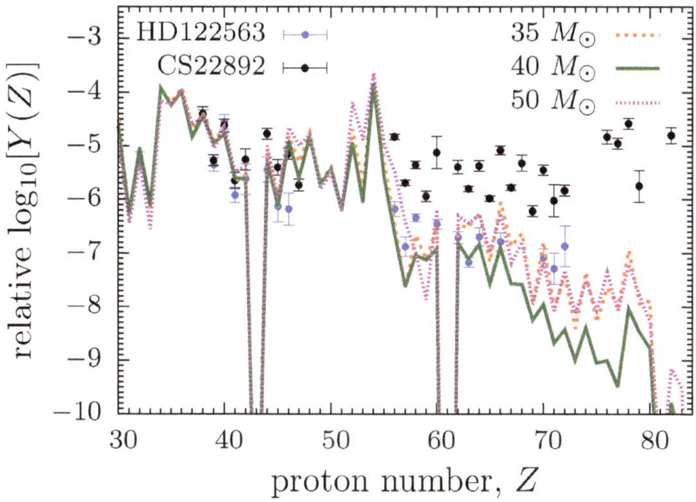

Fig. 8. Integrated elemental yields from three QCD driven core-collapse supernova explosions, launched from 35, 40 and 50 $M_\odot$ progenitors. For comparison, observations are shown for two metal-poor stars, HD122564 from Roederer *et al.* (2012) and CS22892 from Sneden *et al.* (2003). Figure taken from Fischer *et al.* (2020b).

3. Quark Matter in Binary Neutron Star Mergers

For the astrophysics community the very first unambiguous detection of a neutron star coalescence in August 2017 was a seminal event as the observations revealed a "proto-type" neutron star merger, which settled many questions that had been open or tentative before (Abbott *et al.*, 2017). Furthermore, the detection highlights the potential to learn about the presence of deconfined quark matter in neutron stars, i.e., at densities of a few time nuclear saturation density. Merging neutron stars lead to a number of observable phenomena, which in principle can carry in imprint of quark deconfinement. This includes foremost the gravitational-wave signal but also electromagnetic counterparts, where various processes generate emission in different parts of the spectrum (LIGO Scientific Collaboration *et al.*, 2017).

The merger of two neutron stars proceeds through different stages, starting with a long-lasting "inspiral" phase, where the binary looses angular momentum and energy as a result of gravitational-wave emission, which leads to a decrease of the orbital separation. The inspiral continuously accelerates and finally leads to the coalescence of the binary components as a highly dynamical process with typical time scales of milliseconds (see

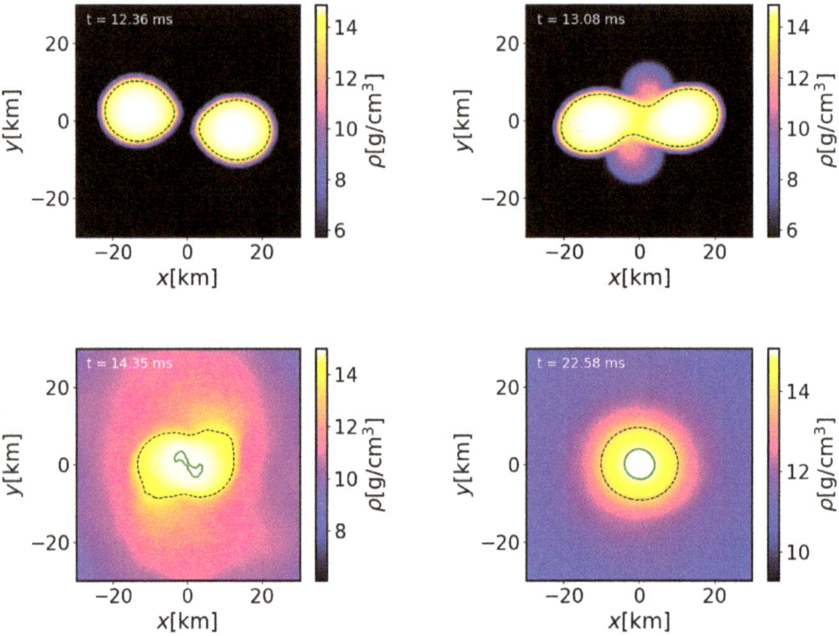

Fig. 9. Density evolution of a neutron star merger with a hybrid EOS with a phase transition to deconfined quark matter. Lines display density contours of $\rho = 10^{14}$ g/cm^3 (dashed) and 8.7×10^{14}g/cm^3 (solid). The latter density roughly corresponds to the onset density of the phase transition at zero temperature. Figure taken from Bauswein *et al.* (2019b).

Fig. 9). For higher total binary masses the merger forms a black hole-torus system because the merger remnant cannot be stabilized against the gravitational collapse. Recall that typical neutron star masses are about 1.3–1.4 $M_\odot$ such that the mass of the merged object will usually exceed the maximum mass of non-rotating neutron stars. If the total binary mass is below a certain threshold mass, the merging binary components create a neutron star merger remnant. Since most neutron star binaries may have a total mass of about 2.7 $M_\odot$ and since the threshold mass for black hole formation is likely in the range between 2.8 and 3.4 $M_\odot$, one may expect that the typical outcome of a merger is the formation of a neutron star remnant. This object is rapidly rotating and initially highly deformed and strongly oscillates reaching a quasi-equilibrium on time scales of several 10 ms to enter a phase of secular evolution.

During the merging process the densities increase as a more massive and more compact object forms. Therefore, the prospects for the occurrence of deconfined quark matter in the "postmerger" stage are generally better compared to the inspiral. Moreover, the temperature in the merger remnant increases and can reach several 10 MeV and in the shock-heated contact interface even about 100 MeV. This further enhances the likelihood for the hadron-quark phase transition to take place. Whether deconfined quark matter is present in the postmerger remnant or even before merging during the inspiral, is currently open and depends on the unknown temperature-dependent onset density of the hadron-quark phase transition. This motivates to identify specific signatures of quark deconfimenent in neutron star merger observables.

3.1. *Gravitational waves*

The gravitational wave signal of a neutron star merger reflects the stages of the coalescence, i.e., the inspiral and the postmerger evolution, and encodes the dynamics of the system, which are dominantly determined by the binary masses and the EOS. In turn, the detailed analysis of the gravitational wave signal reveals the binary masses and information on the EOS (see Abbott *et al.*, 2019). The impact of the binary masses is described by two parameters. The chirp mass, defined as follows:

$$\mathcal{M}_{\rm chirp} = (M_1 M_2)^{3/5}/(M_1 + M_2)^{1/5}, \tag{1}$$

with M_1 and M_2 being the masses of the individual binary components, has the strongest influence on the inspiral gravitational wave emission and is thus measured with high precision. The binary mass ratio $q = M_1/M_2$ affects the signal only during the last orbits before merging and can thus be inferred only with less precision. Combining $\mathcal{M}_{\rm chirp}$ and q yields the physical masses of the system.

EOS effects during the inspiral are described by the so-called combined tidal deformability, which is defined as follows (c.f., Flanagan and Hinderer, 2008; Hinderer, 2008),

$$\tilde{\Lambda} = \frac{16}{13} \frac{(M_1 + 12M_2)M_1^4\Lambda_1 + (M_2 + 12M_1)M_2^4\Lambda_2}{(M_1 + M_2)^5}. \tag{2}$$

Here, Λ_1 and Λ_2 are the tidal deformabilities of the individual stars given by $\Lambda = \frac{2}{3}k_2 \left(\frac{c^2 R}{GM}\right)^5$ with the stellar radius R and the tidal Love number k_2. k_2 is a stellar structure parameter, which is as the radius uniquely determined by the EOS. A larger $\tilde{\Lambda}$ accelerates the inspiral since systems with a finite tidal deformability lead to stronger gravitational-wave emission compared to point-particle binaries of the same masses. Its definition illustrates that Λ scales tightly with the stellar radius, and the constraint which GW170817 put on the tidal deformability excludes neutron stars in the given mass range with radii larger than about 13.5 km (Abbott *et al.*, 2018).

As the stellar radius, the tidal deformability is similarly affected by the presence of a phase transition to deconfined quark matter. Specifically, the hadron-quark phase transition can lead to a kink in the relation between tidal deformability and mass. The kink occurs at the neutron star mass $M_{\rm onset}$, where quark matter starts to appear. Hence, the presence of quark matter may be identified by detecting systems with different masses and measuring the respective tidal deformability (detailed can be found in Chatziioannou and Han, 2020; Chen *et al.*, 2020; Pang *et al.*, 2020, and references therein). However, even for models which predict a rather strong impact of the phase transition on the stellar structure, the kink in $\Lambda(M)$ may not be very pronounced (c.f., Bauswein *et al.*, 2019b; Christian *et al.*, 2019; Han and Steiner, 2019). Therefore, identifying quark matter in neutron stars during the gravitational-wave inspiral would require very precise measurements of the tidal deformability, which seems challenging in regard to statistical and systematic uncertainties in the gravitational-wave data analysis.

The definition of the tidal deformability also shows that Λ strongly decreases with the stellar mass since more massive stars become more compact and thus more similar to point particles. Hence, it is more challenging to measure finite-size effects in high-mass binaries. This implies that the identification of quark matter through the inspiral may be more intricate in particular if the onset density of the hadron-quark phase transition is relatively high. Although the currently very low number of neutron star merger detections does not yield good statistics on the population properties of neutron star binaries, one may expects that there will be fewer systems with high total binary masses.

In this regard the postmerger phase may be more promising since the remnant probes a density regime which is higher than that of the initial stars, and, thus, low-mass binaries may already reveal signals of quark matter. The EOS affects the frequencies of the postmerger oscillations and therefore the frequencies of the postmerger gravitational-wave emission. Typical gravitational-wave spectra are shown in Fig. 10, where one can clearly identify different peaks. The most prominent feature is a pronounced peak in the kHz range, which corresponds to the dominant oscillation mode of the remnant and is generated by the fundamental quadrupolar

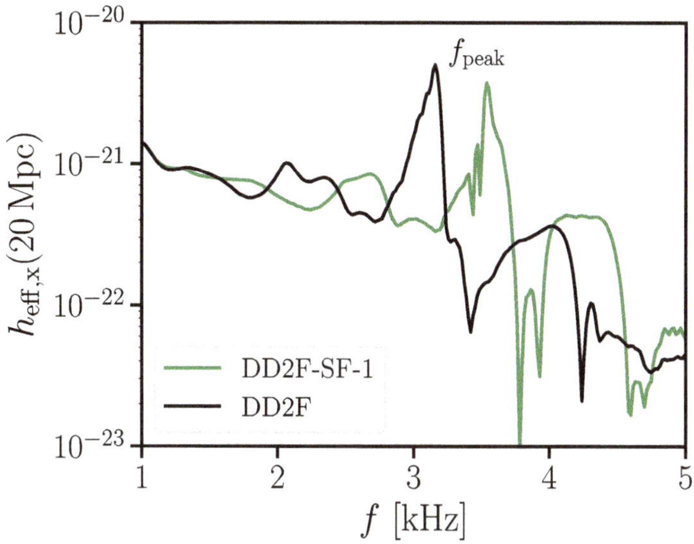

Fig. 10. Gravitational-wave spectrum of two neutron star merger simulations with a total mass of 2.7 $M_\odot$. The black curve adopts a purely hadronic EOS, while the green curve displays the spectrum for a model undergoing a phase transition to deconfined quark matter. Figure taken from Bauswein *et al.* (2019a).

fluid mode. The frequency of this oscillation is determined by the compactness of the remnant, where more compact stellar objects oscillate at higher frequencies. This implies a the strong sensitivity of the frequencies on the EOS since the properties of the EOS shape the stellar structure of the remnant.

Figure 10 shows simulation results for an EOS model with phase transition to quark matter (green) and one model without phase transition (black) for a merger of two stars with 1.35 $M_\odot$. Both models are based on the same hadronic EOS at lower densities and only differ at higher densities, where the model with phase transition features a significant softening. This softening leads to a more compact remnant and thus to a higher postmerger oscillation frequency, while the general morphology of the spectra is similar. In the particular case displayed in Fig. 10 the phase transition takes place right after merging (see Fig. 9) such that the inspiral proceeds identically.

While a frequency shift by a phase transition is intuitive, it is not obvious that a high postmerger frequency alone is indicative of the presence of deconfined quark matter because also purely hadronic EOS models can reach frequencies in the range up to 4 kHz for this binary configuration. However, an increased postmerger frequency is in fact characteristic of a phase transition if one compares $f_{\rm peak}$ to the tidal deformability inferred from the inspiral phase (details can be found in Bauswein *et al.*, 2019a, and references therein). This comparison is displayed in Fig. 11, which compiles simulation results for many calculations with the same binary mass configuration but with different EOS models with and without phase transition to deconfined quark matter.

All purely hadronic models follow a tight correlation between $f_{\rm peak}$ and $\Lambda_{1.35}$ (note that for equal-mass binaries $\tilde{\Lambda} = \Lambda_1 = \Lambda_2$). This scaling is not fully unexpected because $\Lambda_{1.35}$ is a good proxy for the stellar structure of the inspiralling stars and consequently also for the postmerger remnant. Although densities increase during merging, purely hadronic EOS do not change very abruptly such that $\Lambda_{1.35}$ does also describe the stellar structure of objects with densities exceeding those in 1.35 $M_\odot$ neutron stars.

In contrast, the occurrence of deconfined quark matter can change the EOS strongly beyond the onset density of the phase transition. Consequently, the softening of the EOS and the resulting frequency shift of the postmerger gravitational-wave emission is not captured by stellar parameters which are only sensitive to the density regime below the onset density (such as $\Lambda_{1.35}$ in Fig. 11). This explains the characteristic offset of EOS models with quark matter in Fig. 11. Hence, a significant deviation of the postmerger gravitational wave frequency $f_{\rm peak}$ from the empirical relation $f_{\rm peak}(\Lambda_{1.35})$ for purely hadronic models (black line) does provide very strong evidence for the occurrence of deconfined quark matter in neutron stars (see Bauswein *et al.*, 2019a).

Some of the purely hadronic models in Fig. 11 feature a phase transition to hyperonic matter, which apparently does not soften the equation of state sufficiently to yield a strong frequency deviation. Hence, a characteristic frequency shift relative to $\Lambda_{1.35}$ does indeed represent an unambiguous signature of quark matter.

It is conceivable that the frequency shift is a signature which also occurs for other hadronic equations of state describing the density regime below the phase transition (see Bauswein *et al.*, 2019a; Blacker *et al.*, 2020; Hanauske *et al.*, 2021; Liebling *et al.*, 2021; Prakash *et al.*, 2021; Weih *et al.*, 2020, and references therein). The models depicted by plus signs in Fig. 11 are based on the same hadronic EOS but make different assumptions about the properties of the hadron-quark phase transition and the properties of deconfined quark matter in the non-perturbative regime beyond the phase

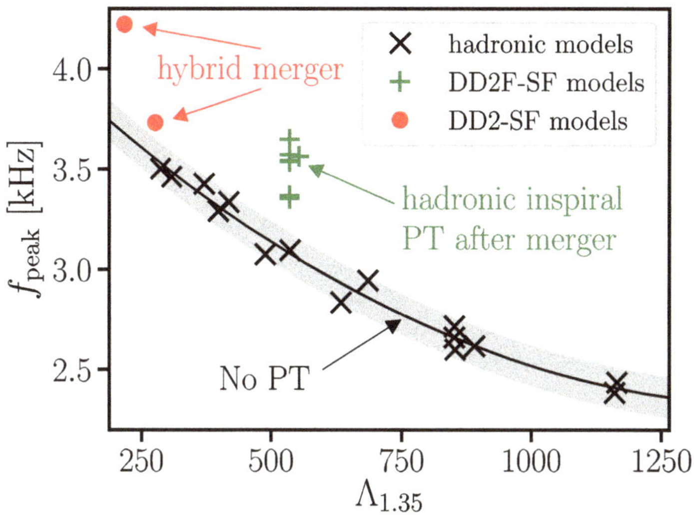

Fig. 11. Postmerger gravitational-wave frequency as function of the tidal deformability for 1.35-1.35 $M_{\odot}$ binaries with many equation of state models. Purely hadronic models (crosses) follow a tight relation fitted by the black curve. Models undergoing a phase transition before merging (dots) or after merging (plus signs) lead to a frequency shift relative to the models without phase transition. Figure taken from Bauswein and Blacker (2020).

transition. Therefore, the models occur at the same tidal deformability $\Lambda_{1.35}$ but result in differently strong frequency shifts.[2] In fact, the magnitude of the frequency increase encodes the "strength" of the phase transition. For instance, for a roughly similar onset density and stiffness of the quark matter phase, the frequency shift scales with the density jump across the phase transition (see Bauswein *et al.*, 2019a). Generally, it is clear that the signature of quark matter will be more pronounced for models which feature a stronger deviation from the corresponding hadronic reference model. This may also imply that for certain hybrid models the aforementioned signature of a relative frequency increase may not be very pronounced or even absent if the EOS closely resembles the behavior of a purely hadronic model. See also the model in Most *et al.* (2019), where quark matter exhibits only a very weak impact on the gravitational-wave frequency but influences the collapse time scale of the remnant.

[2]One model with a slightly higher $\Lambda_{1.35}$ features a small modification of the purely hadronic EOS.

Moreover, the frequency shift may not be very strong if only a small quark matter core forms in the center of the remnant for binary configurations which are barely massive enough to reach conditions in the postmerger remnant to undergo a phase transition (see Blacker *et al.*, 2020). In this case, however, a more massive binary with a sizeable quark matter core may yield a very strong signature. Blacker *et al.* (2020) provides an extensive study of the impact of quark matter in neutron star mergers for different binary mass configuration including unequal-mass systems, which can similarly yield characteristic frequency shifts. Depending on the specific properties of the phase transition and the quark matter EOS, the occurrence of deconfined quark matter during the postmerger evolution may be delayed, as was discused in Weih *et al.* (2020).

The dots in Fig. 11 demonstrate that a frequency increase relative to the empirical $f_{\mathrm{peak}}(\Lambda_{1.35})$ relation may also occur for mergers of hybrid stars, i.e., systems which contain quark matter cores already before the coalescence during the inspiral (Bauswein and Blacker, 2020).

Figure 11 illustrates that a relatively coarse determination of the tidal deformability of about $\Delta\Lambda \approx 100$ may be sufficient to identify the occurrence of the hadron-quark phase transition through a postmerger frequency shift. The dominant postmerger gravitational-wave frequency was not detected in GW170817. However, simulations show that relatively minor upgrades of the sensitivity of existing gravitational-wave instruments in the kHz range are sufficient to detect the postmerger peak of a signal similar to GW170817 (details can be found in Chatziioannou *et al.*, 2017; Torres-Rivas *et al.*, 2019; Wijngaarden *et al.*, 2022). These gravitational-wave data analysis studies reveal that f_{peak} can be measured within a few 10 Hz, which is sufficiently precise to detect sizeable deviations from the $f_{\mathrm{peak}}(\Lambda)$ relation for purely hadronic models.

These considerations imply that any sufficiently precise measurement of f_{peak} and the tidal deformability will be informative about properties of the hadron-quark phase transition. While a significant postmerger frequency shift will provide very strong evidence for the presence of deconfined quark matter, data compatible with the $f_{\mathrm{peak}}(\Lambda)$ relation for purely hadronic models rules out the presence of a strong phase transition in the density regime of the merger remnant. This would thus place a constraint on the onset of quark deconfinement assuming that the presence of quark matter would lead to a *strong* signature as described above.

The probed density regime by the postmerger remnant and thus the concrete constraint that can be put on the onset density of the hadron-quark phase transition, is also accessible by gravitational wave parameters. For purely hadronic EOS models there exist tight empirical relations between the tidal deformability or the postmerger gravitational wave frequency f_{peak}, respectively, and the maximum density in the merger remnant (see Bauswein *et al.*, 2019a; Blacker *et al.*, 2020). Those universal relations hold for a large range of binary masses and can be employed to infer an upper or lower limit on the onset density of the hadron-quark phase transition depending on whether the comparison between f_{peak} and Λ for the given binary masses reveals evidence for the presence or absence of quark matter.

Two complications may arise, which have been stressed in detail in Blacker *et al.* (2020). If the quark matter core is too small, the impact on the stellar structure and thus postmerger gravitational wave emission will not be very pronounced but only occur for more massive binary systems. Moreover, the presence of quark matter may not in any case lead to a sizeable effect if for instance the properties of the quark matter EOS are mimicking those of hadronic matter as in Alford *et al.* (2005).[3] The latter issue requires further investigations to reveal possible other signatures of quark matter that can inform about the presence of quark matter in the case its impact on the stellar structure is relatively weak.

3.2. *Mass ejection and electromagnetic counterparts*

Gravitational waves are not the only messenger from neutron star mergers. Apart from gamma-ray bursts and associated electromagnetic emission in different parts of the spectrum, mass ejection during the merger process can lead to electromagnetic radiation in the infrared, optical and ultraviolet. Material ejected from a neutron star merger expands with high velocities (up to a few 10% of the speed of light) and undergoes the rapid neutron capture process forming heavy elements (Cowan *et al.*, 2021). The freshly synthesized elements are radioactive and the subsequent decays deposit heat in the initially opaque expanding ejecta cloud. The thermal emission of this matter produces an electromagnetic counterpart commonly called kilonova, which evolves on time scales of hours and days and becomes bright enough to be observable with telescopes (Metzger, 2019). In the case of GW170817

---

[3]If values of f_{peak} and Λ are compatible with $f_{\text{peak}}(\Lambda)$ of hadronic matter, either no phase transition occurred or quark matter properties are very similar to hadronic matter. Hence, the sketched procedures can only rule out the presence of a *strong* phase transition up to a given density and thus constrain the onset density of a strong phase transition (Blacker *et al.*, 2020).

the gravitational wave signal provided an estimate of the sky localization, and the corresponding kilonova was found several hours after the initial alert circulated after the detection by the gravitational wave instruments (LIGO Scientific Collaboration *et al.*, 2017).

The properties of the electromagnetic emission yield a coarse estimate of the ejecta parameters like their mass and outflow velocities. The brightness of the kilonova for instance scales approximately with the amount of ejecta, although the quantitative interpretation is very challenging because of various uncertainties, e.g., about atomic data of heavy nuclei, and different ejection mechanisms playing a role on different time scales and leading to different ejecta components. The ejecta characteristics depend on the merger dynamics and therefore on the binary masses and the EOS (c.f., Bauswein *et al.*, 2013; Hotokezaka *et al.*, 2013, and references therein). Hence, kilonova properties in principle contain information on the EOS and thus possibly on the presence of quark matter in neutron stars.

Figure 12 displays the amount of unbound matter as function of time for 1.35–1.35 $M_\odot$ mergers. The calculations run only for a few 10 ms after merging and thus do only provide the initial dynamical ejecta component,

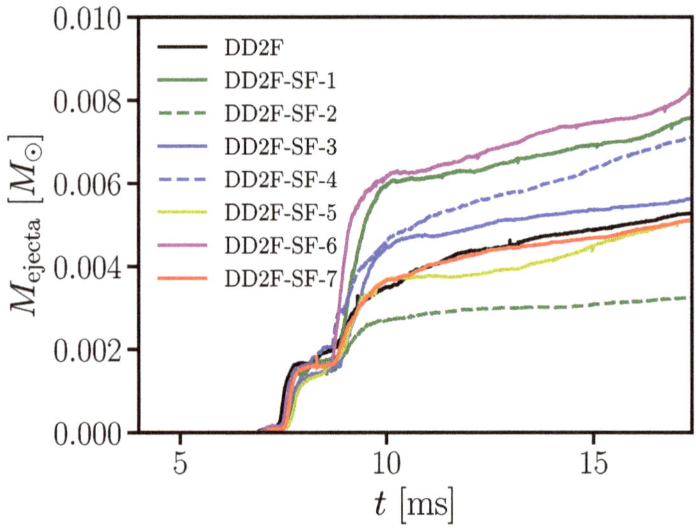

Fig. 12. Amount of unbound matter as function of time in simulations of 1.35–1.35 $M_\odot$ binaries for different hybrid equations of state. The black curve shows the purely hadronic reference model without phase transition. Figure taken from Bauswein *et al.* (2019b).

while matter becoming unbound during the subsequent secular evolution is not captured. The black curve shows the result for a purely hadronic EOS model, while all other curves exhibit the ejecta mass for hybrid equations of state undergoing a phase transition to deconfined quark matter. This set of models demonstrates that there is no very distinct and unique impact of the hadron-quark phase transition on mass ejection in neutron star mergers, as some models lead to enhanced ejecta masses while others yield a comparable or even reduced amount of unbound material in comparison to the purely hadronic reference EOS (black line).

The scatter among the different models visible in Fig. 12 is to be compared to ejecta mass variations of various purely hadronic equations of state. Neutron star merger simulations for a larger sample of purely hadronic EOS show roughly comparable variations of the ejecta mass even for models which result in similar stellar parameters (see Bauswein *et al.*, 2013). This implies that it is very difficult to unambiguously deduce the occurrence of quark matter from kilonova observations alone since bulk ejecta properties apparently do not carry unambiguous signatures of quark matter (Bauswein *et al.*, 2019b; Prakash *et al.*, 2021). Moreover, nuclear network calculations postprocessing merger simulations do not reveal significant differences in the resulting elemental abundance patterns comparing models with and without phase transition (Prakash *et al.*, 2021), although much more work is required to properly address the full complexity of this point.

3.3. *Black-hole formation*

Albeit kilonova observations may not be immediately informative about the occurrence of deconfined quark matter, they do contain information about the merger outcome. All models displayed in Fig. 12 yield a neutron star merger remnant since their total binary mass of 2.7 $M_\odot$ is below the respective threshold mass M_{thres} for prompt black-hole formation. If the total binary mass exceed the threshold for prompt black-hole formation, the ejecta mass and thus the kilonova brightness is strongly reduced (Bauswein *et al.*, 2013). Hence, combining the binary masses measured by the gravitational wave inspiral and the interpretation of the kilonova emission can yield an estimate of the threshold mass M_{thres} for direct black hole-formation. Along these lines, it has been argued that GW170817 did not result in a prompt collapse. Also, the presence or absence of postmerger gravitational-wave emission informs about a gravitational collapse after merging. In essence, it is conceivable that M_{thres} can be determined or at least constrained from future observations.

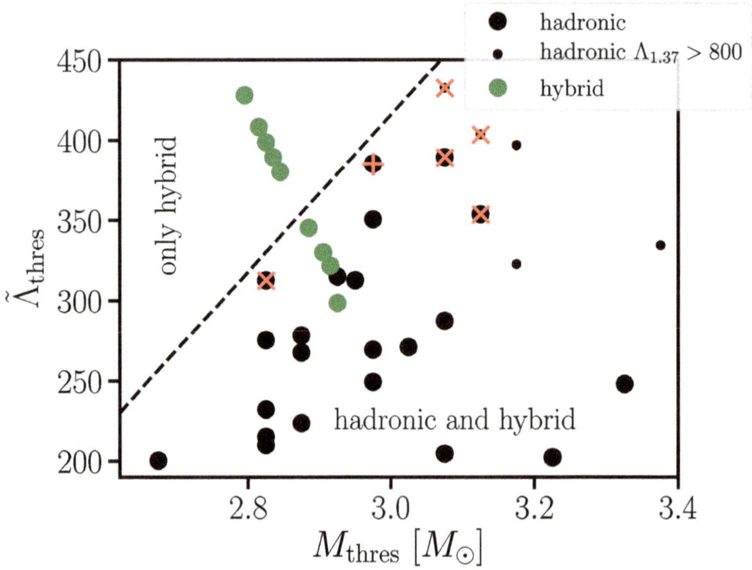

Fig. 13. Collapse behavior of purely hadronic models (black) and hybrid equations of state with phase transition to deconfined quark matter (green). $M_{\rm thres}$ is the threshold mass for prompt black-hole formation. $\Lambda_{\rm thres}$ is the tidal deformability of the binary with a total mass of $M_{\rm thres}$, i.e., the tidal deformability of the system with a marginally stable remnant for the given EOS. Purely hadronic equations of state lead to threshold properties below the black dashed line.

This poses the question whether the occurrence of deconfined quark matter affects the threshold mass $M_{\rm thres}$ for black-hole formation. It is intuitively clear that the stability of a neutron star system in general and of $M_{\rm thres}$ in particular will strongly depend on the properties of the EOS. Numerically $M_{\rm thres}$ can be determined by performing several simulations for different binary masses for a given EOS and by checking the respective outcome. Such estimates for a large sample of different purely hadronic EOSs yield $M_{\rm thres}$ in a relatively broad range between roughly 2.8 and 3.8 $M_\odot$ depending on the specific properties of the microphysical model, as was discussed in Bauswein *et al.* (2020).

A $M_{\rm thres}$ estimate may yield signatures of quark matter in neutron stars since the occurrence of a phase transition can but does not necessarily need to result in a destabilization of the merger remnant (Bauswein *et al.*, 2020). This is summarized in Fig. 13, which displays $M_{\rm thres}$ versus $\Lambda_{\rm thres}$ for different equations of state. $\Lambda_{\rm thres}$ is the tidal deformability measured in a system with a total mass of $M_{\rm tot} = M_{\rm thres}$ and can be inferred from the gravitational-wave inspiral signal. Hence, future observations will narrow down the area where the true EOS lies in Fig. 13.

It is striking that in Fig. 13 purely hadronic models are located below the dashed line, whereas some hybrid equations of state with phase transition to deconfined quark matter occur in the upper left corner of the diagram. Hence, some hybrid models feature a relatively low M_{thres} in comparison to the tidal deformability. This may not be surprising because in these cases the tidal deformability is determined by the hadronic regime of the EOS below the onset of the phase transition, whereas the appearance of quark matter during the merging softens the EOS and destabilizes the remnant, i.e., reduces M_{thres}. The softening of the EOS and the associated reduction of M_{thres} is not captured by the tidal deformability since it is only sensitive to the EOS at lower densities. This effect is somewhat similar to the shift of the postmerger gravitational-wave frequency as result of a strong compactification of the remnant. Such sudden changes of the EOS do not occur for purely hadronic models, which is why they cannot occur much beyond the dashed line in Fig. 13.

Identifying the presence of quark matter through the collapse behavior in Fig. 13 requires only a moderate precision of a tidal deformability measurement in comparison to resolving directly the kink induced by a phase transition in $\Lambda(M)$. Moreover, a single measurement may already provide evidence for quark matter if Λ_{thres} and M_{thres} are correspondingly constrained. As apparent from Fig. 13 the comparison between Λ_{thres} and M_{thres} may not in any case reveal the presence of quark matter as some hybrid models can mimic the behavior of purely hadronic EOSs.

4. Summary

In this contribution, we have illustrated that there exist pronounced and unambiguous features which reveal the occurrence of deconfined quark matter in supernova explosions and in neutron star mergers.

In supernova simulations, the quark matter deconfinement phase transition leads to an explosion mechanism for massive blue supergiant stars which without the phase transition would evolve into black holes and represent failed supernovae. We have demonstrated that as a result of the deconfinement-driven explosion mechanism neutron stars with high masses $\gtrsim 2\ M_\odot$ are born, while the population of this mass segment within the standard neutrino-driven supernova explosion is quite unlikely, see Müller *et al.* (2016). As observational signatures we have discussed a second neutrino signal which appears about 1.2 ms after bounce and would be detectable by Super-Kamiokande if the supernova goes off in our galaxy, as well as a gravitational wave signal in the spectrogram of an axially-symmetric core-collapse supernova about 400 ms post bounce. It has also been found that a supernova exploding by the quark deconfinement mechanism must be a relatively rare event because it cannot account for the bulk part of the r-process nucleosynthesis in the galaxy due to a strong suppression of the synthesis of rare-earth elements and those of the third r-process peak.

Most of the signals of quark deconfinement in binary neutron star mergers are connected to the postmerger evolution, which may not be unexpected since the merger remnant probes higher densities than the inspiralling stars and thus naturally carries a stronger imprint of the EOS at higher densities. We have focussed on the discussion of two deconfinement signals: the shift of the peak frequency of postmerger gravitational wave signal (occurring for both merging hadronic and merging hybrid stars) and the correlation between binary mass and tidal deformability at the threshold for prompt black-hole formation. The strength of these signatures depend on details of the phase transition and the EOS of quark matter. This implies that under certain circumstances it may be challenging to reveal or exclude the occurrence of quark deconfinement. This highlights the necessity of further research investigating more hybrid models and details of the various signals from neutron star mergers.

The prospects to identify signals of quark deconfinement through the various features described above emphasizes the outstanding importance of future measurements and sensitivity upgrades of existing instruments. This includes in particular the ability to detect postmerger gravitational-wave emission in the kHz range but also electromagnetic follow-up observations to infer kilonova properties.

5. Epilogue

We want to take up the discussion of the Introduction that the simulations of core-collapse SN explosions and BNS mergers with model EOS are a unique tool to investigate the QCD phase diagram in the region of low temperatures and high baryon densities ($T \lesssim 60$ MeV at $1 \lesssim n/n_0 \lesssim 5$) which is inaccessible to lattice QCD simulations and HIC experiments. Not only that the detection of the above-described signals of a strong first-order phase transition in BNS mergers and/or supernovae would provide support for the existence of a CEP in the phase diagram that has so far been unsuccessfully been sought for in HIC experiments, there is theoretical evidence for a crossover transition at very low temperatures that suggests the existence of a second CEP or even a crossover-all-over situation. This arises from the observation (Abuki *et al.*, 2010; Hatsuda *et al.*, 2006) that a coexistence of chiral symmetry breaking and diquark condensation occurs at low temperatures due to the $U_A(1)$ anomaly-generating triangle diagram which, after Fierz transformation mixes diquark and meson condensates. This effect realizes the concept of quark-hadron continuity (Schäfer and Wilczek, 1999) in a crossover transition. Support for such a picture comes also from recent progress in NS phenomenology.

We are currently witnessing a paradigm change in the interpretation of mass and radius measurements of pulsars that is induced by the observation that from the multi-messenger analysis of typical-mass neutron star radii with $R_{1.4M_\odot} = 11.7^{+0.86}_{-0.81}$ km (Dietrich *et al.*, 2020) (see also Capano *et al.*, 2020) and the recent NICER radius measurement $R_{2.0M_\odot} = 13.7^{+2.6}_{-1.5}$ km (Miller *et al.*, 2021) (see also Riley *et al.*, 2021) follows that $R_{2.0M_\odot} \gtrsim R_{1.4M_\odot}$. The description of such a behavior as solution of the Tolman–Oppenheimer–Volkoff equations requires a soft-stiff transition in the EOS at densities $n \lesssim 2n_0$, just before the hyperon onset. This transition could be the hadron-to-quark matter transition. In recent descriptions one joins a standard nuclear EOS with a constant speed of sound (CSS) model for the high-density phase either by a first-order phase transition (with a vanishing speed of sound in the mixed phase (Somasundaram and Margueron, 2022)) or by directly matching the nuclear and quark matter squared speed of sound c_s^2 at a certain transition density $n_{\rm tr}$ without a density jump, thus mimicking a crossover transition (Drischler *et al.*, 2021). The best phenomenological description fulfilling simultaneously the constraints on both radii $R_{2.0M_\odot}$ and $R_{1.4M_\odot}$ is obtained in this simple picture by $n_{\rm tr} \sim 0.5\, n_0$ and $c_s^2 \sim 0.5$. We would like to remark that a CSS model with $c_s^2 = 0.45\ldots0.54$ provides an excellent fit to a microscopic non-local chiral quark model of the Nambu–Jona-Lasinio (NJL) type with diquark condensation (color superconductivity) and repulsive vector meson mean field (Antić, *et al.*, 2021; Contrera *et al.*, 2022). A direct, one-zone interpolation scheme between the safely known soft nuclear matter EOS (up to about 1.1 n_0 as in Hebeler *et al.* (2013)) and the suitably chosen stiff quark matter EOS (e.g., from a NJL model with coupling to a repulsive vector meson mean field) was pioneered in the works of (Masuda *et al.*, 2013a,b). Such a phase transition construction can be understood as a shortcut for three physical effects as ingredients:

(i) a stiffening of the nuclear matter EOS $P_H(\mu)$ due to the repulsive quark Pauli blocking effect between nucleons (Blaschke *et al.*, 2020; Röpke *et al.*, 1986) which can be effectively accounted for with a nucleonic excluded volume (see, e.g., Alvarez-Castillo *et al.*, 2016),

(ii) a strong reduction of the quark matter pressure $P_Q(\mu)$ at low chemical potentials due to confining forces (which result then in a good crossing of curves at $\mu = \mu_c$ allowing for the Maxwell construction $P_H(\mu_c) = P_Q(\mu_c)$ of a first-order phase transition), and

(iii) a mixed phase construction (e.g., by a parabolic interpolation (Ayriyan *et al.*, 2018)) that mimics the effects of finite-size structures (pasta phases) in the quark-hadron coexistence region.

For more details on the physics background, (see Ayriyan *et al.*, 2021; Baym *et al.*, 2018, and references therein). With this microphysical basis behind the interpolation approach, a two-zone interpolation construction for the hadron-to-quark matter has been developed in Ayriyan *et al.* (2021), where at the matching point μ_c situated between μ_H and μ_Q, one can choose the condition of continuous density ($\Delta n = 0$, crossover) or a finite density jump ($\Delta n \neq 0$, first-order transition). In concluding this epilogue, we want to give an outlook to the generalization of this two-zone interpolation scheme to finite temperatures (and arbitrary isospin densities) as a necessary prerequisite for investigating the consequences of these recent developments in the interpretation of neutron star phenomenology at zero temperature to simulations of supernova explosions and of binary neutron star merger events. The goal is to model the general class of hybrid EOS that corresponds to a phase diagram which has not only one CEP at high temperatures which marks the change from a first-order to a crossover transition regime, but also second CEP at low temperatures that arises from the competition and mixing between dynamical chiral symmetry breaking and color superconductivity. Within the finite-temperature generalization of the two-zone interpolation scheme, this can be achieved by defining the function $\Delta n[\mu_c(T)]$ along the matching line $\mu_c(T)$ between the hadron-like and the quark-like interpolation zone in the phase diagram. The function $\Delta n[\mu_c(T)]$ encodes the position of the CEPs $T_{\mathrm{CEP},1}$ and $T_{\mathrm{CEP},2}$ where $\Delta n = 0$ as well as the strength of thee first-order transition between these points where $\Delta n \neq 0$. For details, see Ivanytskyi and Blaschke (2022).

It is the aim of our ongoing research to investigate the dependence of the above described signals of a strong phase transition in supernova explosions and binary neutron star mergers on the detailed structure of the QCD phase diagram at low temperatures and high baryon densities and thus to be prepared for interpreting the possible observation of signals from such events in the near future.

Acknowledgements

This work was supported by the Polish National Science Centre (NCN) under grant No. 2019/33/B/ST9/03059 (D.B.) and No. 2020/37/B/ST9/00691 (T.F.). A.B. acknowledges support by the European Research Council (ERC) under the European Union's Horizon 2020 research and innovation programme under grant agreement No. 759253, by Deutsche Forschungsgemeinschaft (DFG, German Research Foundation) - Project-ID 279384907 - SFB 1245, by DFG - Project-ID 138713538 - SFB 881 ("The Milky Way System", subproject A10) and by the State of Hesse within the Cluster Project ELEMENTS. The work was performed within a project that has received funding from the European Union's Horizon 2020 research and innovation program under grant agreement STRONG - 2020 - No. 824093.

References

Abbott, B. P. *et al.* (2017). *Physical Review Letters* **119**, p. 161101.

Abbott, B. P. *et al.* (2018). *Physical Review Letters* **121**, p. 161101.

Abbott, B. P. *et al.* (2019). *Physical Review X* **9**, p. 011001.

Abuki, H., Baym, G., Hatsuda, T., and Yamamoto, N. (2010). *Physical Review D* **81**, p. 125010.

Alford, M., Braby, M., Paris, M., and Reddy, S. (2005). *Astrophysical Journal* **629**, p. 969.

Alvarez-Castillo D., Ayriyan, A., Benic, S., Blaschke, D., Grigorian, H., and Typel S. (2016). *European Physical Journal A* **52**, p. 69.

An X., *et al.* (2022). *Nuclear Physics A* **1017**, p. 122343.

Antić S., Shahrbaf M., Blaschke D., and Grunfeld A. G. (2021). arXiv:2105.00029 [nucl-th].

Antoniadis J., *et al.* (2013). *Science* **340**, p. 448.

Arsene I. C., *et al.* (2007). *Physical Review C* **75**, p. 034902.

Ayriyan, A., Bastian, N. U., Blaschke, D., Grigorian, H., Maslov, K., and Voskresensky D. N. (2018). *Physical Review C* **97**, p. 045802.

Ayriyan A., Blaschke D., Grunfeld, A. G., Alvarez-Castillo, D., Grigorian, H., and Abgaryan, V. (2021). *European Physical Journal A* **57**, p. 318.

Bauswein, A. and Blacker, S. (2020). *European Physical Journal Special Topics* **229**, p. 3595.

Bauswein, A., Goriely, S. and Janka, H.-T. (2013). *Astrophysical Journal* **773**, pp. 78.

Bauswein, A., Bastian, N.-U. F., Blaschke, D. B., Chatziioannou, K., Clark, J. A., Fischer, T., and Oertel, M. (2019a). *Physical Review Letters* **122**, p. 061102.
Bauswein, A., *et al.* (2019b). *AIP Conference Proceedings* **2127**, p. 020013.
Bauswein A., *et al.* (2020). *Physical Review Letters* **125**, p. 141103.
Baym, G., Hatsuda, T., Kojo, T., Powell, P. D., Song, Y., and Takatsuka, T. (2018). *Reports on Progress in Physics* **81**, pp. 056902.
Bazavov, A., *et al.* (2017). *Physical Review D* **95**, p. 054504.
Bazavov, A., *et al.* (2019). *Physics Letters B* **795**, p. 15.
Benić, S., Blaschke, D., Alvarez-Castillo, D. E., Fischer, T., and Typel, S., (2015). *Astronomy and Astrophysics* **577**, p. A40.
Bethe, H. A. and Wilson, J. R. (1985). *Astrophysical Journal* **295**, p. 14.
Bionta, R., Blewitt, G., Bratton, C., Casper, D., and Ciocio, A., *et al.* (1987). *Physical Review Letters* **58**, p. 1494.
Blacker, S., Bastian, N.-U. F., Bauswein, A., Blaschke, D. B., Fischer, T., Oertel, M., Soultanis, T. and Typel, S. (2020). *Physical Review D* **102**, p. 123023.
Blaschke, D., Grigorian, H. and Röpke, G. (2020). *Particles* **3**, p. 477.
Bollig, R., Yadav, N., Kresse, D., Janka, H.-T., Müller, B., and Heger, A. (2021). *Astrophysical Journal* **915**, p. 28.
Bruenn, S. W., *et al.* (2013). *Astrophysical Journal* **767**, p. L6.
Burrows, A. and Vartanyan, D. (2021). *Nature* **589**, p. 29.
Capano, C. D. *et al.* (2020). *Nature Astronomy* **4**, p. 625.
Chatziioannou, K. and Han, S. (2020). *Physical Review D* **101**, p. 044019.
Chatziioannou, K., Clark, J. A., Bauswein, A., Millhouse, M., Littenberg, T. B., and Cornish, N. (2017). *Physical Review D* **96**, p. 124035.
Chen, A., Johnson-McDaniel, N. K., Dietrich, T., and Dudi, R. (2020). *Physical Review D* **101**, p. 103008.
Christian, J.-E., Zacchi, A. and Schaffner-Bielich, J. (2019). *Physical Review D* **99**, p. 023009.
Contrera, G. A., Blaschke, D., Carlomagno, J. P., Grunfeld, A. G., and Liebing, S. (2022). *Physical Review C* **105**, p. 045808.
Cowan, J. J., Sneden, C., Lawler, J. E., Aprahamian, A., Wiescher, M., Langanke, K., Martínez-Pinedo, G., and Thielemann, F.-K. (2021). *Reviews of Modern Physics* **93**, p. 015002.
Cromartie, H. T., *et al.* (2020). *Nature Astronomy* **4**, p. 72.
Dasgupta, B., Fischer, T., Horiuchi, S., Liebendorfer, M., and Mirizzi, A., *et al.* (2010). *Physical Review D* **81**, p. 103005.
Dietrich, T., Coughlin, M. W., Pang, P. T. H., Bulla, M., Heinzel, J., Issa, L., Tews, I., and Antier, S. (2020). *Science* **370**, p. 1450.
Dimmelmeier, H., Ott, C. D., Janka, H. T., Marek, A., and Müller, E., (2007). *Physical Review Letters* **98**, p. 251101.
Dimmelmeier, H., Ott, C. D., Marek, A. and Janka, H. T. (2008). *Physical Review D* **78**, p. 064056.
Ding, H. T., *et al.* (2019). *Physical Review Letters* **123**, p. 062002.
Drischler, C., Han, S., Lattimer, J. M., Prakash, M., Reddy, S., and Zhao, T. (2021). *Physical Review C* **103**, p. 045808.

Farouqi, K., Thielemann, F.-K., Rosswog, S. and Kratz, K.-L. (2021). *Astronomy and Astrophysics* **663**, p. A70.
Fischer, T. (2021). *European Physical Journal A* **57**, p. 270.
Fischer, T., *et al.* (2011). *Astrophysical Journal Supplementary* **194**, p. 39.
Fischer, T., *et al.* (2017). *Publications of the Astronomical Society of Australia* **34**, p. e067.
Fischer, T., *et al.* (2018). *Nature Astronomy* **2**, p. 980.
Fischer, T., Guo, G., Dzhioev, A. A., Martínez-Pinedo, G., Wu, M.-R., Lohs, A., and Qian, Y.-Z. (2020a). *Physical Review C* **101**, p. 025804.
Fischer, T., Wu, M.-R., Wehmeyer, B., Bastian, N.-U. F., Martínez-Pinedo, G., and Thielemann, F.-K. (2020b). *Astrophysical Journal* **894**, p. 9.
Flanagan, É. É. and Hinderer, T. (2008). *Physical Review D* **77**, 021502.
Fonseca, E. *et al.* (2016). *Astrophysical Journal* **832**, p. 167.
Fonseca, E. *et al.* (2021). *Astrophysical Journal Letters* **915**, p. L12.
Han, S. and Steiner, A. W. (2019). *Physical Review D* **99**, p. 083014.
Hanauske, M., Weih, L. R., Stöcker, H., and Rezzolla, L. (2021). *European Physical Journal Special Topics* **230**, p. 543.
Hatsuda, T., Tachibana, M., Yamamoto, N., and Baym, G. (2006). *Physical Review Letters* **97**, p. 122001.
Hebeler, K., Lattimer, J. M., Pethick, C. J., and Schwenk, A. (2013). *Astrophysical Journal* **773**, p. 11.
Hempel, M., Heinimann, O., Yudin, A., Iosilevskiy, I., Liebendörfer, M., and Thielemann, F.-K. (2016). *Physical Review D* **94**, p. 103001.
Hinderer, T. (2008). *Astrophysical Journal* **677**, p. 1216.
Hirata, K., Kajita, T., Koshiba, M., Nakahata, M., Oyama, Y., *et al.* (1988). *Physical Review D* **38**, p. 448.
Hotokezaka, K., Kiuchi, K., Kyutoku, K., Okawa, H., Sekiguchi, Y.-i., Shibata, M., and Taniguchi, K. (2013). *Physical Review D* **87**, p. 024001.
Ivanytskyi, O. and Blaschke, D. (2022). *European Physical Journal A* **58**, p. 152.
Janka, H. T., Langanke, K., Marek, A., Martínez-Pinedo G., and Müller, B. (2007). *Physics Reports* **442**, p. 38.
Kaltenborn, M. A. R., Bastian, N.-U. F. and Blaschke, D. B. (2017). *Physical Review D* **96**, p. 056024.
Klähn, T. and Fischer, T. (2015). *Astrophysical Journal* **810**, p. 134.
Kotake, K., Sato, K. and Takahashi, K. (2006). *Reports on Progress in Physics* **69**, p. 971.
Kuroda, T., Takiwaki, T. and Kotake, K. (2014). *Physical Review D* **89**, p. 044011.
Kuroda, T., Kotake, K. and Takiwaki, T. (2016). *Astrophysical Journal* **829**, p. L14.
Kuroda, T., Kotake, K., Hayama, K., and Takiwaki, T. (2017). *Astrophysical Journal* **851**, p. 62.
Kuroda, T., Fischer, T., Takiwaki, T., and Kotake, K. (2021). *Astrophysical Journal* **924**, p. 38.
LIGO Scientific Collaboration Virgo Collaboration *et al.* (2017). *Astrophysical Journal Letters* **848**, p. L12.

LeBlanc J. M. and Wilson J. R. (1970). *Astrophysical Journal* **161**, p. 541.
Lentz, E. J., *et al.* (2015). *Astrophysical Journal Letter* **807**, p. L31.
Liebling, S. L., Palenzuela, C. and Lehner, L. (2021). *Classical and Quantum Gravity* **38**, p. 115007.
Martínez-Pinedo, G., Fischer, T. and Huther, L. (2014). *Journal of Physics G Nuclear Physics* **41**, p. 044008.
Masuda, K., Hatsuda, T. and Takatsuka, T. (2013a). *Astrophysical Journal* **764**, p. 12.
Masuda, K., Hatsuda, T. and Takatsuka, T. (2013b). *Progress of Theoretical and Experimental Physics* **2013** (7), p. 073D01.
Melson, T., Janka, H.-T. and Marek, A. (2015). *Astrophysical Journal* **801**, p L24.
Metzger, B. D. (2019). *Living Reviews in Relativity* **23**, p. 1.
Mezzacappa, A. *et al.* (2020). *Physical Review D* **102**, p. 023027.
Miller M. C., *et al.* (2021). *Astrophysical Journal Letter* **918**, p. L28.
Mirizzi, A., Tamborra, I., Janka, H. T., Saviano, N., Scholberg, K., Bollig, R., Hüdepohl, L., and Chakraborty, S. (2016). *Nuovo Cimento Rivista Serie* **39**, p. 1.
Morozova, V., Radice, D., Burrows, A., and Vartanyan, D. (2018). *Astrophysical Journal* **861**, p. 10.
Most, E. R., Papenfort, L. J., Dexheimer, V., Hanauske, M., Schramm, S., Stöcker, H., and Rezzolla, L. (2019). *Physical Review Letters* **122**, p. 061101.
Mösta, P., *et al.* (2014). *Astrophysical Journal Letter* **785**, p. L29.
Mösta, P., Ott, C. D., Radice, D., Roberts, L. F., Schnetter, E., and Haas, R. (2015). *Nature* **528**, p. 376.
Müller, B., Janka, H.-T. and Marek, A. (2013). *Astrophysical Journal* **766**, p. 43.
Müller, B., Heger, A., Liptai, D. and Cameron, J. B. (2016). *Monthly Notices of the Royal Astronomical Society* **460**, p. 742.
Müller, B., Melson, T., Heger, A., and Janka, H.-T. (2017). *Monthly Notices of the Royal Astronomical Society* **472**, p. 491.
Müller, B., *et al.* (2019). *Monthly Notices of the Royal Astronomical Society* **484**, p. 3307.
Muroya, S., Nakamura, A., Nonaka, C., and Takaishi, T. (2003). *Progress of Theoretical and Experimental Physics* **110**, p. 615.
Murphy, J. W., Ott, C. D. and Burrows, A. (2009). *Astrophysical Journal* **707**, p. 1173.
Mller, B., Janka, H.-T. and Marek, A. (2012). *Astrophysical Journal* **756**, p. 84.
Nagakura, H., *et al.* (2018). *Astrophysical Journal* **854**, p. 136.
Nagakura, H., Burrows, A. and Vartanyan, D. (2021). *Monthly Notices of the Royal Astronomical Society* **506**, p. 1462.
Ott, C. D., Burrows, A., Livne, E., and Walder, R. (2004). *Astrophysical Journal* **600**, p. 834.
Ott, C. D., Burrows, A., Dessart, L., and Livne, E. (2006). *Physical Review Letters* **96**, p. 201102.

Pang, P. T. H., Dietrich, T., Tews, I. and Van Den Broeck C. (2020). *Physical Review Research* **2**, p. 033514.

Prakash A., *et al.* (2021). *Physical Review D* **104**, p. 083029.

Pratt, S., Sangaline, E., Sorensen, P., and Wang, H. (2015). *Physical Review Letters* **114**, p. 202301.

Qian, Y. and Woosley S. (1996). *Astrophysical Journal* **471**, p. 331.

Radice, D., Burrows, A., Vartanyan, D., Skinner, M. A., and Dolence, J. C. (2017). *Astrophysical Journal* **850**, p. 43.

Riley, T. E., *et al.* (2021). *Astrophysical Journal Letter* **918**, p. L27.

Roberts, L. F., Shen, G., Cirigliano, V., Pons, J. A., Reddy, S., and Woosley, S. E., (2012). *Physical Review Letters* **108**, p. 061103.

Roederer, I. U., *et al.* (2012). *Astrophysical Journal Supplementary* **203**, pp. 27.

Röpke, G., Blaschke, D. and Schulz, H. (1986). *Physical Review D* **34**, p. 3499.

Sagert, I., Fischer, T., Hempel, M., Pagliara, G., Schaffner-Bielich, J., *et al.* (2009). *Physical Review Letters* **102**, p. 081101.

Schäfer, T. and Wilczek, F. (1999). *Physical Review Letters* **82**, p. 3956.

Scheidegger, S., Fischer, T., Whitehouse, S. C., and Liebendörfer, M. (2008). *Astronomy and Astrophysics* **490**, p. 231.

Scheidegger, S., Käppeli, R., Whitehouse, S. C., Fischer, T., and Liebendörfer, M. (2010). *Astronomy and Astrophysics* **514**, p. A51.

Senger, P. (2021). *Physica Scripta* **96**, p. 054002.

Shibagaki, S., Kuroda, T., Kotake, K., and Takiwaki, T. (2021). *Monthly Notices of the Royal Astronomical Society* **502**, p. 3066.

Shibata, M. and Sekiguchi, Y.-I. (2003). *Physical Review D* **68**, p. 104020.

Sneden C., *et al.* (2003). *Astrophysical Journal* **591**, p. 936.

Somasundaram, R. and Margueron, J. (2022). *Europhysics Letters* **138**, p. 14002.

Sotani, H. and Takiwaki, T. (2020). *Monthly Notices of the Royal Astronomical Society* **498**, p. 3503.

Stockinger, G., *et al.* (2020). *Monthly Notices of the Royal Astronomical Society* **496**, p. 2039.

Takiwaki, T., Kotake, K. and Sato, K. (2009). *Astrophysical Journal* **691**, p. 1360–1379.

Takiwaki, T., Kotake, K. and Suwa, Y. (2012). *Astrophysical Journal* **749**, p. 98.

Thielemann, F. K., *et al.* (2011). *Progress in Particle and Nuclear Physics* **66**, p. 346.

Torres-Forné, A., Cerdá-Durán, P., Passamonti, A., Obergaulinger, M., and Font, J. A. (2019). *Monthly Notices of the Royal Astronomical Society* **482**, p. 3967.

Torres-Rivas, A., Chatziioannou, K., Bauswein, A., and Clark, J. A. (2019). *Physical Review D* **99**, p. 044014.

Vartanyan, D., Burrows, A. and Radice, D. (2019). *Monthly Notices of the Royal Astronomical Society* **489**, p. 2227.

Weih, L. R., Hanauske, M. and Rezzolla, L. (2020). *Physical Review Letters* **124**, p. 171103.

Wijngaarden, M., Chatziioannou, K., Bauswein, A., Clark, J. A., and Cornish, N. (2022). *Physical Review D* **105**, p. 104019.

Winteler, C., Käppeli, R., Perego, A., Arcones, A., Vasset, N., Nishimura, N., Liebendörfer, M., and Thielemann, F.-K. (2012). *Astrophysical Journal Letters* **750**, p. L22.

Witt, M., *et al.* (2021). *Astrophysical Journal* **921**, p. 19.

Woosley, S., Wilson, J., Mathews, G., Hoffman, R., and Meyer, B. (1994). *Astrophysical Journal* **433**, p. 229.

Woosley, S., Heger, A. and Weaver, T. (2002). *Reviews of Modern Physics* **74**, p. 1015.

Wu, M.-R., Qian, Y.-Z., Martínez-Pinedo, G., Fischer, T., and Huther, L. (2015). *Physical Review D* **91**, p. 065016.

Zha, S., O'Connor, E. P., Chu, M.-C., Lin, L.-M., and Couch, S. M. (2020). *Physical Review Letters* **125**, p. 051102.

Lightning Source UK Ltd.
Milton Keynes UK
UKHW050751251122
412828UK00002B/22

9 789811 220937